ADVANCES IN THE BIOCHEMISTRY AND PHYSIOLOGY
OF PLANT LIPIDS

DEVELOPMENTS IN PLANT BIOLOGY
VOLUME 3

Other volumes in this series:

Volume 1 Plant Mitochondria
G. Ducet and C. Lance editors, 1978

Volume 2 Chloroplast Development
G. Akoyunoglou and J.H. Argyroudi-Akoyunoglou editors, 1978

ADVANCES IN THE BIOCHEMISTRY PHYSIOLOGY OF PLANT LIPIDS

Proceedings of the Symposium on Recent Advances in the Biochemistry and Physiology of Plant Lipids held in Göteborg, Sweden, August 28-30, 1978.

Editors

Lars-Åke Appelqvist

and

Conny Liljenberg

1979

ELSEVIER/NORTH-HOLLAND BIOMEDICAL PRESS
AMSTERDAM · NEW YORK · OXFORD

©1979 Elsevier/North-Holland Biomedical Press

All rights reserved. No part of this publication may be reproduced, stored in a retrieval system, or transmitted, in any form or by any means, electronic, mechanical, photocopying, recording or otherwise, without the prior permission of the copyright owner.

ISBN for this volume: 0-444-80129-4
ISBN for the series: 0-444-80081-6

Published by:
Elsevier/North-Holland Biomedical Press
335 Jan van Galenstraat, P.O. Box 211
Amsterdam, The Netherlands

Sole distributors for the USA and Canada:
Elsevier North-Holland, Inc.
52 Vanderbilt Avenue
New York, N.Y. 10017

Library of Congress Cataloging in Publication Data

```
Symposium on Recent Advances in the Biochemistry
   and Physiology of Plant Lipids, Gothenburg,
   Sweden, 1978.
   Advances in the biochemistry and physiology
of plant lipids.

   (Developments in plant biology ; 3)
   Bibliography: p.
   Includes index.
   1.  Plant lipids--Congresses.  I.  Appelqvist,
Lars-Åke, 1931-      II.  Liljenberg, Conny.
III.  Title.  IV.  Series.
QK898.L56S94   1978       581.1'9247      79-11642
ISBN 0-444-80129-4
```

PRINTED IN THE NETHERLANDS

PREFACE

This volume contains the lectures and short communications presented at the international symposium "Recent Advances in the Biochemistry and Physiology of Plant Lipids", held in Göteborg, Sweden, August 28-30, 1978.

It was the third in an informal series, which was initiated (inadvertently!) by some British scientists, Drs. T. Galliard, M.I. Gurr, E.I. Mercer and M.J.C. Rhodes who organized a meeting in Norwich, England, April 8-10, 1974 - "Recent Advances in the Chemistry and Biochemistry of Plant Lipids". That meeting was sponsored jointly by the Phytochemical Society and the Lipid Group of the Biochemical Society of Great Britain and was originally considered as unique.

The second meeting was held in Karlsruhe, F.R.G., in 1976. It was organized by four German scientists, Professors E. Heinz, H.K. Lichtenthaler, H.K. Mangold and M. Tevini. Some of them had been in Norwich and found that meeting so valuable that a continuation of this type of symposium was considered very essential. Then the Göteborg meeting was organized by us and some other Swedish scientists at the suggestion of those attending the Karlsruhe meeting.

The two previous publications from Norwich and Karlsruhe and this from Göteborg, although all three dealing with the general theme *Lipids in Higher Plants*, each have their specific character. In Norwich only 12 invited lecturers presented papers on the *Chemistry and Biochemistry* of Plant Lipids and essentially only acyl lipids, since at other, then recent, British symposia polyisoprenoids had been covered.

In Karlsruhe the theme was *Lipids and Lipid Polymers* in Higher Plants. The publication contained only the "main papers" which covered e.g. both acyl lipids, polyisoprenoids and polymers such as cutin.

In the invitation to the Göteborg meeting we emphasized especially the relation between structure and function of lipids in the membranes of higher plants. However, a rather wide span of papers within the general theme: *Biochemistry and Physiology* of Plant Lipids was presented orally or as posters in Göteborg. We are happy to present both lectures and short communications from the Göteborg meeting in this volume.

Lipids in plants, notably higher plants, is indeed a minor topic which receives poor attention at general biochemical meetings as well as at general meetings on botany or plant physiology. This was clearly pointed out also in the preface to the book from the Norwich meeting :"Recent Advances in the Chemistry and Biochemistry of Plant Lipids" (T. Galliard and I. Mercer, eds., A.P. 1975). The considerable interest in the Göteborg meeting, with its possibility for cross-fertilization of ideas from scientists which by classical nomenclature could be considered e.g. as physicists, physical chemists, geneticists, cytologists, plant physiologists or biochemists is taken to mean that this type of symposium deserves a "niche" of its own. The meeting attracted some 130 scientists from 17 countries in the world.

It is the hope of the editors that the outcome of this volume will be to contribute further to the advancement of science in this special field which ought to attract more attention in view of the key importance of plant lipids for such different areas as the structure and function of the photosynthetic apparatus and the production of vegetable oils in increased quantity and with improved quality.

The editors are indeed grateful to the authors, who with very short notice have submitted their manuscripts.

We gratefully acknowledge the financial sponsorship of the Ministry of Education of the Swedish government, The University of Göteborg, the City of Göteborg and the Anna Ahrenberg Fund, Göteborg.

We also wish to express our gratitude to the staff at the Department of Plant Physiology, University of Göteborg, notably to its chairman, Professor Hemming Virgin and to Ms Anna-Stina Sandelius and Ms Eva Selstam, whose support and enthusiastic assistance made the symposium, and thereby this volume, possible.

                            Stockholm and Göteborg, January, 1979
                            Lars-Åke Appelqvist
                            Conny Liljenberg

CONTENTS

Preface   v

Genetics and biosynthesis of plant epicuticular waxes
   P. von Wettstein-Knowles   1

The aqueous system of monogalactosyl diglycerides and diagalactosyl
   diglycerides - Significance to the structure of the thylakoid
   membrane
   K. Larsson and S. Puang-Ngern   27

MEMBRANE LIPIDS : STRUCTURE AND FUNCTION

Chlorophyll-lipid interactions
   J.J. Katz   37

Occurence and function of prenyllipids in the photosynthetic
   membrane
   H.K. Lichtenthaler   57

The chloroplast envelope: An unusual cell membrane system
   R. Douce and J. Joyard   79

Investigations on the origin of diglyceride diversity
   in leaf lipids
   E. Heinz, H.P. Siebertz, M. Linscheid, J. Joyard and R. Douce   99

The enzymic degradation of membrane lipids in higher plants
   T. Galliard   121

Plastid development and tocochromanol accumulation in oil seeds
   H. Beringer and F. Nothdurft   133

Soluble, isomeric forms of glycerolphosphate acyltransferase
   in chloroplasts
   M. Bertrams and E. Heinz   139

The influence of phytohormones on prenyllipid composition and
   photosynthetic activity of thylakoids
   C. Buschmann and H.K. Lichtenthaler   145

Heterogeneity of lipids in wheat freeze-dried chloroplast lamellae
   C. Costes and R. Bazier   151

Lipoxygenase activity distribution in young wheat chloroplast
   lamellae
   R. Douillard and E. Bergeron   159

Tracer kinetic analysis of chloroplast pigments in *hlorella pyrenoidosa*
   K.H. Grumbach, G. Britton and T.W. Goodwin   165

Ultrastructure and lipid composition of chloroplasts of shade and
   sun plants
   T. Guillot-Salomon, C. Tuquet, M. de Lubac, M.F. Hallais and M. Signol   169

Correlation between photosynthesis and plant lipid composition
 K.-P. Heise and G. Harnischfeger   175

Is the chloroplast envelope a site of galactolipid synthesis? Yes!
 J. Joyard, M. Chuzel and R. Douce   181

Building units of prolamellar bodies from etioplasts of *Avena sativa* L.: Saponin content and reaggregation experiments
 J. Kesselmeier and H.G. Ruppel   187

Endogenous lipoxygenase control and lipid-associated free radical scavenging as modes of cytokinin action in plant senescence retardation
 Y.Y. Leshem, S. Grossman, A. Frimer and J. Ziv   193

Interactions of chlorophyll a with chloroplast lipids in mixed monolayer films
 C. Liljenberg and E. Selstam   199

The galactolipid and pigment composition of the thylakoid membranes from naturally differentiating chloroplasts of *Avena sativa* L.
 R.O. Mackender   205

Changes in the mitochondrial lipids of mango fruits at storage temperatures inducing "chilling injury"
 P. Mazliak and O. Kane   211

A proposed model for sudden fluidity changes
 V. McMahon   215

Has *trans*-3-hexadecenoic acid a role in granal stacking?
 M.P. Percival, J. Wharfe, P. Bolton, A.O. Davies, R. Jeffcoat, A.T. James and J.L. Harwood   219

Lipids and lipolytic activities in spinach plastids during seasonal development
 R. Frey and M. Tevini   225

The conversion of 9-D and 13-L-hydroperoxylinoleic acid by soybean lipoxygenase-1 under aerobic conditions
 J. Verhagen, G.A. Veldink, J.F.G. Vliegenthart and J. Boldingh   231

Lipid metabolism during the regreening of the chaetophoralean green alga *Fritschiella tuberosa* in axening culture
 M. Wettern   237

A phospholipid exchange protein from the endosperm of germinating castor bean seeds
 T. Tanaka and M. Yamada   243

LIPID COMPOSITION AND ANALYSIS

Changes in phospholipid levels during cold stratification and germination of *Pinus pinea* seeds
 M.T. Alsasua and E. Palacios-Alaiz   251

Pyrrolidides in the structural analysis of lipids by mass spectrometry
B.Å. Andersson 257

A practical procedure for the analysis of plant volatiles
B.Å. Andersson, L. Lundgren and G. Stenhagen 263

Water stress, epicuticular wax and cuticular transpiration rate
C. Bengston, S. Larsson and C. Liljenberg 269

Distribution of long-chain fatty acids in the cuticular lipids of hemp during plant development
P.Á. Biacs 275

Mass spectrometry of glycolipids
M.E. Breimer, G.C. Hansson, K.-A. Karlsson, G. Larson, H. Leffler, I. Pascher, W. Pimlott and B.E. Samuelsson 281

Triterpenes, steroids, alkanes and waxes from the peel of grapes
C.H. Brieskorn and G. Blosczyk 287

On phospholipids and fatty acids in cell-free fractions from the Rhodospirillaceae *Rhodopseudomonas palustris* and *Rps. spheroides*
K. Knobloch and F. Gemeinhardt 293

Application of TLC/FID in lipid analysis
B. Herslöf 301

Effects of light on lipids in potato tubers
C. Liljenberg, A.S. Sandelius and E. Selstam 307

Polyprenylphosphate derivatives as intermediates in the biosynthesis of cellulose precursors. Subcellular localization
H.E. Hopp, P.A. Romero, G.R. Daleo and R. Pont Lezica 313

Separation of prenyllipids by high performance liquid chromatography
U. Prenzel and H.K. Lichtenthaler 319

LIPID BIOSYNTHESIS

Lipid biosynthesis in plant cell cultures
M. Kates, A.C. Wilson and A.I. de la Roche 329

Biosynthesis of linoleic and linolenic acids - Substrates and sites
L.-Å. Appelqvist and S. Stymne 343

Some aspects of galactolipid formation in spinach chloroplasts
A. van Besouw, G. Bögemann and J.F.G.M. Wintermans 359

Changes in the photodynamic properties of the chlorophyll(ide) during the early stages of greening
L. Axelsson and E. Selstam 363

On the compartmentation of the biosynthesis of aromatic amino acids
and prenylquinones in higher plants
H. Bickel, B. Buchholz and G. Schultz 369

Regulation of prenylquinone synthesis by shikimate pathway in
higher plants
H. Bickel and G. Schultz 377

Phosphatidylglycerol biosynthesis by isolated chloroplasts of
Euglena gracilis
A. Chammai and R. Schantz 381

Incorporation of [^{14}C]-acetate and [^{32}P]-phosphate into phospholipids
of sycamore cell suspensions
R. Bligny, F. Rebeille-Borgella and R. Douce 387

Biosynthesis of the saturated very long chain fatty acids in higher
plants from exogenous substrates
C. Cassagne and R. Lessire 393

Lipid formation in olive fruit (*Olea europea* L.)
A. Cherif, A. Drira and B. Marzouk 399

Lipid synthesis in imbibed fern spores
A.R. Gemmrich 403

Ultrastructural sites involved in petroselinic acid (C18: 1Δ^6)
biosynthesis during ivy seed (*Hedera helix* L.) development
M. Grosbois and P. Mazliak 409

Glycerolipid synthesis in the leaves of *Vicia faba* and *Hordeum vulgare*
treated with substituted pyridazinones (San 9785, San 9774 and
San 6706)
M. Khan, D.J. Chapman, N.W. Lem, K.R. Chandorkar and J.P. Williams 415

Calcium inhibition of phospholipid biosynthesis in a calcicolous
plant (*Vicia faba*); Comparison with a calcifuge one
(*Lupinus luteus*)
A. Oursel 421

Control of fatty acid and lipid formation in Baltic marine algae
by environmental factors
P. Pohl and F. Zurheide 427

On the fatty acid biosynthesis in carrot root. An indirect
approach
J. Soimajärvi 433

Oleyl CoA and linoleyl CoA desaturase activities and α-linolenic
acid biosynthesis in sub-cellular fractions from young pea leaves
A. Tremolieres, J.P. Dubacq, M. Muller, D. Drapier and P. Mazliak 437

Purification of choline kinase from soya bean
J. Wharfe and J.L. Harwood 443

Synthesis and turnover of molecular species of galactolipids of
Vicia faba leaves
J.P. Williams and S.P.K. Leung 449

Author index 455
Subject index 457

GENETICS AND BIOSYNTHESIS OF PLANT EPICUTICULAR WAXES

PENNY von WETTSTEIN-KNOWLES

Institute of Genetics, University of Copenhagen, Øster Farimagsgade 2A, DK-1353 Copenhagen K (Denmark) and
Department of Physiology, Carlsberg Laboratory, Gl. Carlsberg Vej 10, DK-2500 Copenhagen Valby (Denmark)

INTRODUCTION

On the outer cuticular surfaces of plants, very long chain lipid molecules are often present. These have been called the epicuticular waxes. They are easily isolated by dipping a portion of the plant into a solvent such as chloroform for a few seconds and then evaporating the solvent. The rapid identification and quantification of almost all of the lipids present in only a few mg of wax has been made possible by the development of techniques such as thin layer chromatography, gas liquid chromatography and especially gas chromatography-mass spectrometry. Amid the wide range of relatively non-polar lipids which are constituents of plant epicuticular waxes (see 1), the alkanes, esters, β-diketones, aldehydes, primary and secondary alcohols, ketones, free acids and triterpenes have been the most intensively studied. A great diversity is observed in the number of lipid classes present in a wax and also in their relative abundance among plant species. Even within a single plant, the various cuticle surfaces are often characterized by different wax compositions, for example, in barley β-diketones and hydroxy-β-diketones form 50% of the wax on the spikes and 74% on the uppermost internodes and leaf sheaths, but they are absent in the leaf blade wax which may contain as much as 83% of the primary alcohols [2,3,4]. When secondary alcohols are present they are restricted to the awns[5].

In many wax lipid classes, chains longer than 20 carbons are present and form a series in which either even or odd numbers of carbons predominate. The data in Table 1 show that the chain length composition of a lipid class on different parts of a plant may also vary. In fact, two adjacent epidermal cells may have different wax covers. For instance the gl_1 mutation in *Zea mays* inhibits formation of wax structures over the cells of the leaf lamina, excepting the accessory cells of the stomates[6]. This indicates that the synthesis and deposition of epicuticular waxes is cell specifically determined.

In the following pages selected facets of plant epicuticular wax biosynthesis will be considered. These will include a brief summary of the biosynthetic

TABLE 1

COMPOSITION OF NORMAL ALKANES FROM DIFFERENT ORGANS OF 8-WEEK-OLD BONUS BARLEY (WT %)

	C 23	25	27	29	31	33	35
Leaf blades	1	7	7	9	20	55	2
Total spikes	2	2	3	20	70	4	

relationships among the alkanes, secondary alcohols, ketones, aldehydes, primary alcohols, free fatty acids and esters arising from the latter two. I shall not reiterate the data supporting these relationships found in other reviews (e.g. see 7-9). Instead, my goal is to update, to add overlooked results, to reinterpret some observations and to point out where additional work is needed. Secondly, the *eceriferum* (*cer*) mutants of barley which affect wax synthesis and deposition will be introduced and their contribution to enlarging the biosynthetic pathways to include the β-diketones, hydroxy-β-diketones and alkan-2-ol containing esters detailed. Finally, I will discuss the relationship among the elongating systems and their role in the biosynthesis of long chain lipids present in plant waxes.

THE ELONGATION, ELONGATION-DECARBOXYLATION AND ELONGATION-REDUCTION PATHWAYS

By the mid-1930's several different hypotheses to explain the origin of some wax classes had been put forth on the basis of compositional analyses[10-12]. During the past 14 years, some of the wax biosynthetic relationships have been unravelled. As summarized in Figure 1, C_{16} fatty acids are elongated in the epidermal cells to specific chain lengths by the addition of C_2-units from malonyl-CoA. That two or more elongation systems are involved in the formation of an equivalent number of fatty acyl pools is indicated by the double lines. Before arriving on the cuticle surface, these pools of fatty acyl chains may or may not serve as substrates for different pathways. That is, they can be (i) decarboxylated to yield alkanes. Oxidation of specific carbons along the chain then gives rise to secondary alcohols, some of which may be subsequently oxidized to ketones. (ii) Reduced to aldehydes and alcohols. Certain of the latter may then be esterified to give alkan-1-ol containing esters. (iii) Released from the elongation system(s) as free fatty acids. In practice the latter can be regarded as leftovers, that is, they are the products of the elongation system(s) which did not enter either the decarboxylative or reductive pathways. Beneath the name of the resulting wax classes are noted chain lengths commonly found. These pathways explain the origin of seven wax classes including those that were the subject of the early speculations.

Much of the experimental evidence supporting the relationships shown in Figure 1 was obtained from tracer experiments using intact plant organs or tissue slice systems. In these studies the use of the above mentioned identification techniques proved indispensable. Especially in latter years the use of tracers has been combined with attempts to alter wax composition via genetic manipulation and/or the use of inhibitors. These latter two approaches are

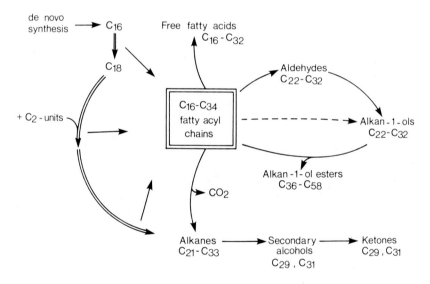

Fig. 1. Biosynthetic relationships among epicuticular wax lipids arising via the elongation-decarboxylation pathway (alkanes, secondary alcohols, ketones), the elongation-reductive pathway (aldehydes, primary alcohols, alkan-1-ol containing esters) and elongation only pathway (free fatty acids)[8,9,13,14].

based on the concept that blocking an enzyme in a metabolic sequence can cause the accumulation of the immediate substrate or an earlier intermediate in that metabolic pathway. However, if alternate pathways for the metabolism of a preceeding intermediate exist, the block may shift the precursors into another pathway. In practice this means that in attempting to pinpoint the site of action of a gene or chemical inhibitor, the effect on all preceeding intermediates and on those compounds in the possible branch pathways should be assayed. In general, while lesions in a structural gene are normally specific for a single enzyme, chemical inhibitors often act on more than one enzyme.

Many mutations have been isolated in higher plants that influence epicuticular waxes. Their effect on wax composition and/or amount, however, has been ascertained in relatively few cases. Of these, those that affect the biosynthesis of the wax lipid classes shown in Figure 1 are listed in Table 2 together with their site of action if known.

Genes affecting synthesis of branched lipids

Branched compounds are normally found in addition to members of the normal series only in the free and ester alcohols as well as ester acids of the *Brassicas*[15,16,18,20,28]. This leads to the inference that either branched aldehydes are not intermediates of branched alcohols or that if synthesized they are not transported to the cuticle surface. Thus in Figure 1 a dashed line is included directly from the elongated fatty acyl chains to the primary alcohols. The primary effect of the gl_6K mutant (Table 2) is the appearance of anteiso members in the alkanes, free acids, aldehydes and possibly the secondary alcohols[15,16]. The question of how the gl_6K mutation brings this about will be considered in the last section. Two deductions were drawn from these experimental results. Firstly, while some of the aldehydes in plant waxes arise via the reductive pathway as indicated in Figure 1, another mechanism more closely allied to that yielding the alkanes may also function. This point certainly deserves investigation. Secondly, more than one parallel elongation system can function in epicuticular wax biosynthesis *in vivo*.

Blocks in elongation and the decarboxylation pathway

Chemical analyses of the waxes on gl_3K, gl_4K, gl_2SC, gl_5C, *wa* and *wsp* plants had been made by 1970. Although unrecognized at the time, these analyses provided *in vivo* evidence for the elongation-decarboxylation origin of the alkanes.

TABLE 2

SITES OF ACTION OF GENES INFLUENCING THE COMPOSITION AND/OR AMOUNTS OF THE WAX CLASSES INCLUDED IN FIG. 1[a]

Plant species	Branching	Decarboxylation pathway	Elongation	Other[b]
Brassica oleracea[c]				
Marrow stem kale	$gl_6K(15,16)$[d]	$gl_4K(15,18,19)$	$gl_3K(15,18)$	
Brussel sprout		$gl_2SI(20)$	$gl_2SC(18)$	$gl_4S(20)$
			$gl_1S(20)$	
			$gl_3S(20)$	
Cauliflower			$gl_5C(18,19)$	
Sprouting broccoli				gl_1B(see 15)
Brassica napus[c]				
Rape	gl_nR[e](17)	$gl_rR(17)$		
Pisum sativum		$wsp(21,22)$	$wa(21,22)$	$was(21)$
				$wlo(22)$
				$wb(21,22)$
Zea mays[c,f]			$gl_2M(24)$	$gl_1M(25)$
				$gl_3M(25)$
				$gl_7M(25)$
Hordeum vulgare		$cer\text{-}^{soh}(5,23)$		$cer\text{-}j^{59}(13,26)$
				$cer\text{-}p^{37}(26)$

[a] Includes those genes whose effect on wax chemistry has been analyzed.
[b] See text.
[c] A capital letter has been appended to each gl gene symbol to indicate which plant it occurs in (see 14). A second capital letter is added to the two gl_2S mutants to distinguish the variety of origin; that is, I for Irish Elegance and C for Cluseed.
[d] References
[e] The glossy mutant in Nilla rape is designated gl_nR.
[f] The 4 gl maize mutants represent different genes (27).

Decarboxylation blocks. Good phenocopies of the decarboxylation mutants gl_4K, gl_2SI, gl_rR and wsp (Table 2) can be obtained by treating with the thiols 1,4-dithioerythritol (DTE) and 1,4-dithiotreitol (DTT)[14,29,30]. In investigating the mechanism by which elongated fatty acyl chains enter the decarboxylative pathway, both the thiols and the genes should prove useful tools. This is an important facet of alkane biosynthesis about which little is known at present. That an α-oxidation mechanism plays a role was suggested in 1974 on the basis of results of experiments using epidermal extracts from pea leaves[31]. The ability of the extracts to convert α-hydroxy C_{32}- and C_{32}-fatty acids to C_{31} alkanes in the presence of various cofactors and inhibitors was studied. Interestingly, the recently reported α-oxidation of very long chain fatty acids in yeast[32] was not inhibited by DTT, whereas alkane biosynthesis was. Whether or not the three decarboxylation mutants identified in *Brassica* represent mutational events of one or more genes is unknown. Making the necessary genetic crosses among these three mutants as well as among additional decarboxylation mutants that could be easily selected for would give an indicator of the complexity of this process.

Only one gene has been identified that acts subsequent to alkane formation in the decarboxylative pathway. This is *cer-*^{soh} which controls introduction of the alcohol group into the carbon chain.

Elongation blocks. The genes listed in Table 2 as functioning in elongation act on one or more of the terminal elongation steps. The inhibition of $C_{30} \rightarrow C_{32}$ elongation is almost complete in gl_2M and *wa*. All the *Brassica* mutants act on the $C_{28} \rightarrow C_{30}$ step except for gl_3K which blocks $C_{26} \rightarrow C_{28}$. While the latter is allelic with gl_5C, the total number of genes represented among the *Brassica* elongation mutants is unknown. By contrast chemical inhibitors have not yet been identified that act only on given terminal elongation steps. Trichloroacetic acid and a number of thiocarbamates (*e.g.*, S-(2,3-dichloroallyl)-diisopropylthiocarbamate and S-ethyldipropylthiocarbamate), however, have been found to inhibit fatty acid elongation in various ways with the total effect being most marked on the longest chains[19,33-37]. Internal lipids are not affected.

Inhibitors and parallel elongation systems

Inhibitors have been used to demonstrate that various elongation systems play a role in lipid biosynthesis. As is well established arsenite effectively blocks the $C_{16} \rightarrow C_{18}$ elongase, but does not inhibit the *de novo* synthesis of palmitate[38-41]. If parallel elongation systems function in wax biosynthesis as was alleged from the study of gl_6K, then an inhibitor should be found to discriminate among them. To draw meaningful conclusions from such experiments necessiates that all the lipid classes are derived from fatty acyl chains having the same length. Thus *Zea mays* or *Bulnesia retamo*, which utilize C_{32}

chains as precursors for both the decarboxylative and reductive pathways[42,43], would be suitable experimental objects to clarify the action of different concentrations of thiocarbamates which are claimed to inhibit the elongation of precursors in *Pisum* giving rise to the C_{31} alkane, but not elongation yielding the C_{26} and C_{28} primary alcohols[36].

Inhibitors and sequential elongation systems

In barley spike waxes at least three sequential elongation systems are involved in the synthesis of the elongated fatty acyl chains used as precursors for the even chain length wax lipids. In these experiments the spikes were allowed to take up various inhibitors before they were fed labelled acetate. As shown in Table 3, the decarboxylation inhibitor DTT caused an accumulation of label in the $C_{28} \rightarrow C_{32}$ chains which must serve as precursors for the alkanes (Table 1). Cyanide inhibited $C_{26} \rightarrow C_{28}$ elongation since the two groups of shorter chains contained increased amounts of label while the longest group had less compared to the control. Feeding of mercaptoethanol or arsenite produced marked accumulations of label in the C_{18} and C_{20} chains indicating that these two chemicals blocked elongation of $C_{20} \rightarrow C_{22}$. These results imply that the $C_{20} \rightarrow C_{22}$ elongating system must be in some respects similar to the $C_{16} \rightarrow C_{18}$ one which is also inhibited by arsenite.

Results of recent experiments[44] with microsomal preparations from leek epidermal cells led to the conclusion that one system was responsible for $C_{18} \rightarrow C_{20}$ elongation and another for $C_{20} \rightarrow C_{30}$. This conclusion was based on the differential effect of various cofactors on the incorporation of label into C_{20} vs $C_{22} \rightarrow C_{30}$ fatty acids from $(1-{}^{14}C)$-stearic acid.

Other genes influencing epicuticular waxes

The sites of action of the ten genes in the last column in Table 2 cannot be pinpointed from the available chemical data, although some suggestions have been made. For example, the elongation process in the $C_{26} \rightarrow C_{32}$ range might be affected by gl_3M. Reduction of available precursors for wax synthesis is mentioned as a mode of action of wlo and $cer-p$[37]. A failure to synthesize and use C_{26} chains in the reductive pathway is suggested for $cer-j$[59], while gl_1B is thought to be defective in the esterification process.

BIOSYNTHESIS OF THE β-DIKETONE LIPIDS AND ESTERS CONTAINING ALKAN-2-OLS

Three of the lipid classes present in some barley waxes whose mechanism of synthesis are not accounted for by the relationships in Figure 1 are the β-

TABLE 3

EFFECTS OF PRE-INCUBATIONS OF INHIBITORS ON THE INCORPORATION OF $(2-^{14}C)$-ACETATE INTO THE EVEN CHAIN LENGTH LIPIDS OF INTACT $cer-u^{69}$ BARLEY SPIKE WAX[a]

Inhibitor	(μmoles)	C_{16-20}	CPM x 10^{-3} C_{22-26}	C_{28-32}
None		82	127	391
Dithiothreitol	(36)	140	122	589
Cyanide	(60)	193	287	178
Mercaptoethanol	(60)	315	141	366
Arsenite	(6)	392	158	193

[a] Includes free acids, primary alcohols, aldehydes and acid moieties of the short chain esters but not alcohol or acid moieties of the long chain esters. Data from reference 14.

diketones, hydroxy-β-diketones and esters containing alkan-2-ols. The major β-diketone is hentriacontan-14,16-dione (97%) while the other 3% is almost entirely tritriacontan-16,18-dione[2,45]. The hydroxyl group of the hydroxy-β-diketones is on carbon-25 of hentriacontan-14,16-dione[3]. The alkan-2-ol containing esters have chain lengths of C_{31}-C_{37} in which C_{16}-C_{22} fatty acyl chains are esterified primarily with the short C_{13} and C_{15} alkan-2-ols. The latter are not found free in the wax[46]. While a number of speculations about the origin of the β-diketones (but not the esterified alkan-2-ols) can be found in the literature[9,47], none of them resemble the mechanism suggested by the experimental evidence from barley. In these studies the availability of a number of mutants affecting the relevant pathways has proved an invaluable tool.

<u>Isolation of genes with blocks in selected steps of epicuticular wax biosynthesis</u>

Ultrastructural examination of epicuticular waxes by transmission or scanning electron microscopy has verified de Bary's conclusions that these waxes exist in a striking diversity of morphological forms[48]. The first technique although time consuming reveals far greater detail of wax body structure than the second which is useful for rapidly scanning large surface areas. Compare

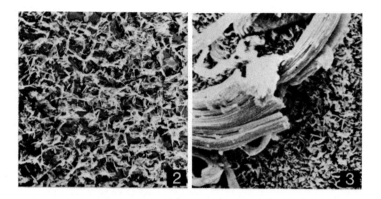

Fig. 2. The characteristic lobed plates forming the leaf waxes on Bonus barley are revealed by carbon replicas examined in a transmission electron microscope[2]. x 7,000.

Fig. 3. A scanning electron microscope (Courtesy Geology Department, University of Copenhagen) was used to prepare this picture of the two types of wax structures present on sugar-cane leaf sheaths. x 8,000.

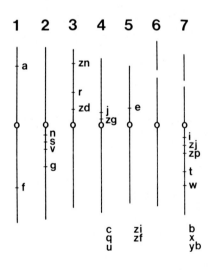

Fig. 4. Location of 17 *cer* genes on the barley chromosomes. Below chromosomes 4, 5 and 7 are listed an additional eight *cer* genes assigned to these chromosomes[60-64].

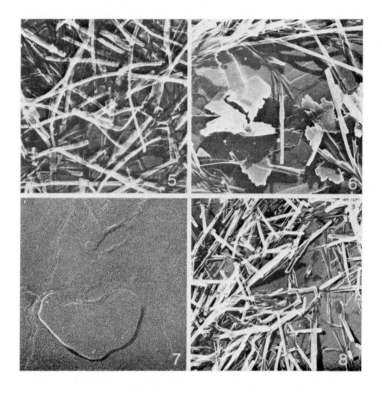

Figs. 5 to 8. Wax coatings on barley spike lemmas as revealed by transmission electron micrographs of carbon replicas[2,55]. 5) Long thin tubes in the wild type Bonus; lightly covered area. x 11,000. 6) Both long thin tubes and very thin, upright plates in $cer-u^{69}$. x 7,000. 7) Thin sheets lying on the cuticle surface in $cer-cu^{108}$. x 16,000. 8) Fewer, shorter tubes in $cer-i^{16}$; densely covered area. x 8,000.

the small plates in Figures 2 and 3. In a number of cases a particular wax structure can be correlated with wax chemistry[49-54]. For instance leaf waxes containing large amounts of primary alcohols are characterized by the presence of lobed plates (Fig. 2), whereas lemma waxes with β-diketones have long thin tubes (Fig. 5). Since changes in wax composition often alter the wax structures which thereby influences light refraction from the plant surface, mutations in genes controlling synthesis and/or deposition of wax lipids can be selected for. In barley 1123 mutations, designated *eceriferum* (*cer*) have been isolated and allocated to 65 loci[56-58]. The chromosomal location of approximately one third of these genes (Fig. 4) has been determined[60-64]. Selection of mutations in 45 of the genes was made because the blue-grey color of the wild type spikes was altered to a yellow-green. At the ultrastructural level various modifications such as shown in Figures 6, 7 and 8 had occurred. Chemical examination of the spike waxes, from the 28 genes studied thus far, revealed quite expectedly that the synthesis of the β-diketones at least was affected. We do not expect to isolate mutants blocked only in the reductive or decarboxylative pathway, as the resulting lipid classes do not contribute significantly if at all to the blue-grey color. The recent demonstration that the internode waxes of *Sorghum* may consist of as much as 90% free fatty acids[65], suggests that this would be a good plant to select for genes involved in elongation.

Exploitation of the *cer* mutants

Barley waxes containing esterified alkan-2-ols were observed to be the same ones having the β-diketone lipids. Since all the other lipid classes are present on all organs, this suggested that the esterified alkan-2-ols and β-diketones were synthesized from a common precursor[3]. This idea was substantiated by examining the amounts of these lipids in waxes from the relevant organs of a number of *cer* mutants. Most of the mutations examined thus far affect synthesis of these lipids in an approximately parallel manner (Table 4). Some such as *cer-q*[42] create a total block, others such as *cer-a*[6] are slightly leaky letting through traces of all three lipids, while the remainder exemplified by *cer-i*[16] allow substanial albeit reduced formation of the three lipids. Two mutants deviate from this pattern (Table 4). A block of hydroxy-β-diketone formation with a proportionate increase in the amount of β-diketones but no alteration of the relative quantities of the two ester types infers that the *cer-u*[69] mutation inhibits insertion of the hydroxyl group into the β-diketone chain[2,3]. A total inhibition of β-diketone lipid synthesis accompanied by a stimulation of alkan-2-ol ester synthesis results from the presence of *cer-c*[36].

TABLE 4

AMOUNTS OF β-DIKETONES, HYDROXY-β-DIKETONES AND ALKAN-2-OL CONTAINING ESTERS IN WAXES OF 5 *cer* MUTANTS AND THEIR WILD TYPE BONUS[a]

Plant part	Genotype	Diketones, % of wax		% of esters containing alkan-2-ols
		β	OH-β	
Total spike	Bonus	24	26	36
	$cer-i^{16}$	15	17	18
	$cer-a^{6}$	tr	tr	2
	$cer-q^{42}$	–	–	0
	$cer-c^{36}$	–	–	59
	$cer-u^{69}$	47	3	35
Spike minus awns	$cer-u^{69}$	+++	–	57
Awn	$cer-u^{69}$	–	–	0

[a] Data from reference 3 where methods used for quantitating the two types of β-diketones and esters containing alkan-2-ols are given. Qualitative observations from TLC separations: tr = trace, – = absent, +++ = more than in the wild type.

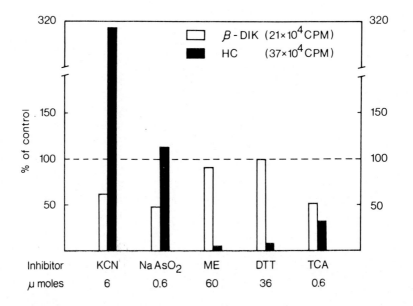

Fig. 9. The amount of $(2-^{14}C)$-acetate label incorporated into the β-diketones and alkanes of $cer-u^{69}$ barley spike epicuticular wax can be changed by preincubations with five inhibitors[30]. ME, 2-mercaptoethanol; TCA, sodium trichloroacetate. Label in control given in upper right corner.

These observations suggest that $cer\text{-}c^{36}$ acts early enough to block synthesis of the β-diketone lipids but late enough to direct precursors destined for the β-diketones toward the alkan-2-ol containing esters. That these are the sites of action of the $cer\text{-}c$ and $\text{-}u$ loci is supported by the observation that no significant changes occur in the distribution of chain lengths within the other wax classes[23].

Simultaneously with the analyses described above, the question of the biosynthetic origin of the β-diketone's carbon skeleton was being probed. $(1\text{-}^{14}C)\text{-}$ and $(2\text{-}^{14}C)$-acetate were fed to spikes of $cer\text{-}u^{69}$ and the β-diketones isolated. Barley plants of this genotype rather than the wild type have been used in all biochemical investigations dealing with β-diketone synthesis, since larger amounts of β-diketones can be isolated more easily. Thereafter, the labelled hentriacontan-14,16-dione was degraded and the specific activity of carbons-6 through 26 determined. The data showed that the most likely mode of synthesis of this molecule was via elongation starting from the C-31 end as follows[66]. C_2-units are added to short precursors to give a chain with 12 carbon atoms. During the subsequent elongation to 16 carbon atoms, the four added carbons were unexpectedly uniformly labelled. Finally C_2-units are added to give the complete molecule.

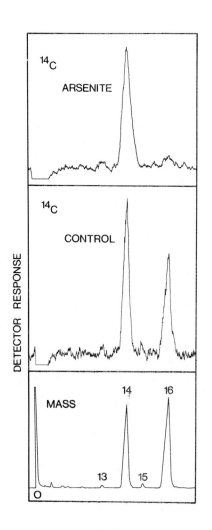

Fig. 10. Effect of a pre-incubation with arsenite on the incorporation of label from $(2\text{-}^{14}C)$-acetate into the two ends of hentriacontan-14,16-dione[30]. Alkaline degradation yields a C_{14} fatty acid from carbons-1 to 14 and a C_{16} fatty acid from carbons-16 to 31 which are then separated via radio gas liquid chromatography.

TABLE 5

INCORPORATION OF LABELLED ACIDS INTO β-DIKETONES AND ALKANES BY TISSUE SLICES PREPARED FROM cer-u^{69} SPIKES MINUS AWNS[a]

(1-^{14}C)-	dpm x 10^{-3}	
	β-diketones	Alkanes
Acetate	30	50
Lauric acid	208	31
Myristic acid	284	47
Palmitic acid	236	40
Stearic acid	tr	41

[a] Data from references 30 and 45

Combined with the chemical analyses of Bonus and cer mutant waxes, the above results infer that the β-diketones arise via a different elongation system than that giving rise to the alkanes[3]. Corroboration of this hypothesis comes from two types of experimental results. Firstly, preincubations with inhibitors influence the ability of cer-u^{69} spikes to incorporate (2-^{14}C)-acetate into the various epicuticular lipid classes. Figure 9 illustrates the markedly different responses of the β-diketones and alkanes to four of the five tested inhibitors[30]. Secondly, when the β-diketones from spikes fed arsenite followed by (2-^{14}C)-acetate were isolated and degraded, label was found in the C-1 end but not the C-31 end of hentriacontan-14,16-dione (Fig. 10). This is expected given that arsenite blocks $C_{16} \rightarrow C_{18}$ elongation and that the β-diketone is made via elongation starting from the C-31 end and proceeds toward the C-1 end. The label incorporated into the β-diketone chain comes from addition of labelled C_2-units to C_{18} or longer chains synthesized before addition of the arsenite[30].

To determine at what point the elongation systems leading to the β-diketones and alkanes diverge, long chain fatty acids were fed to tissue slices prepared from the spikes minus awns of cer-u^{69}. The results in Table 5 imply that the last common intermediate is a C_{16} acid[30,45]. When the CoA derivatives of palmitic and stearic acid were used as precursors, the same results were obtained except that the efficiency of labelling was increased several-fold[45].

Figure 11 summarizes schematically all the available experimental results pertaining to the biosynthetic relationships among the β-diketones, hydroxy-β-diketones and esterified alkan-2-ols. The elongation systems leading to the other wax classes are indicated, as in Figure 1, by the double arrows extending

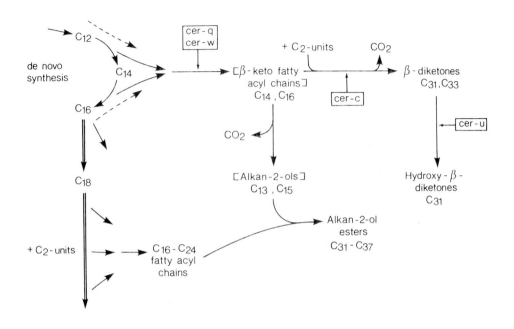

Fig. 11. Biosynthetic relationships among the β-diketones, hydroxy-β-diketones and alkan-2-ol containing esters present in barley epicuticular waxes[3,14]. The sites of action of four *cer* genes affecting only these wax classes are shown. Brackets indicate hypothetical intermediates.

downward from the C_{16} acid, the apparent point of divergence of the β-diketone and alkane elongation systems. Two types of evidence indicate that a C_{16} acid need not always be the starting point for the β-diketone elongation system. Firstly, the carbonyl groups can occur at different distances from the ends of the carbon chains; for example, hentriacontan-10,12-dione occurs in *Poa colensoi*[49], tritriacontan-12,14-dione in *Festuca glauca*[67], hentriacontan-8,10-dione in *Buxus sempervirens*[68] and nonacosan-8,10-dione in many *Rhododendrons*[69]. Secondly, more than 50% of the label in β-diketones synthesized from (1-^{14}C)-myristic acid by tissue slices of *cer-u*[69] spikes minus awns was found in nonacosan-14,16-dione[45]. Apparently, therefore, the barley enzyme can also channel C_{14} fatty acyl chains if available into the β-diketone elongation system.

A proposed mechanism is diagrammed in Figure 12 by which a fatty acyl chain channeled into the β-diketone elongation system eventually becomes a β-diketone[30]. For illustrative purposes a C_{16} fatty acid may be chosen as the start-

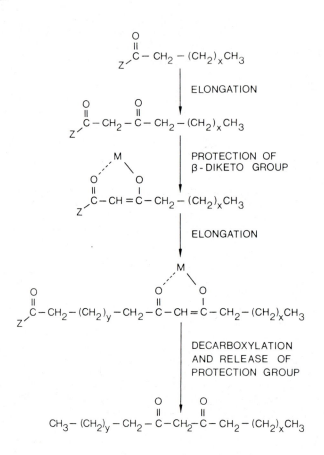

Fig. 12. A proposed pathway for the biosynthesis of β-diketones[30]. X and Y = odd numbers; for hentriacontan-14,16-dione X = 13 and Y = 11. Z = ACP or CoA. M = a metal ion such as Cu or Fe.

ing compound for elongation via the addition of C_2-units. Twice in succession the mechanism of the C_2-unit addition is thought to be modified at the β-keto reductive step. Such variations of *de novo* fatty acid synthesis are known. For example, a change at the dehydrase step in *E.coli* results in synthesis of unsaturated fatty acids. In the present case the β-keto group is envisaged as complexing with the prosthetic group of a metal containing enzyme. Two carbonyl groups can thereby be protected and retained until chain elongation is finished. Release of the protection group and a decarboxylation would yield a completed β-diketone molecule. The decarboxylation mechanism can not be identical to that forming the alkanes as β-diketone synthesis is not inhibited by DTT or ME. This proposal for the origin of the carbonyl groups in the β-diketone molecule, which is consistent with all available experimental data, is in marked contrast to the insertion of an hydroxyl group into a preformed chain as occurs in secondary alcohol synthesis (Fig. 1). The most pertinent areas for investigation arising from the proposal of Figure 12 include (i) the nature of the C_2-unit donor, (ii) the mechanism of the decarboxylation and (iii) the mechanism by which the carbonyl groups are protected and retained in the chain during elongation.

The introduction of the hydroxyl group into the β-diketone chain to form an hydroxy-β-diketone (Fig. 11) may proceed via a mechanism similar to that described for the secondary alcohols of the *Brassicas*[70,71]. Likely, however, more than one mechanism for hydroxyl introduction into a preformed β-diketone chain exists[3,72]. Certainly this is true if one includes the recognition of specific carbons in the chain as part of the mechanism. In different *Gramineae* the hydroxyl group has been found in the carbon-1 end of the molecule as a mixture of isomers, in the carbon-31 end on position 25, and in the carbon-31 end as a mixture of the 25 and 26 positional isomers (see 3, 72). Similar conclusions can be drawn from the knowledge that the hydroxyl group of the secondary alcohols occurs at various positions along the chain in different waxes[1,54]. Specific oxidation of given hydroxyl groups in the β-diketone chain would explain the biosynthetic origin of recently identified compounds such as 25-oxohentriacontan-14,16-dione[72].

In Figure 11 an alternative fate for the β-keto fatty acyl chains about to enter the β-diketone elongation system is shown as a decarboxylation accompanied or followed by a reduction of the original β-keto group to an alcohol group. The hydroxyl group is then esterified to fatty acyl chains to form the esters containing alkan-2-ols[3]. Whether or not the fatty acyl chains come from the same pool as that used for synthesis of the alkan-1-ol containing esters (Fig. 1) is unknown. The acyl moieties of both types of esters are of the same lengths although the chain length distributions are quite different[46]. The most suitable material for experiments to test the validity of the proposed synthetic origin of the esterified alkan-2-ols (Fig. 11) are spikes of *cer-c*[36] in which esters are the major wax class and the β-diketone lipids are absent (Table 4). Initial results of tracer experiments are in accord with the hypothesis, for example, $(1-^{14}C)$-myristic acid is a very efficient precursor of pentadecan-2-ol[23].

The *cer-cqu* region

Three of the genes included in Figure 11, namely *cer-c*, *-q* and *-u*, are very interesting. Genetic analyses[73] have shown that they lie within .03 map units of one another on chromosome 4. Their linear order is still unknown. Further-

TABLE 6

ORIGIN OF THE SINGLE AND MULTIPLE MUTATIONAL EVENTS AT THE *cer-cqu* REGION[a]

Mutagenic agent	Number of events involving *cer*-						
	c	*q*	*u*	*c+u*	*q+u*	*c+q*	*c+q+u*
Neutrons	7	5	1	2	2	1	6
γ-rays	8	5	5				1
Ethyl methanesulfonate	51	32	48	1			
Ethyleneimine	28	20	24				
Other	49	52	28				
Total	143	114	106	3	2	1	7

[a] Data from references 50, 59 and 74

more, the only apparent double or triple mutational events of *cer* loci that have been uncovered during the allele testing of the 1123 *cer* mutants all involve these genes[56-59]. As summarized in Table 6, all possible pairwise as well as triple mutational events have been identified. Combined they account for 3.5% (13/376) of the mutational events involving the *cer-cqu* region. Two possible interpretations of these multiple phenotypic changes are possible. Firstly, they may indeed be the result of two or three simultaneous mutations. This would be supported by the fact that 11 of the 13 multiple events were induced by neutrons. Secondly, they may be the result of single mutational events in a cluster-gene. That is, the *cer-cqu* region codes for a single polypeptide having at least three enzymatic functions or domains. Single mutational events would then have the possibility of affecting one or more domains (see 75). Two such cluster-genes, *fas*-1 and *fas*-2, determine the fatty acid synthetase complex in yeast. Each codes for the synthesis of a multifunctional polypeptide[76-78]. The fatty acid synthetase consists of a dimer made from these two proteins. This is illustrated schematically in Figure 13. The only feasible genetic method in barley to determine whether the multiple phenotypic changes involving *cer-c*, *-q* and *-u* are the result of single or multiple mutational events is via back mutation experiments using sodium azide. Such experiments have been initiated.

BIOSYNTHESIS OF WAX LIPID CLASSES

De novo fatty acid synthesis can be described as the process whereby to an activated primer a number of short carbon chains donated by an activated

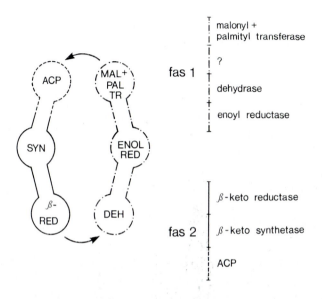

Fig. 13. The cluster genes *fas*-1 and *fas*-2 of the yeast *Saccharomyces cerevisiae* code for two multifunctional polypeptides which combine into the fatty acid synthetase complex for *de novo* synthesis of fatty acids. The linear order of the genetic information for the enzyme activities within the *fas*-1 polypeptide, as shown on the gene map, do not correspond to the known sequence of enzymatic reactions. Adapted from 76-78.

elongating agent are successively added by a group of enzymes until a given chain length, most frequently C_{16}, is attained. Numerous variations of the process are known. They have been identified in the systems contributing the primer, donating the short carbon chains, carrying out the actual elongation and terminating the process. Furthermore, two distinct types of molecular organization are known for soluble enzymes carrying out the elongation, namely, aggregated and non-aggregated. The enzymes may also be membrane bound. Despite all this diversity, the total chemical result is remarkably similar among all investigated organisms or parts thereof. While considerably less is known about the systems which elongate preformed chains, these systems most likely have many facets in common with *de novo* synthesis, especially with regard to the group of enzymes doing the elongating. Thus, given that a number of parallel as well as sequential elongation systems take part in synthesis of wax lipid classes, these systems can be expected to share some of their enzymes or subunits thereof. Experimental evidence to support this contention can be derived from the analyses of waxes on the *cer* mutants. For example, the three mutants *cer*-968, -a^6 and -b^4 not only block synthesis, either completely or almost so, of the β-diketone lipids and the esterified alkan-2-ols, but they also affect the chain length distributions of varying numbers of the other wax classes. Data for the alkanes and primary alcohols, which represent two lipid classes whose predominant chain lengths arise from different sequential elongation systems, are presented in Tables 7 and 8, respectively. In wax from *cer*-968 the primary alcohols are affected but the alkanes are not. The significant alterations in the alkane distribution caused by *cer*-b^4 and -a^6 are accompanied by lesser changes in the alcohols.

Since a number of different elongating systems contribute the fatty acyl chains found in the wax lipids, the question can be asked where these systems are located within the epidermal cells which are known to be the major site of wax synthesis[79,80]. Results from studies summarized in Table 9 have been used to support the hypothesis that these elongating systems are microsomal. This deduction is reasonable for the results with leek. Whether or not the elongating systems detected in the pea and potato, however, are those giving rise to the wax lipids is questionable, since the synthesized fatty acyl chains are shorter than those characteristic for wax lipids, excepting the acid moieties of the esters. In addition, whether or not wax lipids other than free and esterified acids were synthesized simultaneously was not investigated. The microsomal preparation from the germinating peas, however, was also shown to carry out *de novo* synthesis of C_{16} and its elongation to C_{18}. Thus, at least three enzyme systems synthesizing saturated fatty acids were present in this microsomal preparation.

Recently the analysis of the leek system has been extended[44,86]. The crude microsomal pellet was subdivided into fractions enriched in either endoplasmic reticulum or plasmalemma membranes. While both fractions synthesized C_{20}-C_{30} fatty acids using (1-^{14}C)-stearyl CoA as a primer and malonyl CoA as donor of C_2-units, the endoplasmic reticulum was much more efficient. Similar results were obtained with yeast[87]. In the yeast study the long chain alkanes (C_{22}-C_{33}), however, were specifically located in the plasmalemma enriched fraction. It is tempting to speculate from these experiments that all the elongating and decarboxylating systems involved in alkane synthesis are located in the plasmalemma.

In this connection the results of a study of a microsomal preparation from dark grown *Euglena* are of interest[88-90]. All the enzymatic machinery necessary for *de novo* synthesis as well as for reduction of the acids and for ester formation was present in the microsomes. Since the reductase and transacylase

TABLE 7

COMPOSITION OF NORMAL ALKANES ON SPIKES OF BONUS BARLEY AND THREE cer MUTANTS (WT %)

Genotype	C 23	25	27	29	31	33
Bonus (+)	2	2	2	20	70	4
$cer-^{968}$	3	3	3	14	69	7
$cer-b^4$	2	3	8	51	35	1
$cer-a^6$	27	20	16	23	12	1

TABLE 8

COMPOSITION OF PRIMARY ALCOHOLS ON SPIKES OF BONUS BARLEY AND THREE cer MUTANTS (WT %)

Genotype	C 22	24	26	28	30	32
Bonus (+)	1	7	53	17	12	10
$cer-^{968}$	1	3	88	6	1	1
$cer-b^4$	2	6	66	19	7	1
$cer-a^6$	5	16	56	11	9	2

TABLE 9

SYNTHESIS OF LONG CHAIN FATTY ACYL CHAINS FROM ^{14}C-MALONYL CoA PLUS ENDOGENOUS PRIMERS BY PLANT MICROSOMES

Plant material		Chain length	References
Peas:	cotyledons	20-28	(81)
	germinating	20-24	(38,82)
	epidermal cells of young leaves	20-26	(83)
Potatoes:	tubers	20-24	(84)
Leeks:	epidermal cells of young leaves	20-30	(85)

were not easily solubilized, the conclusion that the fatty acid substrate for the reductase must come from the microsomal rather than the cytoplasmic or mitochondrial *de novo* synthesizing systems was strengthened. Although the chain lengths of the esters are shorter in *Euglena* than those characteristic for plant wax esters, the interesting point is that the enzymes of a reductive pathway are apparently coupled with the system producing its precursor chains. On the other hand, a 100,000 x g supernatant from epidermal cells of pea leaves was capable of converting a ^{3}H-C_{32} fatty acid to a C_{31} alkane with the aid of

Fig. 14. Non-uniform distribution of epicuticular wax structures illustrated with a carbon replica of tubes from a lemma of $cer-i^{16}$. x 9,000.

an α-oxidation system[31]. These results suggested that the elongation and decarboxylation systems forming alkanes are in separate compartments. Other work with α-oxidation systems in plants, however, revealed that they could be either tightly bound or loosely associated with the microsomes[91].

From observations on the diversity of wax bodies, from attempts to reform isolated waxes into the structures found in nature and from the establishment of structural-chemical correlations, the conclusion has been drawn that the wax reaching the cuticle surface may not necessarily have a uniform composition at all the sites on a single cell[5,52,53]. Figure 14 clearly demonstrates a non-random distribution of the β-diketone containing tubes on $cer-i$[16] spike lemmas.

On the basis of the experimental evidence presented and/or discussed hereto, I would like to make the following proposal. Grouped together at sites in the plasmalemma adjacent to the outer wall of plant epidermal cells are a *de novo* fatty acid synthesizing system, all the sequential elongating systems as well as the systems giving rise to a given set of wax lipid classes. Thus one such group yields alkanes and the other lipids of the decarboxylative pathway while another group yields the β-diketones and lipid classes derived therefrom. The relative number of the different groups would in part be responsible for the varying proportions of the wax lipid classes on any given plant surface. This proposal permits a visualization of the action of the gl_6K mutation. That is, the mutation either makes available to the *de novo* system of the group yielding alkanes a branched primer or the *de novo* system is modified to accept a branched primer.

REFERENCES

1. Tulloch, A.P. (1976) in Chemistry and Biochemistry of Natural Waxes, Kolattukudy, P.E. ed., Elsevier, Amsterdam, pp. 235-287.
2. Wettstein-Knowles, P. von (1972) Planta (Berl.), 106, 113-130.
3. Wettstein-Knowles, P. von (1976) Molec. gen. Genet., 144, 43-48.
4. Giese, B.N. (1975) Phytochem., 14, 921-929.
5. Wettstein-Knowles, P. von and Netting, A.G. (1976) Carlsberg Res. Commun., 41, 225-235.
6. Lorenzoni, C. and Salamini, F. (1975) Maydica, 20, 5-19.
7. Martin, J.T. and Juniper, B.E. (1970) The Cuticles of Plants, Edward-Arnold, Edinburgh, pp. 156-171.
8. Kolattukudy, P.E. and Walton, T.J. (1972) Progr. Chem. Fats Other Lipids, 13, 119-175.
9. Kolattukudy, P.E., Croteau, R. and Buckner, J.S. (1976) in Chemistry and Biochemistry of Natural Waxes, Kolattukudy, P.E. ed., Elsevier, Amsterdam, pp. 289-347.

10. Clenshaw, E. and Smedley-McLean, I. (1929) Biochem. J., 23, 107-109.
11. Channon, H.J. and Chibnall, A.C. (1929) Biochem. J., 23, 168-175.
12. Chibnall, A.C. and Piper, S.H. (1934) Biochem. J., 28, 2209-2219.
13. Wettstein-Knowles, P. von (1974) FEBS Letters, 42, 187-191.
14. Mikkelsen, J.D. (1978) Carlsberg Res. Commun., 43, 15-35.
15. Netting, A.G. (1971) Ph.D. Thesis, Univ. New South Wales, Sydney, Australia.
16. Netting, A.G., Macey, M.J.K. and Barber, H.N. (1972) Phytochem., 11, 579-585.
17. Holloway, P.J., Brown, G.A., Baker, E.A. and Macey, M.J.K. (1977) Chem. Phys. Lipids, 19, 114-127.
18. Macey, M.J.K. and Barber, H.N. (1970) Phytochem., 9, 13-23.
19. Macey, M.J.K. (1974) Phytochem., 13, 1353-1358.
20. Baker, E.A. (1974) New Phytol., 73, 955-966.
21. Macey, M.J.K. and Barber, H.N. (1970) Phytochem., 9, 5-12.
22. Holloway, P.J., Hunt, G.M., Baker, E.A. and Macey, M.J.K. (1977) Chem. Phys. Lipids, 20, 141-155.
23. Wettstein-Knowles, P. von, unpublished data.
24. Bianchi, G., Avato, P. and Salamini, F. (1975) Maydica, 20, 165-173.
25. Bianchi, G., Avato, P. and Salamini, F. (1977) Maydica, 22, 9-17.
26. Giese, B.N. (1976) Hereditas, 82, 137-148.
27. Neuffer, M.G., Jones, L. and Zuber, M.S. (1968) The Mutants of Maize, Crop Science Society of America, Madison, pp. 1-74.
28. Baker, E.A. and Holloway, P.J. (1975) Phytochem., 14, 2463-2467.
29. Buckner, J.S. and Kolattukudy, P.E. (1973) Arch. Biochem. Biophys., 156, 34-45.
30. Mikkelsen, J.D. and Wettstein-Knowles, P. von (1978) Arch. Biochem. Biophys., 188, 172-181.
31. Khan, A.A. and Kolattukudy, P.E. (1974) Biochem. Biophys. Res. Commun., 61, 1379-1386.
32. Blanchardie, P. and Cassagne, C. (1976) C.R. Acad. Sc. Paris Ser. D., 282, 227-230.
33. Still, G.G., Davis, D.G. and Zander, G.L. (1970) Plant Physiol., 46, 307-314.
34. Harwood, J.L. (1975) in Recent Advances in the Chemistry and Biochemistry of Plant Lipids, Galliard, T. and Mercer, E.I. eds., Academic Press, London, pp. 43-93.
35. Kolattukudy, P.E., Croteau, R. and Brown, L. (1974) Plant Physiol., 54, 670-677.
36. Kolattukudy, P.E. and Brown, L. (1974) Plant Physiol., 53, 903-906.
37. Bolton, P. and Harwood, J.L. (1976) Phytochem., 15, 1507-1509.
38. Harwood, J.L. and Stumpf, P.K. (1971) Arch. Biochem. Biophys., 142, 281-291.
39. Harwood, J.L. and Stumpf, P.K. (1972) Arch. Biochem. Biophys., 148, 282-290.
40. Kannangara, C.G. and Stumpf, P.K. (1972) Plant Physiol., 49, 497-501.

41. Jaworski, J.G., Goldschmidt, E.E. and Stumpf, P.K. (1974) Arch. Biochem. Biophys., 163, 769-776.
42. Bianchi, G. and Salamini, F. (1975) Maydica, 20, 1-3.
43. Brenner, R.R. and Fiora, J. (1960) Industria Quim. (B. Aires), 20, 531-532.
44. Cassagne, C. and Lessire, R. (1978) Arch. Biochem. Biophys., 190, in press.
45. Mikkelsen, J.D., unpublished data.
46. Wettstein-Knowles, P. von and Netting, A.G. (1976) Lipids, 11, 478-484.
47. Kolattukudy, P.E. (1968) Plant Physiol., 43, 1466-1470.
48. Bary, A. de (1871) Bot. Zeitung, 29, 128 ff.
49. Hall, D.M., Matus, A.I., Lamberton, J.A. and Barber, H.N. (1965) Aust. J. Biol. Sci., 18, 323-331.
50. Lundqvist, U., Wettstein-Knowles, P. von and Wettstein, D. von (1968) Hereditas, 59, 473-504.
51. Hallam, N.D. and Chambers, T.C. (1970) Aust. J. Bot., 18, 335-386.
52. Wettstein-Knowles, P. von (1974) J. Ultrastruct. Res., 46, 483-498.
53. Jeffree, C.E., Baker, E.A. and Holloway, P.J. (1975) New Phytol., 75, 539-549.
54. Holloway, P.J., Jeffree, C.E. and Baker, E.A. (1976) Phytochem., 15, 1768-1770.
55. Wettstein-Knowles, P. von (1971) in Barley Genetics II. Proc. 2nd Int. Barley Genet. Symp. 1969, Nilan, R.A. ed., Washington State Univ. Press, Pullman, pp. 146-193.
56. Lundqvist, U. and Wettstein, D. von (1973) Barley Genetics Newsletter, 3, 110-112.
57. Lundqvist, U. and Wettstein, D. von (1975) Barley Genetics Newsletter, 5, 88-90.
58. Lundqvist, U. and Wettstein, D. von (1977) Barley Genetics Newsletter, 7, 92-96.
59. Lundqvist, U. and Wettstein, D. von, unpublished data.
60. Søgaard, B. (1973) Barley Genetics Newsletter, 3, 57-61.
61. Søgaard, B. (1977) Carlsberg Res. Commun., 42, 35-43.
62. Fester, T. and Søgaard, B. (1969) Hereditas, 61, 327-337.
63. Søgaard, B., unpublished data.
64. Jensen, J. (1978) Barley Genetics Newsletter, 8, 59-60.
65. Bianchi, B., Avato, P., Bertorelli, P. and Mariani, G. (1978) Phytochem., 17, 999-1001.
66. Netting, A.G. and Wettstein-Knowles, P. von (1976) Arch. Biochem. Biophys., 174, 613-621.
67. Horn, D.H.S. and Lamberton, J.A. (1962) Chem. Ind. (London), 2036-2037.
68. Dierickx, P.J. (1973) Phytochem., 12, 1498-1499.
69. Evans, D., Knights, B.A., Math, V.B. and Ritchie, A.L. (1975) Phytochem., 14, 2447-2451.

70. Kolattukudy, P.E. and Liu, T.J. (1970) Biochem. Biophys. Res. Commun., 41, 1369-1374.
71. Kolattukudy, P.E., Buckner, J.S. and Liu, T.J. (1973) Arch. Biochem. Biophys., 156, 613-620.
72. Tulloch, A.P. and Hoffman, L.L. (1976) Phytochem., 15, 1145-1151.
73. Søgaard, B. and Wettstein, D. von, unpublished data.
74. Lundqvist, U. and Wettstein, D. von (1962) Hereditas, 48, 342-362.
75. Fincham, J.R.S. (1977) Carlsberg Res. Commun., 42, 421-430.
76. Tauro, P., Holzner, U., Castorph, H., Hill, F. and Schweizer, E. (1974) Molec. gen. Genet., 129, 131-148.
77. Knobling, A. and Schweizer, E. (1975) Eur. J. Biochem., 59, 415-421.
78. Knobling, A., Schiffmann, D., Sickinger, H.D. and Schweizer, E. (1975) Eur. J. Biochem., 56, 359-367.
79. Kolattukudy, P.E. (1968) Plant Physiol., 43, 375-383.
80. Cassagne, C. (1972) Qual. Plant. Mater. Veg., 21, 257-290.
81. Macey, M.J.K. and Stumpf, P.K. (1968) Plant Physiol., 43, 1637-1647.
82. Bolton, P. and Harwood, J.L. (1977) Biochem. J., 168, 261-269.
83. Kolattukudy, P.E. and Buckner, J.S. (1972) Biochem. Biophys. Res. Commun., 46, 801-807.
84. Bolton, P. and Harwood, J.L. (1976) Phytochem., 15, 1501-1506.
85. Cassagne, C., Lessire, R. and Blanchardie, P. (1976) Abst. Symp. Lipids and Lipid Polymers in Higher Plants, Karlsruhe, pp. 13-14.
86. Cassagne, C., Lessire, R. and Carde, J.P. (1976) Plant Sci. Letters, 7, 127-135.
87. Blanchardie, P., Carde, J.P. and Cassagne, C. (1977) Biol. Cellulaire, 30, 127-136.
88. Khan, A.A. and Kolattukudy, P.E. (1973) Biochemistry, 12, 1939-1948.
89. Khan, A.A. and Kolattukudy, P.E. (1973) Arch. Biochem. Biophys., 158, 411-420.
90. Khan, A.A. and Kolattukudy, P.E. (1975) Arch. Biochem. Biophys., 170, 400-408.
91. Shine, W.E. and Stumpf, P.K. (1974) Arch. Biochem. Biophys., 162, 147-157.

© 1979 Elsevier/North-Holland Biomedical Press
Advances in the Biochemistry and Physiology of Plant Lipids,
L.-Å. Appelqvist and C. Liljenberg, editors

THE AQUEOUS SYSTEM OF MONOGALACTOSYL DIGLYCERIDES AND DIGALACTOSYL DIGLYCERIDES - SIGNIFICANCE TO THE STRUCTURE OF THE THYLAKOID MEMBRANE

K. LARSSON and S. PUANG-NGERN
University of Lund, Department of Food Technology, Box 740, S-220 07 Lund (Sweden)

ABSTRACT

The ternary system monogalactosyl diglycerides (MGDG) - digalactosyl diglycerides (DGDG) - water has been analysed in order to determine phase equilibria and characterize the structures. The MGDG and DGDG were isolated from wheat endosperm flour, and the binary systems MGDG-water and DGDG-water showed the same behaviour as the corresponding systems from lipids isolated from pelargonia reported earlier. MGDG forms a reversed hexagonal liquid crystalline phase with water, and DGDG forms a lamellar liquid-crystalline phase. A surprising feature of the ternary system is that these two phases coexist over a wide concentration range. The relevance of these phase relations to the structure of the thylakoid membrane is discussed. It is proposed that the chlorophyll molecule must be organized in the lipid bilayer of this membrane in the same way as the polar lipids in order to form a stable lipid bilayer.

INTRODUCTION

The reason why lipid molecules spontaneously associate in an aqueous environment and form bimolecular membranes is the amphiphilic nature of the molecules, i.e. the molecules possess one region with affinity for water and another tending to avoid water contact. Knowledge on the physical structures occurring in simple lipid-water systems is therefore important in order to understand the molecular organization of biological membranes, and general discussions on lipid systems have been reviewed[1-3]. It is surprising to find that almost all physical studies of membrane lipids concern phospholipids, whereas almost no similar work on galactolipids has been reported. Binary systems, monogalactosyl diglycerides (MGDG) -water and digalactosyl diglycerides (DGDG)-water, have been reported by Shipley et al.[4], and aqueous systems of lipids have also been described[5].

In connection with studies of the phase behaviour of wheat lipids[6] we have isolated MGDG and DGDG from wheat endosperm grains. The structures formed in the ternary system MGDG-DGDG-water will be described below, and consequences with regard to the structural organization of the chloroplast thykaloid membrane are discussed.

MATERIALS AND METHODS

Isolation of the polar lipids from wheat flour and separation of MGDG and DGDG by column chromatography has been described earlier[5]. The MGDG and DGDG samples were of about 90% (w/w) purity according to thin layer chromatography. The ternary mixtures MGDG-DGDG-water were prepared in the following way. First MGDG and DGDG were mixed in the actual proportions and solved in chloroform-methanol (2:1). This is done in order to get molecular mixing of the lipid components. After evaporation of the solvent water was added. Phase equilibria were followed by ultracentrifugation, and the phases formed were examined in the polarizing microscope and by X-ray low angle diffraction in order to determine the liquid-crystalline phase structures.

RESULTS

The phase diagram of the system MGDG-DGDG-water is shown in Figure 1.

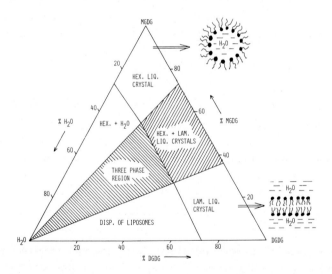

Fig. 1. Phase diagram of the ternary system MGDG-DGDG-water. The structures of the liquid crystallin phases are indicated to the left.

The phases were identified by the X-ray diffraction lines. The hexagonal phase of MGDG swells in a way resemling a lamellar structure (close to linear relation between the repitition distance and the water volume fraction), but

the diffraction symmetry is clearly hexagonal. The MGDG phase at different
water compositions can solubilize about 1:4 in molar ratio of DGDG in this
structure. The solubilization of MGDG in the lamellar liquid crystalline
phase formed by DGDG is about the same. It is quite surprising that there is
not a critical molar ratio when the hexagonal phase is transformed into the
lamellar phase and vice versa, as the molecules are so closely related in
structure. In most known ternary amphiphile-water systems showing these
phases, there is an only narrow gap between them[7]. The reason why a two-phase
region exists over a wide concentration range in this system might therefore
be related to the structure in the sugar group region. The structure of the
hexagonal phase is of the reversed type, with MGDG molecules oriented at the
interphase of water cylindres (see Figure 1). Due to the liquid nature of the
hydrocarbon chains, these cylindres are hexagonally arranged with the hydrocarbon chains forming a continuous matrix.

DISCUSSION

There is hardly any reason to expect that the structural properties shown
above are influenced by the difference in fatty acid pattern between thylakoid
membrane galactolipids, with the strong dominance of linolenic acid, and wheat
endosperm galactolipids where linolic acid dominates. Above the hydrocarbon
chain melting transition such differences, as known from sample model systems,
do not give different types of liquid crystalline structures. If we compare
the binary systems MGDG-water and DGDG-water from pelargonia lipids[3], and
the corresponding binary systems shown above, there are minor differences in
swelling behaviour of the hexagonal and lamellar phases, respectively. These
differences might reflect these differences in fatty acid composition.

The lamellar liquid crystalline phase swells with water to a maximal
lipid bilayer separation, which as in other systems of neutral lipids is
about 15 Å. Above this limit there is not a two-phase region of the lamellar
liquid crystalline phase and water, which often is stated even in recent
literature. Instead a closed spherical structure consisting of lipid bilayers
alternating with water layers is formed[8]. These particles are spontaneously
dispersed in the bulk water, (they have later been called liposomes). The
liposomes (or the single-walled visicles) are closely related to the lipid
bilayer structure of a biomembrane. The liposomes exist in equilibrium with
excess of water; the monomer concentration of lipid molecules is thus extremely small (about 10^{-10} M). In the same way membrane lipids should be expected
to give liposomes in excess of water in order to form a stable membrane.

The membrane proteins can be neglected in this connection, as the generally accepted 'fluid mosaic' membrane model requires a stable and continuous lipid bilayer.

If, on the other side, we consider the hexagonal liquid crystalline phase formed by MGDG and water, it coexists with excess of water without formation of any dispersed particles or membrane-like aggregates. A membrane lipid mixture can hardly form this hexagonal liquid crystalline phase in water, as this structure is not compatible with the stable lipid bilayer corresponding to the membrane.

The ratio MGDG to DGDG and the other polar lipids (phospholipids and sulfolipids) is about 1.4. The values are taken from data on bean-leave chloroplasts[9] and spinach chloroplasts[10]. It should be emphasized that DGDG dominates in relation to sulfo- and phospholipids, and as both phospholipids and sulfolipids give a lamellar liquid crystalline phase with excess of water like DGDG, it is reasonable to compare this polar mixture of the thylakoid membrane with DGDG in this phase diagram. The ratio between MGDG:DGDG of 1.4 in the phase diagram is in the center part of the two-phase region between the liquid crystalline phases, or if the excess water region is considered this ratio corresponds to a dispersion of liposomes in equilibrium with the hexagonal phase. A lipid mixture corresponding to the galacto-, phospho- and sulfolipids of the thylakoid membrane can therefore hardly be expected to form stable bilayer lamellae in excess of water unless other amphiphilic molecules are present to modify the packing conditions.

The molecular geometry favouring the lamellar phase and the reversed hexagonal phase are indicated in Figure 2. Each hydrocarbon chain in liquid crystalline phases has a cross-section area of about 30 $Å^2$. The polar head group, and eventually water molecules bound laterally, must therefore have a cross-section area of about 60 $Å^2$ in this type of di-acyl lipids in a lamellar liquid crystalline phase. If the polar head groups tend to adopt a closer lateral packing, this is only possible if the molecules diverge from the polar groups towards the chain tails, which is the case in the hexagonal liquid crystalline phase. There are also amphiphilic molecules with a very large polar head group, resulting in the opposite type of structure, where the hydrocarbon chain tails occupy a smaller cross-section than the polar head group. An example of such a lipid is lysolecithin. If lysolecithin is added to the hexagonal MGDG phase, it is possible to change the structure from the hexagonal to the lamellar type. The compensating geometrical effect of such an amphiphile resulting in a lamellar structure is also demonstrated in Figure

2. In the thylakoid membrane there is one additional lipid, which should be expected to have the same effect, and that is chlorophyll.

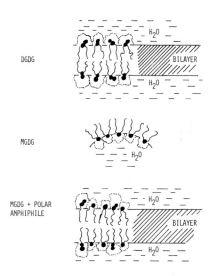

Fig. 2 Schematic illustration of the geometric conditions favouring lamellar and hexagonal structures of DGDG and MGDG, respectively. The water associated to the polar head group is indicated by broken lines. The effect of a strongly polar amphiphile on MGDG is shown below.

When etioplasts change into chloroplasts in connection with formation of chlorophyll, it is known that the lamellar body is broken up into lamellar units, which form the chloroplast membranes[11]. We have recently analyzed a cubic phase in lipid-water systems, which appears to be formed by a network of lamellar units[12]. This phase is remarkably similar in dimensions and structure to ultrastructural pictures of the prolamellar body of etioplasts, and the cubic model of the prolamellar body by Gunning[13] is in general agreement with this lipid-water structure. In order to break up the fused lamellar units of this cubic lipid-water phase into lipid bilayers (such as liposomes) it is necessary to add a strong amphiphile with the same molecular geometry as discussed above, i.e. a large polar head group compared to the hydrocarbon chain region.

An X-ray analysis of maize chloroplast lipids has been reported by Rivas and Luzzati[14]. They separated the lipids into two groups; one fraction consisting of phosphatidyl glycerol, lecithin and phosphatidylinositol, and another fraction consisting of MGDG, DGDG and sulfolipids. Examination of binary systems of these two lipid fractions and water respectively showed that both form a lamellar liquid crystalline phase at high water content. It is surprising that the galactolipid fraction gave a lamellar phase with regard to the results presented above. A possible explanation can be that the fractionation gave other proportions MGDG/DGDG than that occurring in the original chloroplast membranes.

It should be mentioned in this connection that X-ray diffraction studies of intact thylakoid membrane systems have been reported by Kreutz[15], and on this basis he has presented the following membrane model. Each membrane consists of a lipid bilayer in the center. This bilayer is, however, reversed with the polar head group in the middle and the hydrocarbon chain tails directed outwards. The chlorophyll molecules have their phytol tail anchored in the hydrocarbon chain layers and the porphyrin rings outside this reversed bilayer near the protein molecules. As all later work on membrane lipid structure has confirmed the general feature of a lipid bilayer with a hydrocarbon chain core it is hard to accept this model. The phases of the X-ray reflexion could not be determined ambigously, and the electron density profile used in Kreutz´ analysis is therefore uncertain. It is therefore reasonable to assume that the lipids of the thylakoid membranes form the usual type of bilayer. In order to give a stable bilayer, the phase equilibria of the ternary system MGDG-DGDG--water presented here indicate that the chlorophyll molecules are organized as other amphiphiles in the bilayer.

ACKNOWLEDGEMENTS

The separation of the lipids was done by Y. Miezis, and discussions with K. Fontell on phase identification were most valuable.

REFERENCES

1. Luzzati, V. (1968) in Biological Membranes, Chapman, D. ed., Academic Press, New York, pp. 71-123.
2. Shipley, G.G. (1973) in Biological Membranes, Chapman, D. and Wallach, D.F.H. eds., Academic Press, New York, vol. 2, pp. 1-89.
3. Larsson, K. and Lundström, I. (1973) Adv. in Chem. Series Lyotropic Liquid Crystals and the Structure of Biomembranes, Academic Press, 152, 43-70.
4. Shipley, G.G., Green, J.P. and Nichols, B.W. (1973) Biochim. Biophys. Acta 311, 531-544.
5. Carlson, T., Larsson, K. and Miezis, Y. (1978) Cereal Chem. 55, 168-179.
6. Abrahamson, S., Pascher, I., Larsson, K. and Karlsson, K.-A. (1972) Chem. Phys. Lipids 8, 152-179.

7. Ekwall, P. (1975) in Adv. in Liquid Crystals, Brown, G.H. ed., Academic Press, New York, vol. 1, pp. 1-142.
8. Larsson, K. (1967) Z. phys. Chem. 56, 173-189.
9. Novitskaya, G.V. and Rutskaya, L.A. (1976) Sov. Plant Physiol. 23, 757--762.
10. Lichtenthaler, H.K. and Park, R.B. (1963) Nature 198, 1070-1072.
11. Virgin, H.I., Kahn, A. and von Wettstein, D. (1963) Photochem. Photobiol. 2, 83-91.
12. Forsén, S., Fontell, K., Larsson, K. and Lindblom, G. to be published.
13. Gunning, B.E.S. (1965) Protoplasma 60, 111-130.
14. Rivas, E. and Luzzati, V. (1969) J. molec. Biol. 41, 261-275.
15. Kreutz, W. (1966) in Biochemistry of Chloroplasts, Goodwin, T.W. ed., vol. 1, pp. 83-88.

MEMBRANE LIPIDS: STRUCTURE AND FUNCTION

CHLOROPHYLL-LIPID INTERACTIONS

JOSEPH J. KATZ

Chemistry Division, Argonne National Laboratory, Argonne, Illinois 60439, USA

ABSTRACT

The chlorophyll molecule is in a sense a surfactant by virtue of its polar macrocycle and its long, nonpolar, esterifying alcohol. Some properties of chlorophyll that derive from its lipid moiety are described. Chlorophyll may also interact with the lipid components of the photosynthetic membrane by coordination interactions involving the central magnesium atom of the chlorophyll and nucleophilic groups present in the lipids. Proton magnetic resonance has been used to explore possible coordination interactions between chlorophyll and galactolipids and between chlorophyll and xanthophylls. As reasonably detailed models of both photo-reaction center chlorophyll and green plant antenna chlorophyll have been developed recently, various ways in which these can be incorporated into the photosynthetic membrane are examined.

INTRODUCTION

It is very generally accepted that chlorophyll is the most important photoreceptor in photosynthesis. In green plants, in blue-green algae, and in photosynthetic bacteria, chlorophyll is the principal agent in the capture of light energy, in subsequent energy transfer, and in the light energy conversion act itself. All photosynthetic organisms that produce molecular oxygen possess chlorophyll a (Chl a) (Figure 1). The principal chlorophyll in photosynthetic bacteria, bacteriochlorophyll a, is a related chlorophyll in which two addition hydrogen atoms are present at positions 3 and 4, and the vinyl group at position 2 is replaced by an acetyl group, $CH_3-\overset{O}{\underset{}{C}}-$. Only a small fraction of the Chl a or BChl a molecules are involved in photosynthesis as the primary electron donor, while the bulk of the Chl a and BChl a molecules participate in photosynthesis in the collection of light energy and in energy transfer to the photo-reaction center. The structures of photo-reaction center and of light harvesting or antenna chlorophyll have been central problems in photosynthesis. Reasonable models for antenna and especially photo-reaction chlorophyll have been advanced, and it is now timely to consider these model structures in an in vivo context.

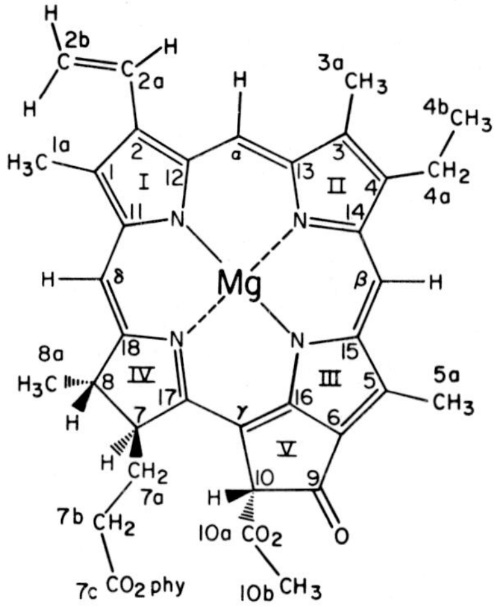

Fig. 1. Structure and numbering of chlorophyll a.

By far the largest effort in the study of chlorophyll function has been devoted to the macrocycle moiety of these substances. Associated with the macrocycle is a highly aromatic delocalized π-system. It is here that light energy is absorbed and converted to electronic excitation energy. Removal of the phytyl chain of Chl a and its replacement by a methyl group does not affect the visible absorption spectrum to any significant extent, indicating that the phytyl chain is only very weakly coupled to the π-system of the chlorophyll. It is this part of the molecule, the chromophore, which is also involved in transfer of electronic excitation energy to the photo-reaction center. The phytyl chain also does not appear to have significant involvement in the light energy conversion step itself as chlorophyll derivatives containing only the chlorophyll macrocycle but not the aliphatic moiety are fully capable of charge separation. The very widely applied spectroscopic techniques of visible absorption, fluorescence and electron paramagnetic resonance spectroscopy as applied to photosynthesis report only on events taking place in the chlorophyll macrocycle. Interest in the lipid portion of the chlorophyll molecule has therefore been confined almost entirely to students of the biosynthesis of chlorophyll.

It would, however, be incorrect to consider the lipid moiety to be without interest in chlorophyll function. Chlorophyll in living photosynthetic organisms, it is generally agreed, is situated in a photosynthetic membrane whose major constituents are lipids and proteins. The interactions of chlorophyll with the other membrane components is clearly a matter of great interest when such questions as the site or sites of the chlorophyll in the membrane are considered. Chlorophyll-protein complexes in the photosynthetic membrane are now under very active study, as evidenced by the large and rapidly growing literature on the subject.[1] This is, to be sure, a voluminous literature on chlorophyll-lipid monolayers[2] and on the fluorescence and charge transfer properties of chlorophyll in lipid matrices[3] or liposomes.[4] Many of these systems use polar phospholipids. However, polar phospholipids are a relatively minor constituent of the photosynthetic membranes of most photosynthetic organisms.[5] It is the neutral lipids, mono- and di-galactosyl diglycerides (MGDG and DGDG) the sulfonolipid sulfoquinovsyl diglyceride (SQDG) that are the characteristic photosynthetic membrane components. (It should be noted that the lipid composition of the photosynthetic membrane in bacteria contains little or no MGDG, DGDG or SQDG.[5] This point is discussed further below.) While such studies are of undoubted interest intrinsically, they are not further considered here.

For chlorophyll-lipid interactions, however, the subject is far less well developed, and little can be said about chlorophyll interactions at the molecular level with the lipid components of the photosynthetic membrane. Two categories of chlorophyll interactions with lipids need to be considered. The first includes the interactions of chlorophyll, which involve its lipid moiety. Interactions between the phytyl chain with the long-chain aliphatic fatty acids of lipids could well be important because of chemical and structural similarity. The second category of chlorophyll-lipid interactions that must be considered are coordination interactions between chlorophyll and lipid, which involve the central magnesium atom of chlorophyll and a nucleophilic or donor group in the lipid.

<u>Chlorophyll as a Lipid</u>. The esterifying alcohol at the propionic acid side chain (position 7) of chlorophyll has a significant effect on the geometry or shape of the chlorophyll molecule. The macrocycle of the chlorophyll molecule is about 1.3 nm across the porphyrin ring. The phytol chain is about 3.6 nm long, making the overall length of the chlorophyll molecule measured from the central Mg atom about 3.5 nm. The chlorophyll molecule is thus comprised of a polar region (the macrocycle) and a non-polar region (the phytyl chain), and is thereby endowed with the fundamental prerequisites for surfactant action. However, as the chlorophyll molecule is completely insoluble in water, it does not form the usual aqueous-type micelle in which the polar regions of the

molecule are on the outside of the micelle and the non-polar regions are on the inside. (In polar solvents, such as alcohol or ethanol, under conditions where micelles might be formed, it is expected that the micelle would have the usual structure.) In non-polar solvents that have little capacity for solvation of the chlorophyll macrocycle, i.e., aliphatic or cycloaliphatic hydrocarbons, inverted micelles can be expected to form in which the aliphatic phytyl tails of the chlorophyll are in the solvent phase and the macrocycle head-groups are in the interior of the micelle. The self-aggregation of chlorophyll in non-polar (or better, non-nucleophilic solvents) arising in coordination interactions is as much a consequence of exclusion of the macrocycles from the solvent as it is the result of particularly strong coordination interactions. Chlorophyll-chlorophyll aggregates themselves can be considered as inverted micelles. Interactions between chlorophyll and a lipid based on the mutual solubility of the phytyl chains and the fatty acid chains of the lipid are thus in competition with the tendency of the chlorophyll to form inverted chlorophyll-chlorophyll micelles. The same forces that lead to micellization in water of ordinary surfactants are also operative when chlorophyll is dissolved in non-nucleophilic solvents because of the lipid moiety of the chlorophyll.

Until recently, phytol was the only esterifying alcohol thought to be present in Chl a and BChl a. It has been known for some time now that BChl a from *Rhodospirillum rubrum* is esterified not with phytol, but with all-trans geranylgeraniol.[6] Chlorophyll a esterified with geranylgeraniol has been observed in newly greened leaves.[7] Geranylgeraniol had long been considered to be the precursor of phytol[8] even before BChl a esterified with geranylgeraniol was identified, and the detection of geranylgeraniol in chlorophyll provides support for this view. Tetrahydrogeranylgeraniol (2 double bonds) also a possible precursor of phytol which is intermediate in unsaturation between geranylgeraniol (4 double bonds) and phytol (one double bond) has recently been identified in newly synthesized Chl a[9] and stearyl alcohol has been very recently reported as the esterifying alcohol in the chlorophyll of a photosynthetic green bacteria. While the predominant esterifying alcohol in Chl a is phytol, esterifying alcohols in which the degree of unsaturation varies between 4 as in geranylgeraniol to the fully saturated C_{20} alcohol[6] may occur. Indeed, there is a possibility that high molecular weight polyprenols (to C_{55}) are also present as esterifying alcohols in chlorophyll.[6] The degree of unsaturation or chain length of the esterifying alcohol cannot be important in antenna chlorophyll, because the non-phytol components are present only in

small amount, and phytylated Chl a is by far the most prominent component of
the antenna. However, chlorophylls esterified with other alcohols (especially
high molecular weight polyprenols) might well be important for the much more
specialized photo-reaction chlorophylls that are present in the plant only in
small amounts. The lipophilic properties of the esterifying alcohol could be
a sensitive function of its degree of unsaturation, which affects the polarizability, rigidity, and geometry of the aliphatic hydrocarbon chain. It should
be noted, however, that it is easy by conventional mass spectroscopy to produce
artifacts in the mass spectrometer source; the possibility of an active chloro-
phyllase producing post-mortem artifacts by transesterification must also be
considered before definitive statements about the esterifying alcohol(s) of
chlorophyll in vivo can be made.

Arguing against the view that the photo-reaction center chlorophyll is differentiated from the bulk antenna chlorophyll by a different esterifying alcohol
is the remarkable finding that the esterifying alcohol in the bacteriochlorophyll of the isolated photo-reaction center of a R. *rubrum* mutant (G-9$^+$) is
geranylgeraniol, the same alcohol as in the bulk BChl a.[11] The bacteriopheophytin a in these photo-reaction centers, however, is reported to be phytol.
The phytylated bacteriopheophytin a_p is exclusively present in the reaction
centers. Taken at face value, these findings suggest the possibility of a
much more important role for the esterifying alcohols in the organization of
bacterial photo-reaction center than has previously been considered. It is
conceivable that the phytol in the bacteriopheophytin a of the isolated reaction
centers is formed by reduction of geranylgeraniol by the reducing power generated in the bacterial photo-reaction center, but no evidence bearing in this
point is available.

Rosenberg[12] has advanced an interesting hypothesis for the nature of the
interaction between the phytyl chain of chlorophyll and the fatty acid chains in
galactolipids. According to Rosenberg, the methyl groups at positions 3, 7,
11, and 15 of the phytyl chain protrude from the chain and interact with
depressions in the acyl chains where double bonds occur in a lock and key
arrangement. The evidence for this is based on molecular models and the high
degree of unsaturation of the acyl chains in the galactolipids. Unfortunately
for this view, the structural features required for the lock and key arrangement are observed only in one particular molecular model. Godfrey molecular
models show the phytyl chain with protruding methyl groups and the double bonds
in the acyl chains as depressions. But this is not true for Courtauld, Stuart-
Briegleb or CPK space-filling models. Constructed from any of these molecular

models, the phytyl chain is smooth, and there are no visible depressions in the unsaturated acyl chains. Rosenberg's hypothesis thus seems to be valid only for a particular kind of molecular model, and thus cannot claim general validity. While the unsaturation in the acyl chains of the galactolipids must certainly affect the geometry and polarizability of the fatty acid residues, it is unlikely that sufficiently strong and specific forces occur between the unsaturation centers and the methyl groups in the esterifying alcohols to form definite structures.

It is evident from the above that new and interesting chemical and biosynthetic problems are emerging as attention becomes focussed on chlorophyll as a lipid. These and no doubt many other problems as well will need resolution before the lipophilic interactions of chlorophyll with both lipids and proteins in the photosynthetic membrane can be described on the molecular level.

Chlorophyll as Donor-Acceptor. The key to the molecular interactions of chlorophyll is to be found in the donor and acceptor properties inherent in the molecular structure of chlorophyll. The central Mg atom of chlorophyll as shown in the structural formula of Figure 1 has the coordination number of 4, i.e., it is bonded to the 4 nitrogen atoms of the pyrrole rings. Abundant and convincing evidence from infrared[13] and nuclear magnetic resonance[14] spectroscopy indicates that Mg with coordination number 4 in chlorophyll is coordinatively unsaturated and has strong electrophilic (electron-seeking) properties. As a result the Mg axial positions must be occupied by either one or two nucleophilic (electron-donor) groups. Depending on the electron-donor strength (basicity) of the nucleophile, the Mg atom in chlorophyll can have a coordination number of 5 (one axial position filled) or 6 (both axial positions occupied). Chlorophyll dissolved in weak donor solvents such as acetone or ethyl acetate has Mg with a coordination number of 5, but in pyridine, a much stronger base, the Mg has a coordination number of 6.[15]

The chlorophylls themselves have functional groups that enable them to act as electron donors in a coordination interaction. Chl a has two ester C=O groups and a keto C=O function that in principle can act as donor, and BChl a has in addition an acetyl C=O that could have donor properties. The keto C=O groups appears to be the best of the donor functions of the chlorophyll molecule.[16] Under conditions where extraneous nucleophiles are not available, coordination interactions between the central Mg atom of one chlorophyll molecule and the keto C=O function of another forms chlorophyll dimers or oligomers. With bifunctional nucleophiles that can also form hydrogen bonds such as water or alcohol, chlorophyll aggregates can be formed. Chlorophyll self-aggregates and chlorophyll-water aggregates generated by coordination interactions have served as paradigms for antenna in green plants[17] and for photo-reaction center chlorophyll in both green plants and photosynthetic bacteria.[18]

When considering possible coordination interactions between chlorophyll and lipids it is the acceptor properties of the chlorophyll and the donor properties of the lipid that are primarily involved. For a lipid to compete successfully for coordination to the Mg of chlorophyll it must displace the nucleophile that is already present in the Mg axial position. Nuclear magnetic resonance spectroscopy is a useful tool for establishing the competitive power for the potential donor groups in the lipid molecule.

Chlorophyll-Sulfonolipid Coordination Interactions. The coordination interactions between Chl a and green plant sulfonolipid has been examined by [1]Hmr. Sulfoquinovosyl diglyceride (SQDG) is present in all green plants. It is primarily localized in the lamellae of the chloroplast and is an important structural component of the photosynthetic membrane.[19] From its structural formula (Figure 2), it might be inferred that SQDG could act as a nucleophile for the Mg atom of Chl a as it contains ester C=O groups, hydroxyl groups in the carbohydrate moiety, and a sulfonic acid group.

Fig. 2. Structure of sulfoquinovosyl digyceride (SQDG).

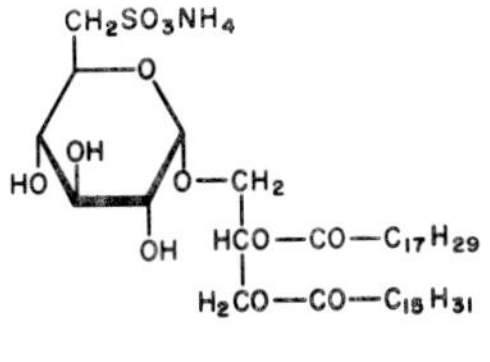

Chlorophyll dissolved in carbon tetrachloride occurs as a dimer formed by a keto C=O---Mg interaction between the Ring V keto C=O group of one chlorophyll acting as donor to the central Mg atom of the other Chl a molecule. A nucleophile that successfully competes for coordination to the Mg will necessarily disrupt the dimer. The disaggregation of the dimer can readily be followed by [1]Hmr. The chlorophyll macrocycle is a highly aromatic system, and as a consequence exhibits some rather unusual properties in a magnetic resonance experiment. In a magnetic field, the π-electron system of the macrocycle produces local magnetic fields that have a strong effect on the resonance frequencies of protons situated in the plane of the macrocycle and an equally strong but opposite effect on protons situated above or below the plane of the macrocycle. A proton positioned directly above the Mg atom experiences a large up-field shift to an extent determined by its distance from the Mg. Thus, in a chlorophyll dimer, the ^{1}Hmr spectrum characteristic of Chl a monomer is strongly distorted by ring current effects, and disaggregation of the dimer can be followed by changes in the ^{1}Hmr spectrum of the system as the Chl a dimer is converted to monomer.

The interpretation of the ^{1}Hmr spectral data is considerably facilitated by appropriate adjustments in the isotopic composition of the system. As it is possible to grow photosynthetic organisms (as well as many other microorganisms) in fully deuterated form,[20] it is practical to procure either chlorophyll or lipid in fully deuterated form. Thus, a system for ^{1}Hmr can be prepared in which one of the components is transparent, making it possible to observe changes in ligation at Mg without the complication of two sets of spectral lines.

In the experiment described here, the interaction between Chl a dimer and SQDG has been examined with Chl a of ordinary isotopic composition and fully deuterated SQDG. The latter is obtained from the blue-green alga *Synechococcus lividus* grown in 99.7% D_2O. It is characteristic of monomeric Chl $a \cdot L_1$ spectra to observe resonances from the four methyl groups at positions 1, 3, 5, and 10b in the molecule as sharp, well-resolved lines. In $(Chl\ a)_2$ the methyl region of the ^{1}Hmr spectrum at room temperature is the lines are broad and very poorly resolved. Disaggregation of the dimer to monomer by an effective competitor for coordination at Mg converts the braod and ill-defined methyl resonances of the dimer to the sharp spectrum of the monomer.

Figure 3 shows the methyl region of Chl a in the presence of increasing concentrations of ^{2}H-SQDG. Whereas two equivalents of a base such as methanol will essentially completely convert $(Chl\ a)_2$ to Chl $a \cdot L_1$, two equivalents of ^{2}H-SQDG has only a small disaggregating effect. Even a 10 mole excess of ^{2}H-SQDG has not converted all of the dimer to monomer, although from the spectrum it can be inferred that disruption of the dimer has proceeded to an appreciable extent.[21]

The data suggests that despite its numerous potential donor groups SQDG is not nearly as effective a coordinating or disaggregating agent as are nucleophiles such as acetone, methanol, or ethylacetate. Probably the most important reason for the relative weakness of SQDG as a donor is the formation of inverted micelles of SQDG.

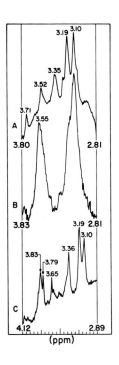

Fig. 3. ^{1}Hmr of chlorophyll a in carbon tetrachloride solution in the presence of A, 2 moles of ^{2}H-SQDG per mole of Chl a; B, 3 moles of ^{2}H-SQDG/mole Chl a; and C, 10 moles of ^{2}H-SQDG/mole Chl a.

^{13}Cmr investigations on SQDG indicate quite unequivocally that SQDG in chloroform has an inverted micellar structure.[22] This conclusion is arrived at principally from correlation times for the carbon atoms in the molecule deduced from relaxation data. The correlation time is a measure of the freedom for the carbon atoms of the sulfoquinovosyl moiety are consistent with a decrease of two orders of magnitude in the freedom of motion of the SQDG in chloroform as compared to a solution in methanol. Molecular tumbling in chloroform is thus extremely slow. The correlation times of carbon atoms in the acyl chains even in chloroform, however, are very short. These results can be rationalized by the assumption that the polar head groups of SQDG associate in chloroform to form a micelle in which the sulfoquinovosyl head portion is in the interior of the micelle and essentially immobilized, whereas the acyl chains extend into the solvent and move freely. The inverted micelle can reasonably be expected to impose substantial steric obstacles to interaction between the nucleophilic groups of SQDG and the Mg atom of chlorophyll.

The high molar ratio of SQDG to Chl a required to produce a significant concentration of monomeric Chl a species suggests that at lower ratios the

Chl a-SQDG interaction may to a large extent involve the phytyl chains of the chlorophyll and the acyl chains of the SQDG. The chlorophyll dimers could be bound to the inverted SQDG micelle by lipophilic interactions, much as pins in a pin-cushion. The monomeric chlorophyll that ultimately is generated at very high SQDG concentrations could then be interpreted as arising from mass action considerations that compensate for the intrinsic weakness of the donor functions of SQDG.

The carbon tetrachloride solvent, facilitating as it does the formation of inverted micelles, introduces factors that might not be present in an in vivo lipid membrane, and the chlorophyll-lipid interaction may be quite different in the laboratory system. With conventional apparatus, a high resolution NMR study of systems containing only chlorophyll and SQDG or mono- or di-galactosyl glycerides was impossible because of the immobilization of the components. However, new developments in techniques for recording high resolution spectra on immobilized systems in which the molecules of interest cannot tumble freely have been developed, and the prospects for successful investigation of such systems now seem bright. Particularly useful will be ^{1}Hmr with magic angle sample spinning on systems in which the chlorophyll is magnetically dilute.

Magic angle sample spinning is a newly developed technique for securing well-resolved NMR spectra on solids or otherwise immobilized molecules.[23] This technique is particularly useful when the observed nucleus is present at low concentration (magnetically dilute) and the dipolar interactions responsible for line-broadening in solids are produced by deuterons rather than protons. Deuteron dipolar perturbations are weaker than those of protons and are therefore more easily cancelled by magic angle spinning. Thus, a system containing ^{2}H-Chl a with a few percent ^{1}H uniformly distributed in it in a fully deuterated lipid will be particularly suitable for ^{1}Hmr experiments with magic angle spinning. The needed components are readily accessible by biosynthesis.

Chlorophyll-Galactolipid Interactions. Only a small amount of information is available about chlorophyll-galactolipid interactions. Trosper et al.[24] have measured fluorescence depolarization and the concentration dependence of Chl a fluorescence in mixed monolayers containing galactolipids or SQDG. The fluorescence depolarization and fluorescence yield decrease in both MGDG and SQDG mixed films. However, the polarization of the emitted fluorescence, which falls to zero with increasing Chl a concentration in MGDG films, never reaches zero in SQDG films. Trosper et al.[24] account for their observations by assuming a random distribution of Chl a in MDGD but partial ordering in SQDG. Pressure-area behavior of the mixed films indicates the Chl a is fully miscible with both SQDG and MGDG with no evidence for segregation or special interactions between chlorophyll molecules. Liljenberg and Selstam[2] have stacked chlorophyll-galactolipid monolayers with a surface balance technique. The miscibility of Chl a with MGDG is strongly dependent on unsaturation in the acyl groups of MGDG.

Because in vivo green plant antenna is only slightly fluorescent, Beddard and Porter[25] have advanced an explanation for the concentration quenching of Chl a fluorescence that postulates the creation of non-fluorescent traps at high Chl a concentrations. These traps are not considered to be true $(Chl\ a)_2$ dimers, however, because there appears not to be any concentration-dependent changes of the Chl a absorption spectrum in polar solvents such as ethanol that indicate dimer formation. The trap, rather is a pair of monomeric Chl a molecules the separation of which is less than a critical distance (about 1.0 to 1.2 nm). Based on this concept, a model for antenna chlorophyll was proposed in which monomer chlorophyll molecules, even though closely spaced, are separated from each other by strongly coordinated molecules that prevent formation of a quenching trap. Galactolipids were suggested for the separating molecules.

In a continuation of this work, Beddard et al.[26] studied the concentration quenching of Chl a fluorescence in bilayer lipid vesicles and liposomes. Of the systems investigated, a lipid matrix composed of a mixture of MGDG and DGDG in the ratio 3:1 was the most effective in preventing concentration quenching of any of the lipid matrices that were explored. Beddard et al.[26] interpreted these results to indicate that the bulky carbohydrate groups in the galactolipids hold the chlorophyll molecules farther apart, presumably by specific interactions with the Mg atom or by hydrogen-bonding interactions to the keto C=O groups of the chlorophyll. It is evident that the chlorophyll is largely in monomeric form in these galactolipid matrices. This would seem to preclude identification of these systems with in vivo antenna. Although no visible absorption spectra for these systems are given by Beddard et al.,[26] it is reasonable to assume that all of the chlorophyll absorbs at wavelengths below 670 nm. In vivo antenna chlorophyll, however, absorbs at 680 nm.

From the foregoing it can be concluded that considerably more information of a structural nature is needed about chlorophyll-galactolipid interactions before a role in the organization of antenna chlorophyll can be assigned to the galactolipid.

<u>Chlorophyll-Cartenoid Interactions</u>. The carotenes and their oxygenated derivatives the xanthophylls are other prominent lipid components of the

photosynthetic membrane. β-Carotene is universally present in chloroplasts. Coordination interactions between β-carotene and chlorophyll involving the chlorophyll macrocycle can readily be studied by ^{1}Hmr taking advantage of the ring current effect exerted by the π-system of the macrocycle. Remembering that the resonances of proton positioned above or below the plane of the macrocycle are substantially shifted to higher fields, possible interactions between β-carotene and Chl a can be detected by upfield shifts in the resonances of the β-carotene protons. β-carotene-Chl a interaction that puts some of the protons of the β-carotene above the macrocycle can easily be detected by an upfield shift in those protons. Again, the experiment can be simplified by suitable adjustments in the isotopic composition of the interactants. When ^{2}H-Chl a is one of the partners, only the carotene or xanthophyll is observed by ^{1}Hmr, and the interpretation of the experimental results if facilitated.

Figure 4 shows the ^{1}Hmr spectra of β-carotene in the absence and in the presence of ^{2}H-Chl a. The spectra are identical, indicating that any interaction that does occur between β-carotene and Chl a cannot involve the chlorophyll macrocycle. This is scarcely surprising as β-carotene has no donor functions capable of acting as a nucleophile for the Mg atom of Chl a.

The situation is very different for the xanthophylls. Lutein, a dihydroxy β-carotene, shows marked changes in its ^{1}Hmr spectrum in the presence of ^{2}H-Chl a (Figure 5). It is clear from the spectra that coordination of Mg to the

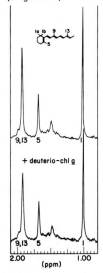

Fig. 4. ^{1}Hmr of β-carotene in CDCl$_3$ (top) and (bottom) in the presence of excess ^{2}H-chlorophyll a. Chemical shifts are relative to TMS.

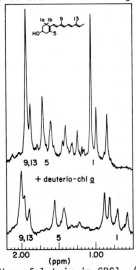

Fig. 5. ^{1}Hmr of lutein in CDCl$_3$ (top) and in the presence of ^{2}H-Chl a (bottom). Only one-half of the lutein molecule is shown. Chemical shifts are relative to TMS.

oxygen functions of lutein occurs, and the protons closest to the nucleophilic center in the lutein are subjected to an up-field ring current shift. The methyl groups at positions 9 and 13 are scarcely affected being remote from the coordination center. The methyl groups at position 5 experience only a slight upfield shift because they are insulated so to speak by the alicyclic ring of the xanthophyll from the macrocycle ring current. The methyl groups at position 1, however, are brought directly over the macrocycle by the coordination interaction of the OH group and the Chl a Mg atom. The assumption that the upfield shifted protons belong to the methyl groups at position 1 appear to be compatible with the xanthophyll ^{1}Hmr assignments.

These kinds of interactions are also seen with β-carotene epoxide and other xanthophylls containing hydroxyl and keto groups. Indeed, ^{2}H-Chl a acts as an NMR shift reagent and can be used for making chemical shift assignments.

These coordination interactions also furnish a direct explanation for the electrochromic shifts in the visible absorption spectrum observed to occur in a lutein monolayer in contact with a chlorophyll monolayer when the chlorophyll is irradiated. The shift in the lutein absorption maximum has been interpreted to be the consequences of complexing of the OH group of lutein with the Mg atom of chlorophyll.[27] An ^{1}Hmr examination of a ^{2}H-Chl a-^{1}H-lutein system irradiated by light, which is now in progress, should help clarify the nature of the interactions in the light excited states that produce the shift in the optical spectrum of the xanthophyll.

<u>Chlorophyll Special Pair-Lipid Interactions</u>. There is now good evidence that the primary electron donor in photosynthesis is a special pair of chlorophyll molecules. This conclusion is based primarily on magnetic resonance observations on photo-reaction center chlorophyll in both green plants and photosynthetic bacteria, and on the properties of in vitro chlorophyll-water adducts that serve as paradigms for photo-reaction center chlorophyll.[28] Several models for the chlorophyll special pair, Chl_{sp}, which is so designated to differentiate it from the true chlorophyll dimer $(Chl\ a)_2$. These models have been described previously in some detail,[18,29] and are not further discussed here. A model that is currently receiving much attention has been suggested by Shipman et al.[30] The Shipman et al. model (Figure 6) cross-links two chlorophylls in a parallel configuration by two molecules of a bifunctional ligand. Water, alcohol, or nucleophilic groups, i.e., -OH, $-NH_2$, or -SH such as are characteristically present in protein side-chains are suitable bifunctional ligands. Combinations of nucleophiles can also be contemplated for forming the chlorophyll special pair. Chl_{sp} have been

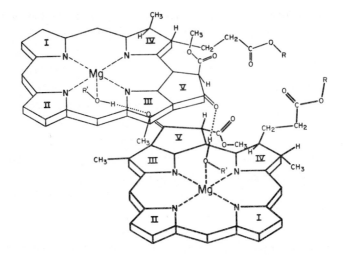

Fig. 6. Chlorophyll special pair model of Shipman et al.[30] The two Chl a macrocycles are cross-linked via their keto C=O groups by two molecules of a bifunctional nucleophile such as water or ethanol. Nucleophiles of the general structure R'XH, where X = O, NH or S and R' = H or alkyl can act as cross-linking agents.

prepared in the laboratory by a variety of techniques. A synthetic Chl_{sp} is shown in Figure 7. Here two Chl a macrocycles have been linked covalently by

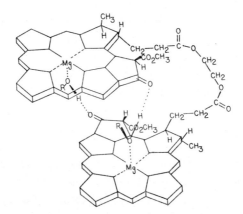

Fig. 7. Covalently linked chlorophyll molecules in the folded configuration. The covalent link prevents the two halves of the chlorophyll special pair from diffusing away from each other. Boxer and Closs[42] have used a similar strategy with a chlorophyll a derivative, pyrochlorophyll a.

an ethylene glycol molecule esterified at the propionic acid side chains of the two macrocycles. The linked "dimer" is induced to fold into the desired configuration by two molecules of water or ethanol. The synthetic Chl_{sp} has the visible absorption spectrum, redox properties, and the ability to share the unpaired spin when it is oxidized to Chl_{sp}^+ characteristic of P700. A linked $BChl_{sp}$ has also been synthesized.[31] The $BChl_{sp}$ has the required spin sharing properties, but does not mimic the visible absorption spectrum of P865.

We can now briefly discuss the question of where in the photosynthetic membrane the synthetic Chl_{sp} ought to be inserted. It is immediately evident that regardless of what else the synthetic Chl_{sp} may mimic, it does not have anything resembling the lipophilic properties of natural P700. In the process of linking the two macrocycles, the phytyl chains are lost, and the synthetic Chl_{sp} without phytyl chains is not at all comfortable in an aliphatic hydrocarbon environment. The covalently linked dimer, for example, is not soluble to an appreciable extent in any hydrocarbon solvent. Consequently it will be necessary to synthesize Chl_{sp} that retain their phytyl chains, which can in principle be accomplished by linkage of the two macrocycles through their vinyl groups, or by linking two chlorophyll b molecules through their aldehyde functions; other synthetic procedures also suggest themselves.

With the information now available, the idea that the Chl_{sp} is organized by nucleophilic groups in protein side chains seems attractive. But Chl_{sp}-protein interactions between phytyl chains and the hydrophobic regions of a protein also must be considered, as well as the possible role that polar regions of a lipid might play in the organization of a Chl_{sp}. Experimental study of these possibilities must wait on the synthesis of special pairs with appropriate lipophilic properties.

Antenna Chlorophyll-Lipid Interactions. As there is no general agreement as to the nature of antenna chlorophyll in either green plants or photosynthetic bacteria, it is obvious that any discussion about antenna-lipid interactions must be very preliminary. Nevertheless, it may be useful to try in a very speculative way to define some of the problems without entering into detail.

Antenna chlorophyll in green plants has its absorption maximum in the red at or near 680 nm. Spectral observations at 77°K have been interpreted to indicate four major forms of chlorophyll with peaks at 662, 670, 677 and 684 nm and two minor forms with absorption maxima at about 693 nm and 704 nm.[32] The peak maxima were obtained by deconvolution of the absorption curve into Gaussian components.

There are several questions that arise about these conclusions. The implicit assumption is made that cooling to 77°K only sharpens the spectra but does not generate new chlorophyll species. However, there is abundant data that shows chlorophyll coordination interactions are subject to equilibrium considerations and are often strongly temperature dependent in a reversible way.[33] Consequently low temperature spectra may not be easily related to room temperature spectra because of the transformation of room temperature species by cooling. The effect of temperature cycling in the integrity of the membrane must also be considered, for an increase in permeability to water could lead to the formation of chlorophyll-water species that absorb at 695 and 705 nm.[34] And finally, a solution of $(Chl\ a)_2$ dimer in carbon tetrachloride has a red absorption band that deconvolutes into Gaussian components at 667 and 681 nm. In this system there is only one chlorophyll species, namely, the dimer, yet there is a multiplicity of Gaussian components, which cannot be assigned to different forms or structures of chlorophyll.

The straightforward way to account for the red-shift in antenna chlorophyll is by chlorophyll-chlorophyll interactions. It has long been known that films or concentrated solutions of chlorophyll in nonpolar solvents have spectral properties quite similar to those of antenna chlorophyll. Boardman et al.[35] however prefer "to consider that specific interactions of the chlorophyll molecules in vivo are influenced to a large extent by some coordination of chlorophyll molecules to amino acid side chains of protein." In support of this view is the crystal structure determination of a bacteriochlorophyll-protein complex by Fenna and Matthews.[36] This complex has red-shifted BChl a with no apparent BChl-BChl coordination interactions that could account for the red-shift. The optical properties of the BChl a-protein complex are still an enigma. Despite serious efforts, no really satisfactory explanation of the BChl red shift in the complex in terms of exciton interactions has been advanced. Unless a convincing explanation for chlorophyll red-shifts on the basis of transition density-transition density interactions is forthcoming, direct interactions seem a more satisfactory approach.

The monomer chlorophyll antenna model of Beddard et al.[25,26] likewise appears to the writer to have problems in accounting for the optical properties of in vivo green plant antenna. Shipman[37] has recently concluded that for any antenna composed of monomer units to have its absorption maximum shifted to ~680 nm by transition density coupling requires that the local chlorophyll concentration must be much greater than the average chlorophyll concentration of ~0.1 M, perhaps as high as 1 M. At such a high concentration fluorescence quenching would be complete, and the end for which the model was developed would not be attained.

An antenna model that invokes chlorophyll interactions with both protein and lipid has been suggested by Anderson.[38] In this model, the phytyl chains of the chlorophylls are associated with the hydrophobic exterior of the major intrinsic proteins of the thylakoid membrane. The chlorophyll thus forms part of the boundary lipids of these proteins. The edge of the chlorophyll macrocycle adjacent to the phytyl chain is considered to be hydrophilic and is postulated to interact at the membrane surface with the exposed hydrophilic region of the intrinsic protein; the more hydrophobic portions of the macrocycle would be buried in the protein. It is not clear whether the chlorophyll in this model is monomeric, i.e., the chlorophylls do not experience coordination interactions with each other. If such is the case, then it may be difficult to account for a red-shift in the absorption spectrum. The proposed arrangement of the chlorophyll at the protein-lipid interface also seems to suggest that there are regions in the protein surface more hydrophobic than those of the fatty acid tails of the lipid, which seems doubtful. Nevertheless, the idea of chlorophyll aggregates as boundary lipids would seem to be an interesting concept worth exploring.

A model for antenna chlorophyll that appears to be able to account for the general features of in vivo antenna chlorophyll has been advanced by Katz et al.[29] In this model, the antenna is considered to consist of Chl a oligomers, possibly of different lengths. The structure of such an oligomer is shown in Figure 8. In this structure, the chlorophyll molecules are orthogonal to each

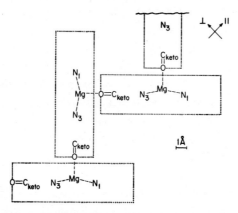

Fig. 8. Structure of chlorophyll a oligomer projected onto a plane containing the oligomer axis. All distances are drawn to scale. Arrows in the upper right hand corner indicate directions of parallel (||) and perpendicular to the oligomer axis.[39]

other. This chlorophyll species is formed in aliphatic hydrocarbon solvents at high concentrations of chlorophyll in the absence of extraneous nucleophiles. Equating chlorophyll oligomers with in vivo antenna chlorophyll is largely on the basis of comparison of the visible absorption spectra by deconvolution methods.[39] Both the red band of Chl a oligomer and the red band of antenna chlorophyll in various photosynthetic organisms can be deconvoluted with precisely the same Gaussian components.[39]

Arguments in support of such an antenna model have been given in detail elsewhere.[29] At least two arguments have been raised against this model. The oligomer is for all practical purposes non-fluorescent. If the ability to fluoresce is essential for energy transfer by a Förster mechanism, then the oligomer would not serve the purpose. However, because energy transfer takes place on a much faster time-scale than fluorescence, the inability of the oligomer to fluoresce may not be a serious handicap in providing efficient energy transfer. A second objection to oligomer as antenna is the sensitivity of such a species to disruption by nucleophiles such as water. It has been questioned, therefore, whether there are any regions in the photosynthetic membrane that are sufficiently free of water or nucleophilic groups to assure the integrity of a chlorophyll oligomer. However, in vitro, the partial pressure of water must exceed 10 mm Hg at room temperature for water to compete successfully for coordination at Mg in competition with a Chl a keto C=O function. It is quite possible that there exist regions in a lipid bilayer in which the activity of water is below the critical threshold required for disruption of Chl a oligomer. But, again, experimental evidence will be essential before a decision can be taken on this question.

Various ways in which a Chl a oligomer could be accomodated in a lipid bilayer have been considered.[40] The most reasonable location for a chlorophyll oligomer appears to be in the annular region of the lipid bilayer, i.e., in the fatty acid region in the interior of the bilayer. As Chl a oligomer is known to be soluble in aliphatic hydrocarbon solvents, this should pose no problem. The macrocycles of the oligomer are then oriented in such a way that the phytyl chains protrude into the lipid chains. The ends of the antenna could be attached to or wrapped around a membrane protein, where the Chl_{sp} is presumed to be located.[29]

It has been noted above that there are fundamental differences in the lipid composition of green plant and photosynthetic bacterial membranes.[5] The bacterial membrane contains no sulfoquinovosyl diglyceride or galactolipid. Although green plant and bacterial photo-reaction centers appear to be very similar, if not identical, in structure and function this may not be true for their antenna chlorophylls. From their absorption spectra, none of the bacterial antenna correspond to the Chl a oligomers. All of the bacterial chlorophyll is red-shifted to an extent that suggest the BChl a is hydrated or otherwise aggregated by nucleophilic interactions. Such a difference in the organization of BChl a would be consistent with the marked differences in the lipid composition of the two photosynthetic membranes.

It is evident that much more needs to be done before the many questions about the way chlorophyll fits into the photosynthetic membrane, and the relationships of chlorophyll with the protein and lipid components of the membrane are resolved. However, these are very important questions that must be answered if plant photosynthesis is to become the basis for a biomimetic technology of solar energy conversion.[41] New tools, particularly high resolution NMR of immobilized chlorophyll, can be expected to make important contributions to the solution of these questions.

ACKNOWLEDGEMENTS

This work was performed under the auspices of the Division of Basic Energy Sciences of the Department of Energy.

REFERENCES

1. For a review, see Boardman, N. K., Anderson, J. M. and Goodchild, D. J. (1978) Curr. Top. Bioenergetics, 8, 35-109.
2. Liljenberg, C. and Selstam, E. (1978) This Volume.
3. Kelly, A. R. and Porter, G. (1970) Proc. Roy. Soc. London, A315, 149-161.
4. For example, see Oettmeier, W., Norris, J. R. and Katz, J. J. (1976) Z. Naturforsch., 31C, 163-168.
5. Kates, M. (1970) Adv. Lipid Res., 8, 225-265.
6. Katz, J. J., Strain, H. H., Harkness, A. L., Studier, M. H., Svec, W. A., Janson, T. R. and Cope, B. T. (1972) J. Am. Chem. Soc., 94, 7938-7939.
7. Wellburn, A. R. (1976) Biochem. Physiol. Pflanzen, 169, 265-271.
8. Wellburn, A. R., Stone, K. J. and Hemming, F. W. (1966) Biochem. J., 100, 23c-25c.
9. Schoch, S. and Schäfer, W. (1978) Z. Naturforsch., 33c, 408-412.
10. Gloe, A. and Risch, N. (1978) Arch. Mikrobiol., 118, 153-156.
11. Walter, E. (1978) Dissertation (ETH No. 6106), E.T.H., Zurich.
12. Rosenberg, A. (1967) Science, 157, 1191-1196.
13. Katz, J. J., Dougherty, R. C. and Boucher, L. J. (1966) in The Chlorophylls L. P. Vernon and G. R. Seely, Eds., Academic Press, N. Y., pp. 185-251.
14. Scheer, H. and Katz, J. J. (1976) in Porphyrins and Metalloporphyrins, K. M. Smith, Ed., Elsevier, Amsterdam, pp. 399-524.
15. Evans, T. A. and Katz, J. J. (1975) Biochem. Biophys. Acta, 396, 414-426.

16. Shipman, L. L., Janson, T. R., Ray, G. J. and Katz, J. J. (1975). Proc. Natl. Acad. Sci. USA, 72, 2873-2876.
17. Katz, J. J., Norris, J. R. and Shipman, L. L. (1976) Brookhaven Symp. Biol., 28, 16-55.
18. Katz, J. J., Norris, J. R., Shipman, L. L., Thurnauer, M. C. and Wasielewski, M. R. (1978) Ann. Rev. Biophys. Bioeng., 7, 393-434.
19. Hames, T. H. (1973) in Lipids and Biomembranes of Eukaryotic Microorganisms, J. A. Erwin, Ed., Academic Press, N. Y., pp. 200-213.
20. Katz, J. J. and Crespi, H. L. (1970) in Isotope Effects in Chemical Reactions, C. J. Collins and N. S. Bowman, Eds., Van Nostrand Reinhold Co., New York, pp. 286-363.
21. Katz, J. J. (1968) Dev. Appl. Spectroscopy, 6, 201-218.
22. Johns, S. R., Leslie, D. R., Welling, R. I. and Bishop, D. G. (1978) Aust. J. Chem., 31, 65-72.
23. Griffin, R. G. (1977) Anal. Chem., 49, 951A-962A.
24. Trosper, T., Park, R. B. and Sauer, K. (1968) Photochem. Photobiol., 7, 451-469.
25. Beddard, G. S. and Porter, G. (1976) Nature, 260, 366-367.
26. Beddard, G. S., Carlin, S. E. and Porter, G. (1976) Chem. Phys. Lett., 43, 27-32.
27. Sewe, K.-U. and Reich, R. (1977) Z. Naturforsch., 32C, 161-171.
28. Katz, J. J. and Norris, J. R. (1973) Curr. Top. Bioenergetics, 5, 41-75.
29. Katz, J. J., Norris, J. R. and Shipman, L. L. (1976) Brookhaven Symp. Biol., 28, 16-55.
30. Shipman, L. L., Cotton, T. M., Norris, J. R. and Katz, J. J. (1976) Proc. Natl. Acad. Sci. USA, 73, 1791-1794.
31. Wasielewski, M. R., Smith, U. H., Cope, B. T. and Katz, J. J. (1977) J. Am. Chem. Soc., 99, 4172-4173.
32. Brown, J. S., Alberte, R. S., Thornber, J. P. and French, C. S. (1974) Carnegie Inst. Washington Yearbook, 73, 694-706.
33. Cotton, T. M., Loach, P. A., Katz, J. J. and Ballschmiter, K. (1978) Photochem. Photobiol., 27, 735-749.
34. Shipman, L. L. and Katz, J. J. (1977) J. Phys. Chem., 81, 577-581.
35. Reference 1, page 82.
36. Fenna, R. E. and Matthews, B. W. (1975) Nature, 258, 573-577.
37. Shipman, L. L. (1977) J. Phys. Chem., 81, 2180-2184.
38. Anderson, J. M. (1975) Nature, 253, 536-537.
39. Shipman, L. L., Cotton, T. M., Norris, J. R. and Katz, J. J. (1976) J. Am. Chem. Soc., 98, 8222-8230.
40. Katz, J. J., Oettmeier, W. and Norris, J. R. (1976) Phil. Trans. Roy. Soc. London, B273, 227-253.
41. Katz, J. J., Janson, T. R. and Wasielewski, M. R. (1978) in Energy Chemical Sciences, S. D. Christian and J. J. Zuckerman, Eds., Pergamon Press, Oxford, In Press.
42. Boxer, S. G. and Closs, G. L. (1976) J. Am. Chem. Soc., 98, 5406-5408.

OCCURENCE AND FUNCTION OF PRENYLLIPIDS IN THE PHOTOSYNTHETIC MEMBRANE

HARTMUT K. LICHTENTHALER
Botanical Institute (Plant Physiology), University of Karlsruhe,
Kaiserstrasse 12, D-7500 Karlsruhe (FRG)

ABSTRACT

This review summarizes some new aspects on the occurence, function and organization of prenyllipids (chlorophylls, carotenoids, prenylquinones) in the photosynthetic membrane.
1. The appearance of mixed prenyllipids (chlorophylls, prenylquinones) with one or more additional double bonds in the prenyl side chain and their possible function as intermediates in the biosynthetic pathway is discussed.
2. The functional organization of prenylpigments (chlorophylls, carotenoids) is reviewed with respect to the newly developed tripartite model of the photosynthetic membrane.
3. The possible double function of phylloquinone K_1 as a) photosynthetic electron carrier with a position between Q and plastoquinone and b) as a quencher of excitation energy of pigment system II is discussed on the basis of inhibitor studies.
4. The variation of the prenyllipid composition of thylakoids, as seen from the changed prenyllipid ratios in sun and shade leaves, is reported.

INTRODUCTION

The photochemical active biomembranes of chloroplasts, the thylakoids, consist of approximately 50% lipids and protein. The lipid portion is made up by the group of prenyllipids (25 to 30% by weight), the glycolipids (45-60%), and the phospholipids (9-12%)[1]. The group of thylakoid prenyllipids comprises the pure prenyllipids (carotenoids) and the mixed prenyllipids (chlorophylls, prenylquinones). The latter exhibit a prenyl side chain bound to an aromatic nucleus (porphyrin ring, benzo- or naphthoquinone ring). In contrast to mitochondria chloroplasts possess several membrane-bound prenylquinones (plastoquinone-9, phylloquinone K_1, α-tocoquinone + α-tocopherol) which are potential electron carriers of the photosynthetic electron transport chain. As compared to other cell membranes, thylakoids do either not contain sterols or only in trace amounts which might be bound to the chloroplast envelope.

This review deals with certain new aspects of prenyllipid biosynthesis, the possible function and functional organization of prenyllipids and the question of prenyllipid turnover. It is further shown that the prenyllipid composition of thylakoids largely depends on the growth conditions of the plants and can easily be modified by light intensity and by the addition of phytohormones[2-6].

RESULTS

I. BIOSYNTHESIS OF PRENYLLIPIDS

The biogenetic relationships of prenyllipids as seen from the prenyl chain synthesis is shown in figure 1. Though the basic pathways for the formation of chlorophylls, carotenoids and prenylquinones are well established (for literature see[7-9]) there are still some problems in certain details. This refers in particular to the formation of mixed prenyllipids such as chlorophylls and the prenylquinones: phylloquinone K_1, α-tocoquinone and α-tocopherol.

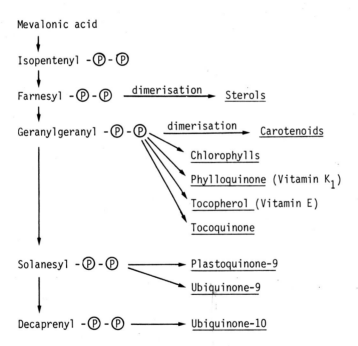

Fig. 1. Biogenetic relationship of plant prenyllipids.

Chlorophyll formation

So far it had generally been accepted that the final step in the chlorophyll formation was the esterification of chlorophyllide with phytol (or phytolpyrophosphate). The detection of geranylgeranyl-protochlorophyll in etiolated barley seedlings which in part was photoconvertible to the chlorophyll[10] and the demonstration of a small amount of geranylgeranyl-chlorophyll a in developing horse chestnut leaves[11] indicates, that a) either two chlorophyll forms (the geranylgeranyl and the phytyl-chlorophyll) exist in green plants or b) that the phytyl-chlorophyll a is formed from geranylgeranyl-chlorophyll a by stepwise saturation of 3 double bonds in the geranylgeranyl side chain. The observation of different chlorophylls with a geranylgeranyl or dihydrogeranyl chain (Fig. 2) in plants treated with the herbicide amitrole is further support for the latter possibility[12]. It appears that amitrole blocks the formation of the enzymes responsible for the saturation of the chlorophyll side chain. On the other hand it can not be excluded that both biosynthetic routes for chlorophyll formation are operating in the plant. The one using geranylgeranylpyrophosphate may be predominant at an early stage of greening, the other one using phytylpyrophosphate at later stage e.g. in the logarithmic phase of chlorophyll accumulation. Further work is required to proof this point.

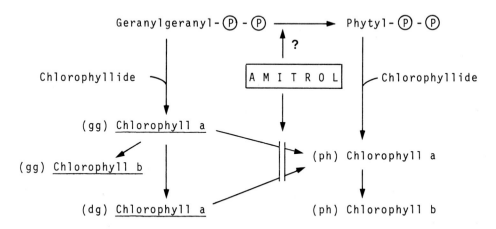

Fig. 2. Alternative pathways of chlorophyll formation. Under the influence of the herbicide amitrole additional chlorophyll forms are found[12]: (gg) and (dg) chlorophyll (=geranylgeranyl and dihydrogeranylgeranyl chlorophyll). (ph) chlorophyll = phytyl chlorophyll.

Another topic in chlorophyll research is the origin of chlorophyll b. Based on the tracer studies of Shlyk[13], Akoyunoglou[14] and others it had been accepted that chlorophyll b derives from chlorophyll a. This is now questioned by new findings of Kasemir et al.[15] who think that not the final phytyl chlorophyll a, but rather an earlier product of the biosynthetic chain may be the ultimate chlorophyll b precursor. As a result several alternative pathways open up for the formation of chlorophyll b (Fig. 3), which have to be checked. Again, also in this case one should look whether there exist differences in alternative pathways between the early and the later stages of leaf greening.

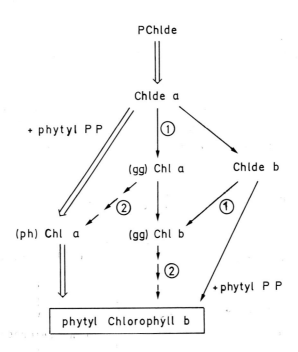

Fig. 3. Hypothetical alternative pathways leading to the formation of chlorophyll b from chlorophyllide a. (gg) and (ph) Chl = geranylgeranyl and phytyl chlorophyll. 1 = + geranylgeranylpyrophosphate, 2 = stepwise saturation of the prenyl side chain.

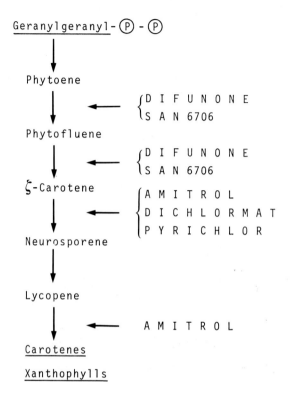

Fig. 4. Inhibition of carotenoid synthesis by herbicides at the level of the mainly colourless C_{40} polyenes. The herbicides apparently block the formation of the enzymes needed for the desaturation of the carotenoid precursors.

Carotenoid formation and herbicides

An interesting present feature of prenyl pigment research is the block of carotenoid and of chlorophyll synthesis by several structurally not related herbicides. These are fluor-containing pyridazinone derivatives (e.g. San 6706 and 9789) as well as amitrole, pyrichlor and difunone. Application of higher herbicide levels result in the formation of albino plants. Depending on the plant and on the herbicide used one obtains an accumulation of the carotenoid precursors such as phytoene, phytofluene or other polyenes (Fig. 4; for literature see[12, 16, 17]). Though their mode action is not yet fully understood it appears that the herbicides block the formation of the enzymes responsible for the desaturation of the early C_{40} carotenoids, and also certain enzymes needed for the formation of chlorophyll. Thus difunone not only blocks carotenoid synthesis but also affects the formation of the enzyme porphobilinogenase[18].

Since chlorophyll and carotenoid accumulation proceed in parallel during the greening phase of leaves it could be possible that the saturation of the prenyl chains of chlorophylls is correlated with the desaturation steps in carotenoid formation. In any case it is of particular interest in this respect that treatment with the herbicide amitrole affects both processes.

Formation of prenylquinones

Plastoquinone-9. The synthesis of the photosynthetic electron carrier plastoquinone-9 from homogentisic acid is clear and shown in figure 5. The nonaprenyl (solanesyl) side chain is unsaturated with one double bond per each isoprene unit. Plastoquinones with either partially or fully saturated side chains have not been found.

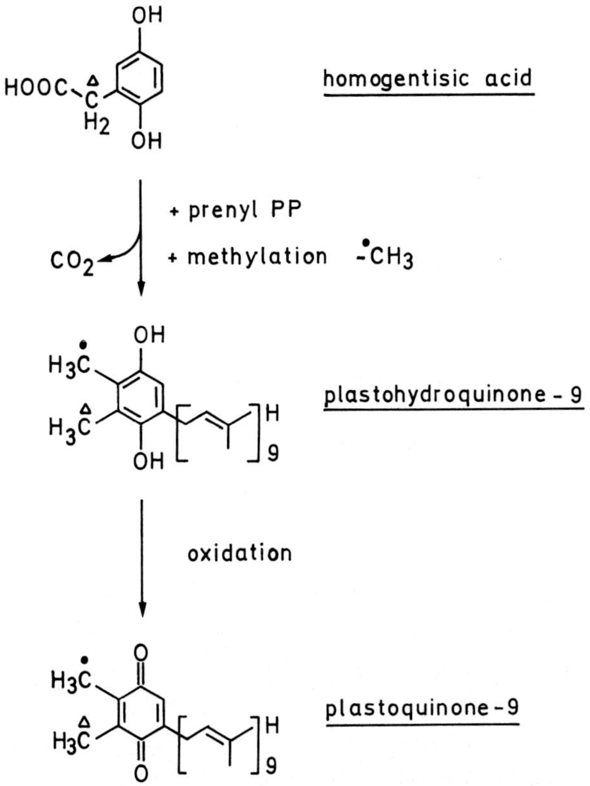

Fig. 5. Formation of plastoquinone-9 from homogentisic acid.

Phylloquinone K_1. The naphthoquinone derivative vitamin K_1 possesses a phytyl side chain like the chlorophylls. This gives rise to two alternative pathways for the phylloquinone formation (Fig. 6). If geranylgeranylpyrophosphate is used, it requires additional steps for the saturation of the side chain. Menaquinone-4, the geranylgeranyl naphthoquinone occuring in some bacteria, has not been found in algae or higher plants.

In the thylakoid-free tissues of several plants a second prenylnaphthoquinone - besides the K_1 - has been detected, its level being decreased upon illumination with a concomitant increase in the concentration of K_1. It was recently shown that this second prenylnaphthoquinone has similar but different chromatographic mobility than the desmethylvitamin K_1[19] and has now been identified as dehydrophylloquinone[20] with an additional double bond in the isoprenoid side chain (=dehydrophytyl chain). The accumulation of this second prenylnaphthoquinone in etioplasts, where thylakoid formation is inhibited, is taken as indication that the phylloquinone biosynthesis, in fact, may proceed via the geranylgeranyl naphthoquinone (Fig. 6) followed by a stepwise saturation of the C_{20} chain.

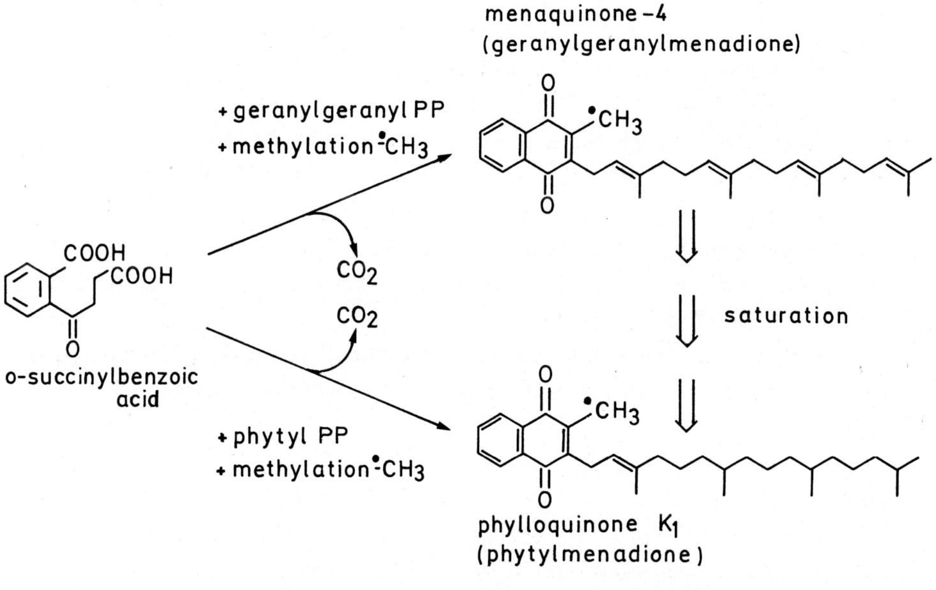

Fig. 6. Alternative pathway for phylloquinone (vitamin K_1) synthesis in higher plants.

α-tocopherol and α-tocoquinone. A C_{20} prenyl chain is used for the formation of the chromanol α-tocopherol and its oxidized form α-tocoquinone which possesses a hydroxyphytyl side chain. As in the biosynthesis of chlorophylls and phylloquinone there may exist alternative biosynthetic pathways which either use geranylgeranylpyrophosphate (Fig. 7). In fully developed chloroplasts only α-tocopherol and α-tocoquinone are found and occasionally some minor amounts of β- and δ-tocopherol, the two isomers with only 2 (instead of 3) methyl groups in the aromatic nucleus.

In the thylakoid-free tissues e.g. in the chloroplasts of fruits or petals and also in the latex of Euphorbiaceae and other plants with milk secretion one may also find the corresponding tocotrienols with an unsaturated prenyl chain. The detection of a dehydroform with one double bond in the side chain for both α-tocopherol and α-tocoquinone in etiolated leaf tissue[20] is further support for the assumption that geranylgeranylpyrophosphate is used for biosynthesis and that α-tocotrienol is the common intermediate.

From the α-tocotrienol to α-tocoquinone there are two routes. Whether α-tocoquinone is formed a) from newly synthesized α-tocopherol by oxidation or b) derives from a direct oxidation of α-tocotrienol followed by a chain saturation step (Fig. 7) can not be decided at present. The observation that the labelling kinetics of α-tocoquinone and α-tocopherol are quite different (Table 3) would, however, favour a separate pathway for the α-tocopherol and α-tocoquinone synthesis from α-tocotrienol.

In any case mixed prenyllipids (chlorophylls, prenylquinones) with differentially saturated side chains are usually found in thylakoid-free tissues or under experimental conditions when the endogenous thylakoid synthesis program is disturbed. It is generally agreed to, that these prenyllipid forms are real biosynthetic intermediates. The possibility, however, that they represent side products because of a block in the main biosynthetic pathway, can at present not been fully excluded.

This as well as the question whether there may exist two different pathways in the early and in the later stages of leaf greening are matter of future research.

Role of the chloroplast envelope

Another current topic in prenyllipid research is the possible function of the chloroplast envelope in the biosynthesis of prenyllipids. It has been shown that the isolated purified envelope contains several carotenoids[21].

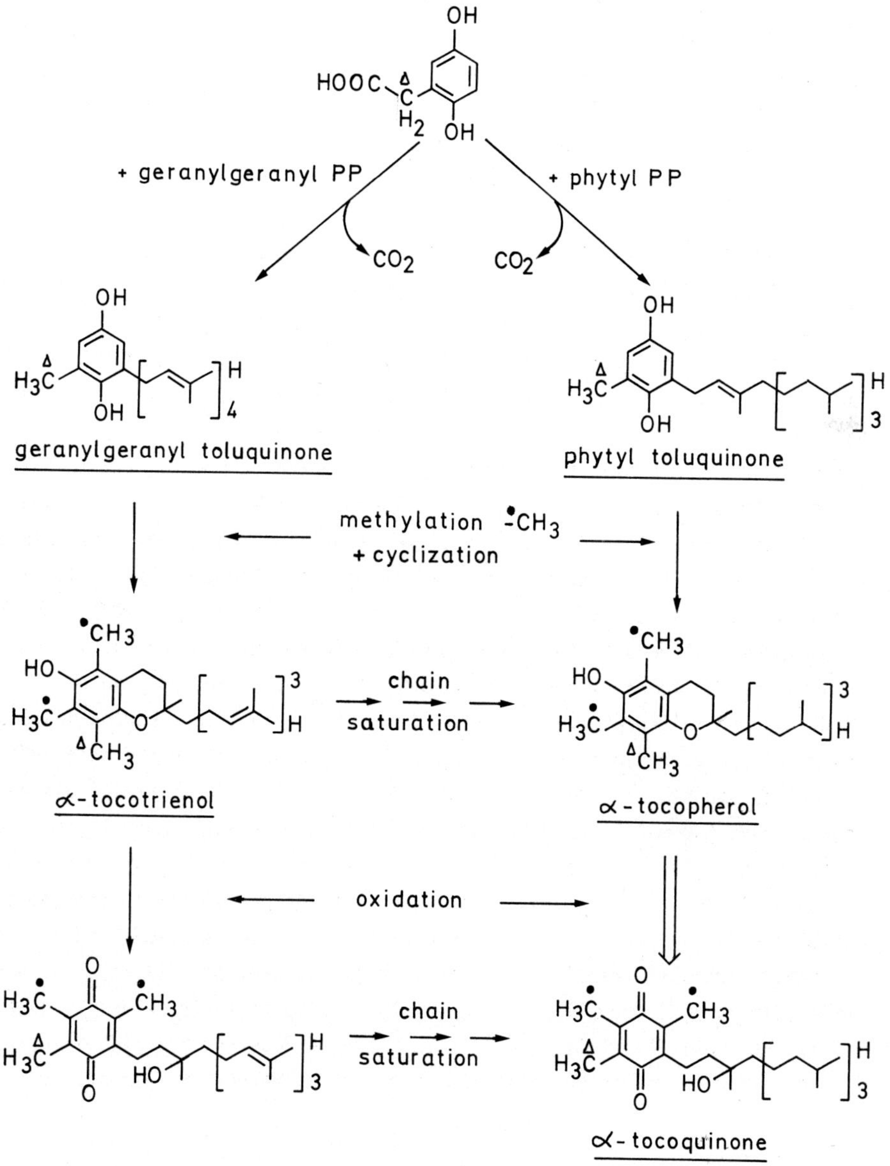

Fig. 7. Alternative biosynthetic pathways leading to the formation of α-tocopherol and α-tocoquinone from homogentisic acid.

We have recently analyzed an envelope preparation, obtained from Dr. Douce, with several methods including high pressure liquid chromatography and found phylloquinone and some plastoquinone-9. The envelope is active in galactolipid synthesis[22,23]. Whether envelope preparations, when provided the proper substrates, are also capable of synthesizing or transforming prenyllipids is matter of current investigation.

II. ORGANISATION AND FUNCTION OF PRENYLLIPIDS IN THE THYLAKOID MEMBRANE

Etiolated leaf tissues are able to synthesize carotenoids and prenylquinones as well as some protochlorophyll(ide), their concentration however, is much lower than in green photosynthetially active leaf tissue from plants of the same age (Table 1). Light not only increases prenyllipid accumulation but also results in an enhanced synthesis promotion for certain prenyllipids. Thus among total carotenoids a higher part of β-carotene is formed and among the prenylquinones a higher proportion of plastoquinone-9, α-tocoquinone and phylloquinone K_1. Therefore the relative prenyllipid composition of chloroplasts is quite different from that of etioplasts. The prenyllipid levels of chloroplasts can be regarded as "functional" prenyllipid concentrations which are representative of a fully functional photosynthetic apparatus.

In contrast to chlorophylls, carotenoids and phylloquinone, which are quantitatively bound to the photochemically active thylakoids, the reduced benzoquinone forms plastohydroquinone and α-tocopherol can be present in the chloroplast in excess amounts which are deposited in the osmiophilic plastoglobuli of the chloroplast stroma[24]. The concentration of the individual prenyllipids of plastoglobuli-free spinach chloroplasts is shown in table 2. By fractionation of chloroplasts or thylakoids with detergents and/or ultrasonic treatment one obtains a "light particle" fraction enriched in pigmentsystem I activity and a "heavy particle" fraction enriched in pigmentsystem II activity. The partition of the individual prenyllipids between the two particle fractions is quite different (Table 2). This indicates that the various prenyllipids are organisized within the photosynthetic membrane in a different functional way.

Prenylpigments

The function of the prenylpigments chlorophylls and carotenoids in the absorption and/or conversion of light seems to be clear. Their functional organisation and distribution within the photosynthetic biomembrane is, however, not yet fully understood. According to recent investigations of

TABLE 1

PRENYLLIPID CONTENT OF 7 D OLD ETIOLATED AND GREEN BARLEY SEEDLINGS µg per 100 shoots (primary leaves).

Prenyllipids	Etiolated	Green	Green/Etiolated
Protochlorophyll(ide)	12	0	
Chlorophylls	0	2190	
β-Carotene	17	90	5.3 x
Xanthophylls	151	198	1.3 x
Plastoquinone-9 [a]	12	85	7 x
α-Tocopherol	18	55	3 x
α-Tocoquinone	1	8	8 x
Phylloquinone K_1	0.7	7	10 x

[a] Oxidized + reduced form

Butler[26] and others the photosynthetic membrane shows a tripartite organization consisting of 1. the pigmentsystem I units and 2. the pigmentsystem II units which are embedded in 3. the light-harvesting chlorophyll a/b protein complex (CP II). Mainly based on detergent fragmentation (e.g. Table 2) and on gelelectrophoresis of thylakoids[27] it is generally assumed that, within the thylakoid-membrane, chlorophyll b is quantitatively associated with the light-harvesting chlorophyll protein complex, wich also contains the larger part of the thylakoid xanthophylls. β-Carotene, in turn, is believed to be primarily associated with antennae system (CPI) of pigment system I. A recent model of this tripartite organization of the photosynthetic unit including the photosynthetic electron transport chain is shown in figure 8.

Prenylquinones

Among the prenylquinones the role of plastoquinone-9 as a terminal electron acceptor of photosystem II and as proton translocator (plastoquinone shuttle) is well established[29]. Fragmentation work shows that it is predominantly associated with the "heavy particle" fraction (Table 2) which consists of pigmentsystem II units including the light-harvesting chlorophyll complex.

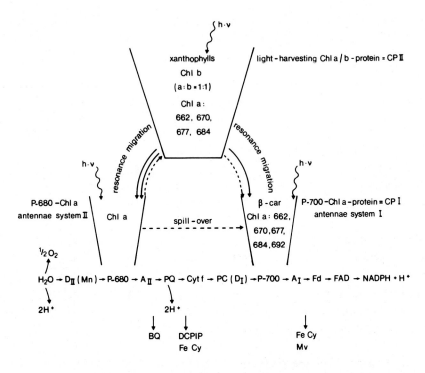

Fig. 8. Tripartite organization of the photosynthetic unit with the photosynthetic electron transport chain showing a different partition of chlorophyll a and b, and of β-carotene and the xanthophylls in the light-harvesting chlorophyll a/b protein complex and the antennae system of pigment system I. (after Wild[28] with permission).

A small part of plastoquinone-9 is found in the "light particle" fraction which mainly contains pigment system I (P 700-chlorophyll a protein complex, CPI).

There is no information on the function of the second chloroplast benzoquinone α-tocoquinone. It is accumulated parallel to the light-induced thylakoid formation and, in contrast to its reduced form α-tocopherol, mainly bound to the thylakoid membrane. α-tocoquinone is a potential electron carrier and present in both the "heavy" and the "light particle" fraction (Table 2).

α-tocopherol, found in both the "light" and the "heavy particle" fraction, apparently functions as a lipid antioxidant in both photosystems and thus helps to protect the photosynthetic membrane lipids against photo-oxidative degradation[30]. The observation that α-tocopherol in liposomes specifically interacts with polyunsaturated phospholipids[31] is consistent with the view

TABLE 2

PARTITION OF CHLOROPHYLLS, CAROTENOIDS AND PRENYLQUINONES BETWEEN TWO THYLAKOID PARTICLE FRACTIONS ENRICHED IN EITHER PIGMENT SYSTEM I OR PIGMENT SYSTEM II ACTIVITY ISOLATED FROM SPINACH CHLOROPLASTS

Values taken from data of Lichtenthaler[25].

Prenyllipid	Chloroplasts and Thylakoids	PS I particle fraction	PS II particle fraction
Chlorophyll a	100	100	100
Chlorophyll b	38	20	54
β-Carotene (c)	13	15	11
Lutein	14	7	21
Violaxanthin } x	5	7	4
Neoxanthin	2	0.5	3.5
Carotenoids (x+c)	34	29.5	39.5
Plastoquinone-9	5	2	8
α-Tocopherol	2.5	2	3
α-Tocoquinone	0.4	0.3	0.5
Phylloquinone K_1	0.7	1.3	0.1
Prenyllipid ratios:			
a / b	2.6	5.0	1.9
x / c	1.6	1.0	2.6
a+b / x+c	4.1	4.1	3.9
a / PQ-9	20	50	12.5
a / α-T	40	50	33.3
a / α-TQ	250	333	200
a / K_1	143	77	1000

that it also possesses a structural role in the photosynthetic membrane. It may exhibit in the sterol-free thylakoids a similar structural function as is assumed for the sterols in other cell membranes[32].

Phylloquinone. Within the photosynthetic membrane the phytylnaphthoquinone K_1 is predominantly associated with the pigment system I particles (Table 2). From this and the fact, that synthetic naphthoquinones can stimulate cyclic electron flow around pigment system I, it was concluded that the phylloquinone K_1 represents the endogenous cofactor for the cyclic electron flow[25].

Recent inhibitor studies[33] point to a role of phylloquinone in the linear electron transport chain near the reaction center of pigment system II. Brom-isopropyl-naphthoquinone (BIN, Fig. 9) blocks the photosynthetic electron transport chain at a position between the quencher Q and the plastoquinone (Fig. 10). The inhibition site is apparently identical with that of the well known herbicide DCMU and is thus quite different from the inhibition point of DBMIB. DBMIB blocks the function of the endogenous benzoquinone plastoquinone-9 (Fig. 9 and 10). In analogy to this it is concluded that bromnaphthoquinones block the function of the endogenous naphthoquinone K_1. The hypothesis that K_1, at its position between Q and plastoquinone-9, may be the re-entry point for cyclic electrons flowing around pigment system I is based on the observation that under photoheterotrophic growth conditions, which only allow cyclic electron flow, the level of the phylloquinone K_1 is considerably increased[33], while that of plastoquinone-9 decreases.

In addition to its role as an electron carrier phylloquinone also seems to have the function of a quencher of excitation energy of pigment system II. Halogenated naphthoquinones and to some extent also exogenous phylloquinone have proved to be good quenchers for the in vivo chlorophyll fluorescence of pigment system II[33].

III. LABELLING DEGREE AND TURNOVER

Recent observations on labelling kinetics of chloroplast prenyllipids from $^{14}CO_2$ in Chlorella[34,35] indicate that the individual prenyllipids of the photosynthetic membrane possess different life times. It is of particular interest that the labelling degree of three prenyl compounds (phylloquinone, α- and β-carotene) is much higher than that of the other chloroplast prenyllipids (Table 3). This means that phylloquinone and the carotenes have a very low half life time in the photosynthetic membrane: In the case of the carotenes the fast turnover may be due to an energy transfer from excited chlorophyll molecules[36,37] which may result in a photo-oxidative degradation of the carotenes. Since phylloquinone can act as quencher of the in vivo chlorophyll fluorescence[33] it is assumed that it may be destroyed in energy dissipation reactions. The high labelling degree, at a constant level of K_1 and the carotenes, indicates that resynthesis of the degraded molecules is very efficient.

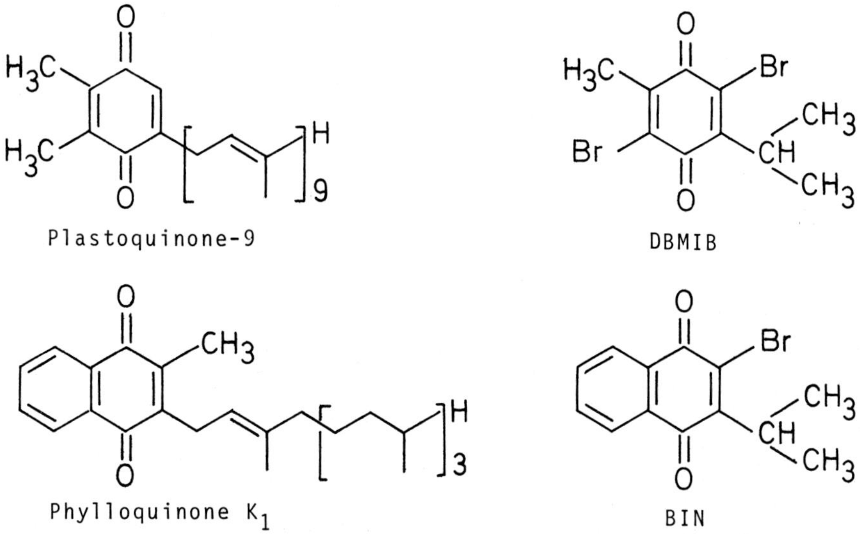

Fig. 9. Structure of the endogenous plastoquinone and phylloquinone and of synthetic bromo-isopropyl analogues. DBMIB = dibromomethylisopropyl benzoquinone. BIN = bromoisopropyl naphthoquinone.

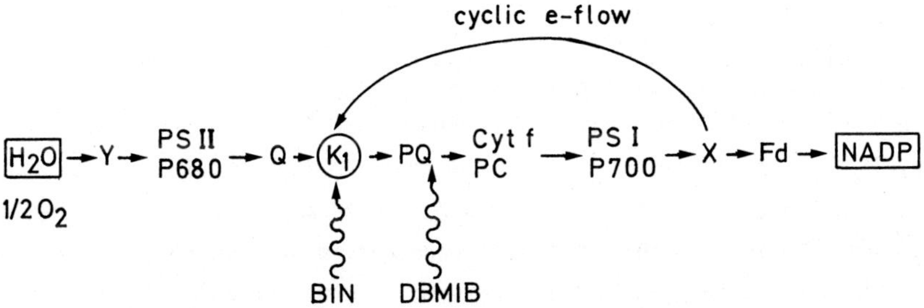

Fig. 10. Photosynthetic electron transport chain from water to NADP showing the possible site of function of the phylloquinone K_1. BIN (bromoisopropylnaphthoquinone) inhibits the electron transport near the reaction center of pigment system II (PS II) presumably by inhibiting the function of the endogenous naphthoquinone K_1.

TABLE 3
^{14}C-LABELLING DEGREE (IN %) OF PRENYLLIPIDS IN THE GREEN ALGA CHLORELLA AFTER 1 AND 2 HOURS OF $^{14}CO_2$ PHOTOSYNTHESIS
The experiment was performed under steady state conditions at constant prenyllipid level (after Grumbach[34, 35]).

Prenyllipid	1 h	2 h
Phylloquinone K_1	12.2	18.0
α-Carotene	13.1	18.8
β-Carotene	13.1	18.5
Plastohydroquinone-9	0.8	5.1
Plastoquinone-9	0.5	1.6
α-Tocopherol	0.1	0.4
α-Tocoquinone	0.5	1.8
Lutein	0.4	1.4
Violaxanthin	2.0	5.1
Antheraxanthin	0.6	1.3
Neoxanthin	0.7	0.7
Chlorophyll a	0.7	2.4
Chlorophyll b	0.5	1.3

In higher plants a breakdown of chlorophylls (Fig. 11) and other thylakoid prenyllipids can be demonstrated if one keeps a green plant in a prolonged darkness which excludes the light-triggered reformation of the degraded prenyl compounds. The loss in chlorophyll is accompanied by a decrease in the photosynthetic activity of the leaf as measured via the variable fluorescence[38] (Fig. 12). Upon re-illumination both the original level of chlorophyll and photosynthetic activity are restorred with time. From the observation that the photosynthetic activity declines much faster than the degradation of chlorophyll it is deduced that the reaction center chlorophyll as well as the plastoquinone-9[38] undergoes a faster turnover than the mass chlorophyll pigments in the antennae or in the light harvesting complex, which gives rise to a faster breakdown of the photosynthetic activity.

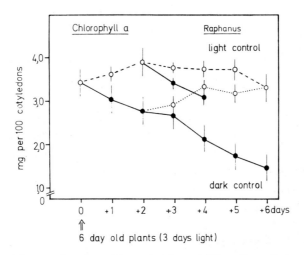

Fig. 11. Breakdown and reformation of chlorophyll a in radish seedlings during prolonged dark phases (●——●) and upon reillumination (o········o). o----o = control in continuous light.
(after Burkhard and Lichtenthaler, unpublished)

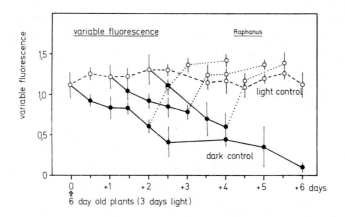

Fig. 12. Loss and reappearance of photosynthetic activity (as measured via the variable fluorescence[38]) in green radish leaves during a prolonged dark phase (●——●) and after reillumination (o········o). o-----o = control in continuous light.
(after Burkhard and Lichtenthaler, unpublished)

IV. VARIATION OF THYLAKOID PRENYLLIPID COMPOSITION

The prenyllipid composition of thylakoids can be changed considerably by application of herbicides[5], phytohormones[6], coloured light[3] and variation of light intensity. The adaptation of the photosynthetic apparatus to high and low light intensity can easily be seen by comparison of the prenyllipid content of sun and shade leaves from the same tree (Table 4). The sun leaves exhibit higher amounts of chlorophyll a, β-carotene and prenylquinones than shade leaves. This is visualized from higher values for the ratio chlorophyll a/b and chlorophylls/carotenoids (a+b/x+c) and from lower values for the ratios xanthophylls/carotene (x/c) and chlorophyll a to prenylquinones (a/PQ-9; a/K_1; a/α-TQ). These changes are the result of a different functional and structural organization of the thylakoids and their pigment protein complexes.

The "suntype" chloroplasts contain fewer light harvesting chlorophyll complexes (CP II) on a chlorophyll basis and accordingly less chlorophyll b and xanthophylls. This also means, when referred to chlorophyll a, more reaction centers and more photosynthetic electron transport chains which gives rise to the observed higher photosynthetic activity on a chlorophyll basis. The higher content of CP II complexes in shade leaves is correlated with higher grana stacks. These relationships between composition, structure and function of the photosynthetic apparatus are summarized in figure 13.

It is of particular interest that the "sun type" situation can be simulated by the application of cytokinins[6]. The accumulation of prenylquinones is strongly enhanced and that of α-tocopherol and carotenoids is markedly decreased (Table 5).

TABLE 4
COMPARISON OF THE PRENYLLIPID CONTENT OF FULLY DEVELOPED SUN AND SHADE LEAVES OF FAGUS SYLVATICA L.

	Sun leaves	Shade leaves
Leaf area	260 cm^2	380 cm^2
Dry weight	2.40 g	1.25 g
Water content	50%	66%
Pigment content (µg) per 100 cm^2 leaf area:		
Chlorophyll a	1340	750
Chlorophyll b	410	290
Chlorophyll a+b	1750	1040
β-Carotene	140	90
Lutein	170	135
Violaxanthin } x	46	22
Neoxanthin	18	13
Carotenoids (x+c)	374	260
Plastoquinone-9 (PQ-9)	520	75
α-Tocopherol (Vit. E; α-T)	320	71
Phylloquinone K$_1$	13	7
α-Tocoquinone	23	10
Prenyllipid ratios:		
a / b	3.3	2.6
x / c	1.6	1.9
a+b / x+c	4.8	4.0
a / PQ-9	2.6	10.0
a / α-T	4.2	10.2
a / K$_1$	103	107
a / α-TQ	58	75

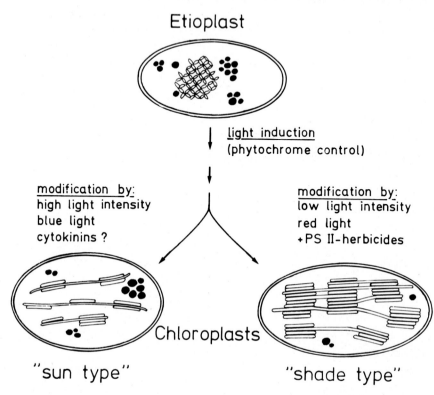

Fig. 13. Light induced formation of two different chloroplasts, the 'sun type' and the 'shade type', which differ in ultrastructure, photosynthetic activity and prenyllipid composition.
(after Lichtenthaler and Buschmann⁺).

TABLE 5
PROMOTION OF PHYLLOQUINONE AND PLASTOQUINONE-9 ACCUMULATION IN
RAPHANUS SEEDLINGS BY THE TREATMENT WITH KINETIN
(after Straub and Lichtenthaler [2]).

Prenyllipid	Control	+ Kinetin	increase or decrease
Chlorophyll a	2700	3100	+15%
Chlorophyll b	900	900	0
Carotenoids	572	480	-16%
Phylloquinone K_1	33	52	+58%
Plastoquinone-9 [a]	299	414	+38%
α-Tocopherol [b]	353	148	-58%

[a] Oxidized + reduced form
[b] α-Tocopherol + α-Tocoquinone

REFERENCES

1. Lichtenthaler, H.K. and Park, R.B. (1963) Nature (London), 198, 1070 - 1072.
2. Straub, V. and Lichtenthaler, H.K. (1973) Z. Pflanzenphysiol., 70, 308 - 321.
3. Buschmann, C. et al. (1978) Photochem. Photobiol., 27, 195 - 198.
4. Lichtenthaler, H.K. and Buschmann, C. (1978) in Chloroplast Development, Akoyunoglou, G. ed., Elsevier, Amsterdam, in press.
5. Lichtenthaler, H.K. (1977) in Lipids and Lipid Polymers in Higher Plants, Tevini, M. and Lichtenthaler, H.K. eds., Springer, Berlin, pp. 231 - 258.
6. Buschmann, C. and Lichtenthaler, H.K. (1977) Z. Naturforsch., 32c, 798 - 802.
7. Schneider, H.A.W. (1975) Ber. Dtsch. Bot. Ges., 88, 83 - 123.
8. Davies, B.H. (1977) in Lipids and Lipid Polymers in Higher Plants, Tevini, M. and Lichtenthaler, H.K. eds., Springer, Berlin, pp. 199 - 217.
9. Threlfall, D.R. and Whistance, G.R. (1971) in Aspects of Terpenoid Chemistry and Biochemistry, Goodwin, T.W. ed., Academic Press, London, pp. 357 - 359.
10. Liljenberg, C. (1977) in Lipids and Lipid Polymers in Higher Plants, Tevini, M. and Lichtenthaler, H.K. eds., Springer, Berlin, pp. 259 - 270.
11. Wellburn, A. R. (1976) Biochem. Physiol. Pflanzen, 169, 265 - 271.
12. Rüdiger, W. et al. (1976) Z. Pflanzenphysiol. 80, 131 - 143.
13. Shlyk, A.A. (1971) Ann. Rev. Plant. Physiol., 22, 169 - 184.

14. Akoyunoglou, G. (1967) Chemica Chronica, 32A, 5 - 8.
15. Kasemir, H. et al. (1979) in Chloroplast Development, Akoyunoglou, G. ed., Elsevier, Amsterdam, in press.
16. Lichtenthaler, H.K. and Kleudgen, H.K. (1977) Z. Naturforsch, 32c, 236 - 240.
17. Urbach, D. et al. (1976) Z. Naturforsch., 31c, 652 - 655.
18. Hampp, R. et al. (1975) Physiol.Plant., 33, 53 - 57.
19. Lichtenthaler, H.K. (1979) in Lipids and Technical Lipid Derivatives, Mangold, H.K. ed., CRC Handbooks of Chromatography, in press.
20. Threlfall, D.R. and Whistance, G.R. (1977) Phytochemistry, 16, 1903 - 1907.
21. Siefermann-Harms, D. et al (1978) Plant Physiol., 61, 530 - 533.
22. Douce, R. and Joyard, J. (1979) in Biochemistry and Physiology of Plant Lipids, Appelquist, L.-A. and Liljenberg, C. eds., Elsevier, Amsterdam, this volume.
23. Van Besouw, A.F.J.M. and Wintermanns, J.F.G.M. (1979) in Biochemisty and Physiology of Plant Lipids, Appelquist, L.-A. and Liljenberg, C. eds., Elsevier, Amsterdam, this volume.
24. Lichtenthaler, H.K. (1968) Endeavor, 27, 144 - 149.
25. Lichtenthaler, H.K. (1969) Progress in Photosynthesis Research, Metzner, H. ed., Vol. 1, pp. 304 - 314.
26. Butler, W.L. and Kitajima, M. (1974) in Proc. 3rd Intern. Congr. Photosynthesis, Avron, M. ed., Elsevier, Amsterdam, Vol. 1, pp. 13 - 24.
27. Thornber, J.P. (1975) Ann. Rev. Plant Physiol., 26, 127 - 158.
28. Wild, A. (1979) Ber. Dtsch. Bot. Ges., in press.
29. Trebst, A. (1974) Ann.Rev. Plant Physiol., 25, 423 - 458.
30. Lichtenthaler, H.K. and Tevini, M. (1969) Z. Naturforsch., 24b, 764 - 769.
31. Lucy, J.A. (1978) in Tocopherol, Oxygen and Biomembranes, de Duve, C. and Hayaishi eds., Elsevier, Amsterdam, pp. 109 - 120.
32. Huang, C. (1977) Chem. Phys. Lipids, 19, 150 - 158.
33. Lichtenthaler, H.K. and Pfister, K. (1978) in Photosynthetic Oxygen Evolution, Metzner, H. ed., Academic Press, London, pp. 171 - 193.
34. Grumbach, K.H. et al. (1978) Planta (Berl.), 141, 37 - 40.
35. Grumbach, K.H. and Lichtenthaler, H.K. (1978) Planta (Berl.), 141, 253 - 258.
36. Witt, H.T. (1971) Quaterly Rev. Biophys., 4, 365 - 477.
37. Mathis, P. (1969) Photochem. Photobiol., 9, 55 - 63.
38. Lichtenthaler, H.K. and Grumbach, K.H. (1974) in Proc. 3rd Intern. Congress on Photosynthesis, Avron, M. ed., Elsevier, Amsterdam, Vol. 3, pp. 2007 - 2015.

ACKNOWLEDGEMENTS

This work was sponsored by a grant from the Deutsche Forschungsgemeinschaft. We wish to thank Mrs. Ursula Prenzel and Miss Eva Schwarz for valuable assistance during the preparation of the manuscript.

THE CHLOROPLAST ENVELOPE : AN UNUSUAL CELL MEMBRANE SYSTEM

ROLAND DOUCE and JACQUES JOYARD
DRF / Biologie Végétale, CENG and USMG, 85 X, F 38041 GRENOBLE-Cedex (France)

SUMMARY

Structure of the chloroplast envelope
Chemical composition of the chloroplast envelope
- Polar lipids
- Pigments
- Polypeptides
Functions of the chloroplast envelope
- Metabolites transport in intact chloroplasts
- Protein transport through the chloroplast envelope
- Role of the chloroplast envelope in lipid synthesis

INTRODUCTION

The structure, chemical composition and function of the internal membrane system of the higher plants chloroplasts (stroma and grana lamellae) have been often described and discussed. Unfortunately, the pair of outer membranes surrounding the chloroplast, or envelope, has received little attention despite its importance in the functional and structural integrity of the chloroplasts. In this article, we describe the structure, chemical composition and functions of the higher plant chloroplast envelope.

STRUCTURE OF THE CHLOROPLAST ENVELOPE

Electron micrographs of higher plant chloroplasts show that the envelope consists of two morphologically and topologically distinct membranes (Figure 1). This applies to all kind of plastids described so far[1].

The total thickness of each envelope membrane of higher plants chloroplasts is reported as 6 nm[1]. The two membranes generally stain with approximately equal electron-density. The envelope membranes are separated by a region about 10-20 nm thick which appears electron-transluscent. Very often, the two envelope membranes

ABBREVIATIONS : MGDG, monogalactosyldiglyceride; DGDG, digalactosyldiglyceride; TGDG, trigalactosyldiglyceride; TTGDG, tetragalactosyldiglyceride; SL, sulfolipid; PC, phosphatidylcholine; PG, phosphatidylglycerol; PI, phosphatidylinositol; PE, phosphatidylethanolamine; PS, phosphatidylserine; DPG, diphosphatidylglycerol or cardiolipin.

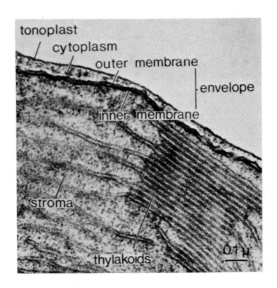

Figure 1 : The chloroplast envelope of higher plants consists of two morphologically and topologically distinct membranes. Note that the two membranes stain with approximately equal electron-density.(portion of a cell of spinach, *Spinacia oleracea* L.). Micrograph provided by Dr. J.P. Carde, University of Bordeaux.

apparently diverge and converge at random, producing small areas of membranes in close contact.

In a classic paper, Singer and Nicolson[2] have postulated that the lipids and the proteins of intact membranes are organized as follows :

a) the polar and ionic head groups of the lipid molecules together with all of ionic chains of the amphipathic globular proteins (integral proteins) are on the exterior surfaces of the membranes ;

b) the non-polar side chains of the integral proteins are in the interior of the membrane, together with the hydrocarbon tails of the polar lipids ;

c) the polar lipids are largely arranged in bilayer form.

The matrix of biological membranes appears to exist as a bilayer of mobile lipids,the relative motion of which determines the fluidity or viscosity of the membrane interior.

Several experiments indicate clearly that the structure of the envelope membranes is consistent with the lipid-globular protein mosaïc model of membranes

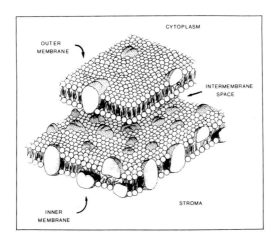

Figure 2 : The structure of both envelope membranes is consistent with the lipid-globular protein mosaic model proposed by Singer and Nicolson.

proposed by Singer and Nicolson[2] :

a) with glutaraldehyde fixation and staining with lead, the envelope membranes of higher plant chloroplasts[3] appear to have globular subunits at high magnification.

b) freeze-cleaving of the outer and inner membranes of the chloroplast envelope membranes from the red alga *Bangia fusco-purpurea*[4], *Euglena*[5], spinach[6] and the amyloplast envelope from root tip cells[7] reveals four fracture faces. The outer membrane of the chloroplast envelope shows a low density (283 particles / μm^2) of 9 nm particles in both complementary fracture faces. The inner membrane of the chloroplast envelope displays a significantly higher particle density. The face of the inner membrane-half in close contact with the chloroplast stroma contains more particles (1820 / μm^2) than that of the half closer to the cytoplasm (980 / μm^2). Their size in the platinium shadowed replica was 9 nm. The particles observed in both envelope membranes represent indubitably the integral proteins described by Singer and Nicolson[2]. It is likely that the differences present in the ultra-structure of inner and outer membranes, as revealed by freeze-fracturing, reflect functionnal and compositionnal differences.

c) Billecocq *et al*[8] and Billecocq[9,10], by means of specific antibodies, have shown that the polar head groups of the sulfolipid and the galactolipids are on the exterior surfaces of the envelope membranes. In other words, the polar heads are exposed to the aqueous phases facing the envelope membranes.

d) Neuburger *et al*[11] have demonstrated that binding of cationic ferricytochrome *c* to the envelope membranes is electrostatic and that the envelope membrane surface are strongly negatively charged. They also provide direct evidence that the outer surface of the outer envelope membrane is highly negatively charged. In accordance with these results, recent studies have shown that the isoelectric point of intact chloroplasts determined by cross-partition is acidic[12]. Therefore, the ionic heads of the acidic lipid molecules (SL and PG) and the acidic polar group of the integral proteins are exposed at the exterior surfaces of the envelope membranes.

For these reasons, we believe that the envelope membrane substructure agrees with the lipid-globular protein mosaic model of membrane substructure proposed by Singer and Nicolson[2] (Figure 2). It is important, however, to bear in mind that a single rigidly defined model may eventually be difficult to reconcile with all biological membrane structure[13].

CHEMICAL COMPOSITION OF THE CHLOROPLAST ENVELOPE

The isolation in large quantities of chloroplast envelope membranes free of intra and extra-chloroplastic contamination[14] allows the complete analysis of their chemical components.

The chloroplast envelope membranes, when compared with other cell membranes analyzed so far, exhibit a most unusual chemical composition. However, the chemical properties attributed to the isolated envelope membranes vary from one group of workers to another (Table 1). Therefore, the reported disparities may

TABLE 1
POLAR LIPID COMPOSITION OF ISOLATED ENVELOPE MEMBRANES
Data are expressed as dry weight percentage of total lipids (excluding pigments).

species	MGDG	DGDG	TGDG	TTGDG	SL	PC	PG	PI	PE	PS	DPG	references
(a)	20	30	4	1	6	20	8	1	tr	0	0	[14]
(a)	27	33	1	-	tr	25	6	1	-	-	tr	[30]
(a)	8	29	5	-	6	27	13	-	tr	-	-	[19]
(b)	27	37	-	-	-	28	8	-	-	-	-	[15]
(c)	24	26	-	-	5	23	9	3	0	0	-	[21]
(d)	22	43	-	-	14	10	11	0	0	0	0	[20]
(e)	31	26	tr	-	1	29	5	1	2	-	0	[18]
(f)	46	18	tr	-	3	7	2	1	1	-	1	[18]

tr : traces (< 0.5 %); - : not recorded.
Species analyzed : (a) *Spinacia oleracea*; (b) *Vicia faba*; (c) *Narcissus pseudonarcissus*; (d) *Triticum sativum*; (e) *Helianthus annuus*; (f) *Zea mays* (mesophyll).

arise due to contamination of envelope preparation by extra and intra-chloroplastic membranes. Up to now, practically all analysis have been carried out on higher plant chloroplast or etioplast envelope membranes. It is probable, however, that there is a high degree of uniformity in the chemical composition of the envelope of various plastids in higher plants.

Polar lipids. Douce et al[14] and Mackender and Leech[15,16] were the first to report the polar lipid and fatty acid composition of the chloroplast envelope membranes from *Spinacia oleracea* and *Vicia faba* leaves. They concluded that qualitatively the polar lipids of both types of chloroplast membranes, thylakoids and envelope, are identical but the proportion in which they are present is different (Table 2). In thylakoid membranes, the two main galactolipids (MGDG and DGDG) are present in a ratio 2:1 and the two main phospholipids (PC and PG) are present in a ratio 1:3. This situation is totally reversed in the envelope fraction. In this case, MGDG and DGDG are present in a ratio of 0.3 to 0.8:1; PC and PG are present in a ratio of 3:1 although the concentration of all polar lipids is greater in the envelope fraction than in the thylakoid fraction, this is particularily true for the relative concentration of PC in both types of membranes. It was concluded[14-17] also that the fatty acids are more saturated in the envelope than in the thylakoids (Table 3). This is particularily true for

TABLE 2

POLAR LIPID COMPOSITION OF ISOLATED PLASTID MEMBRANES AND MITOCHONDRIAL MEMBRANES
Data are expressed as dry weight percentage of total lipids (excluding pigments)

	MGDG	DGDG	TGDG	TTGDG	SL	PC	PG	PI	PE	PS	DPG	references
chloroplast :												
envelope	20	30	4	1	6	20	8	1	tr	0	0	[14]
thylakoids	51	26	-	-	7	3	9	1	0	0	0	[14]
etioplast :												
envelope	22	41	-	-	10	14	9	0	0	0	0	[20]
prolamellar body	36	41	-	-	10	4	9	0	0	0	0	[20]
mitochondria :												
outer membrane	0	0	0	0	0	42	10	21	24	-	3	[26]
inner membrane	0	0	0	0	0	41	3	5	37	-	14	[26]
	0	0	0	0	0	33	4	5	32	-	23	[25]

tr : traces (< 0.5 %); - : not recorded.

TABLE 3

FATTY ACID COMPOSITION OF ISOLATED ENVELOPES AND THYLAKOIDS FROM CHLOROPLASTS
Data are expressed as dry weight percentage of total fatty acids.

	C 16:0	C 16:1	C 16:3	C 18:0	C 18:1	C 18:2	C 18:3	$\frac{\text{unsaturated}}{\text{saturated}}$	references
E	15	3	9	tr	6	10	57	5.7	[14] (a)
T	8	5	13	tr	2	2	70	12.2	[14]
E	41	3	2	10	5	8	29	0.9	[18] (a)
E	13	2	-	4	6	12	63	5.3	[16] (b)
T	6	1	-	2	3	5	83	12.3	[16]
E	19	2	-	1	2	8	67	4.0	[20] (c)
E	21	6	-	4	1	6	62	3.0	[38] (d)
E	36	3	1	5	2	14	40	1.4	[18] (e)
E	61	0	3	14	2	4	13	0.3	[18] (f)

tr : traces (< 0.5 %); - : not recorded; E : envelope; T : thylakoids.
Species analyzed : (a) *Spinacia oleracea*; (b) *Vicia faba*; (c) *Triticum sativum*;
(d) *Phaseolus vulgaris*; (e) *Helianthus annuus*; (f) *Zea mays* (mesophyll).

DGDG (table 4), but the difference found is not as dramatic as reported by Poincelot[18].

All the results have been confirmed by several groups[19-21] and it is interesting to note that the conclusions obtained with mature chloroplasts[14-16] are also valid with etioplasts and greening etioplasts[20] (Table 2). According to Hashimoto and Murakami[19], the high phospholipid content observed in the chloroplast envelope (when compared to that of thylakoids) reflect the chemical nature of the outer membrane of the envelope which could be comparable to that of cytoplasmic membranes[22]. However, there is no convincing evidence for this. As already stated, Billecocq et al[8] and Billecocq[9,10] ,by means of specific antibodies, have clearly shown that the outer membrane of the chloroplast envelope contains galactolipids and sulfolipid. Furthermore PE, which is a major constituent of all the cytoplasmic membranes examined so far[22], is barely detectable in both envelope membranes and in addition, PE is considered now to be a negative marker for purified chloroplasts.

In a review on biomembrane polar lipids in higher plants, Mazliak wrote[22] : "the overwhelming impression is one of the great uniformity of lipid composition among the different cell membranes". This remark is in contradiction to table 2

TABLE 4

FATTY ACID COMPOSITION OF THE DIFFERENT LIPIDS FROM ISOLATED ENVELOPES AND THYLAKOIDS FROM SPINACH CHLOROPLASTS

Data are expressed as dry weight percentage of total fatty acids.

		C 14:0	C 16:0	C 16:1	C 16:2	C 16:3	C 18:0	C 18:1	C 18:2	C 18:3
MGDG	T	tr	tr	tr	tr	26.0	tr	0.6	1.0	72.0
	E	tr	3.9	tr	tr	19.7	tr	0.8	2.0	70.7
DGDG	T	1.0	4.6	tr	tr	4.0	tr	2.5	0.8	87.0
	E	tr	15.3	tr	tr	4.1	1.0	5.3	3.5	69.5
TGDG	E	tr	4.0	tr	tr	19.1	tr	0.8	1.5	74.1
PC	T	tr	13.2	tr	tr	1.1	tr	7.7	19.8	57.2
	E	1.0	22.6	1.0	tr	tr	tr	15.8	21.3	37.3
PG	T	1.0	18.3	39.6	tr	tr	tr	0.5	1.6	38.5
	E	1.4	21.2	38.0	tr	tr	tr	1.6	3.6	32.1
SL	T	tr	47.5	tr	tr	tr	tr	0.8	4.4	46.9
	E	tr	47.1	tr	tr	tr	1.0	2.3	5.9	43.8
PI	T	2.0	42.1	0.5	4.1	2.0	1.0	7.0	14.3	26.3
	E	2.1	42.4	tr	3.0	1.0	1.0	8.0	20.1	22.2
DG	E	4.0	3.3	tr	tr	20.1	1.3	8.7	3.3	58.8

tr : traces (< 0.5 %); E : envelope; T : thylakoids; DG : diacylglycerol.

which shows clearly that the polar lipid content of the envelope membranes is different to that of mitochondrial membranes. Thus the envelope carefully prepared is devoid of PE and DPG but contains large amounts of galactolipids. Conversely, the pure mitochondrial membranes, rich in PE, are practically devoid of galactolipids[23-26]. Consequently, such a result precludes a structural relationship between the plastidial and the mitochondrial structures. It may, however, be noted that the presence of galactolipids in mitochondrial preparations has been reported[16,27,28]. It is important to bear in mind that in a plant cell plastids may represent a large amount of the total membrane area and it is known that their fragmentation during isolation leads to a massive contamination of the mitochondrial fractions by a variable proportion of plastid subfractions (envelope, plastoglubule clusters, intact and broken thylakoids). For these reasons we believe that the galactolipids found in the mitochondrial membranes are entirely attributable to contaminating plastid fragments.

The proportion of lipids and proteins present in the chloroplast envelope is different from other membranes. Douce et al[14] were the first to report a high ratio of acyl lipids to proteins (1.2) in the chloroplast envelope of *Spinacia oleracea*. This value is much higher than that of mitochondrial and "microsomal" membranes[26]. Such a high ratio, which has been confirmed by several groups of investigators[18,21,29], could easily explain the great fragility of higher plant chloroplast envelope and low value of the buoyant density (1.12 g/cm^3) of isolated envelope vesicles.

Pigments. The chloroplast envelope membranes are devoid of chlorophyll[14-16,30] (Figure 3). Consequently, chlorophyll is considered now to be a negative marker for highly purified envelope membranes. The envelope of chloroplast is deep-yellow[11] and "yellowish" when the yield is low[30,31]. Douce et al[14] were the first to report the presence of carotenoids in the envelope membranes (Figure 3).

Qualitatively the carotenoids of both types of chloroplast membranes, thylakoids and envelope, are identical but the proportion in which they are present is different[14,32,33] (Table 5). β-carotene accounted for a higher proportion of the thylakoidal carotenoid content when compared with the envelope fraction. On the other hand, violaxanthin accounted for a higher proportion of the envelope carotenoid content when compared with the thylakoids. These differences may be characterized by the xanthophyll to carotene ratio which is much higher (∼ 6) in the envelope fraction than in the thylakoids (∼ 3).

Criticism, however, has been levelled against the purity of the membranes prepared by the method used. According to Sprey and Laetsch[34] and Goodwin[35], the carotenoids could be present in trapped plastoglobule clusters and are probably absent from the envelope membranes. This criticism is valid only if one starts

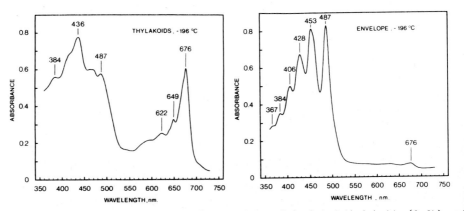

Figure 3 : Low temperature absorption spectrum of isolated thylakoids (left) and envelope (right) membranes. Note that envelope do not contain chlorophylls.

with unpurified chloroplasts. Plastoglobules are generally circular and of 10 to 500 nm in diameter, they are embedded in the stroma between the thylakoids membranes. The broken thylakoïds associated with plastoglobules possess a similar density as the envelope membranes[34]. In this case, envelope fraction prepared from unpurified chloroplasts is contaminated by plastoglobule clusters embedded in swollen grana. So, we believe that carotenoïds are genuine constituents of the envelope membranes for the following reasons :

a) the envelope membranes carefully prepared are devoid of plastoglobule clusters (see the different micrographs in ref. 14,36,37).

b) the concentration ratio of violaxanthin in the envelope is reproducible for various experimental procedure[33].

c) the very low level of chlorophyll present in the spinach envelope fraction is also inconsistent with a significant contribution of carotenoïds from fragments of the thylakoïdal system (Figure 3).

d) the carotenoïd content in envelope membranes is not a contamination by carotenoïds randomly separated from the thylakoïds during osmotic shock and re-dissolved in the highly lipophillic envelope membranes[33] (Figure 4).

These observations rather suggest that the carotenoïds found in unpurified preparations of plastoglobules could be attributable to a contamination by envelope membranes. The presence of carotenoïds in the envelope membranes has been confirmed for spinach leaves[19] and *Phaseolus vulgaris* leaves[38].

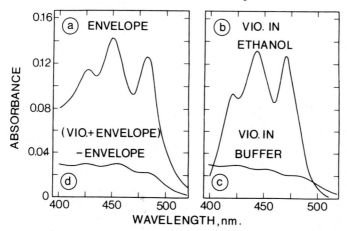

Figure 4 : Comparison of the absorbance spectrum of chloroplast envelopes (a) with the spectrum of purified violaxanthin in ethanol (b), in buffer (c) and in buffer containing envelope membrane vesicles (d). Spectra (d) is a difference spectra : (envelope + violaxanthin) minus envelope. If envelope lipids were able to solubilize violaxanthin, spectra (d) should be the same as spectra (b). In fact, spectra (d) clearly shows that pure violaxanthin cannot be dissolved in envelope membrane lipids and remain as in buffer (spectra c). For experimental conditions, see ref. 33.

TABLE 5

CAROTENOID COMPOSITION

OF THYLAKOIDS AND ENVELOPES FROM DARK- AND LIGHT-TREATED CHLOROPLASTS

For experimental conditions see ref. 33.

	pigment content (μg/mg thylakoid or envelope protein)					
	thylakoids			envelopes		
	dark	light	light - dark	dark	light	light - dark
violaxanthin	4.52	3.16	- 1.36	6.50	4.30	- 2.20
lutein + zeaxanthin	6.05	7.19	+ 1.14	2.50	3.00	+ 0.50
β-carotene				1.42	1.51	--
antheraxanthin		not determined		0.79	0.79	--
neoxanthin				< 0.40	< 0.40	--
violaxanthin decrease in light (% of control)		30 %			34 %	

Jeffrey et al[32] demonstrated that envelope membranes prepared from dark-treated leaves had a violaxanthin content up to 3.5 times the lutein plus zeaxanthin content, whereas in chloroplast envelopes from illuminated leaves, this ratio was only 0.75. Based on this data, the authors suggested that the light-induced pigment changes in envelopes were probably due to de-epoxidation of violaxanthin to zeaxanthin. Carotenoids changes in envelope membranes have been confirmed with intact chloroplasts[33] but the light-induced violaxanthin decrease in the envelope (Table 5) is not caused by an envelope or a stroma de-epoxidase but results probably from a violaxanthin exchange between envelope and thylakoids[33].

It is likely that the envelope membranes could be a site of carotenoid synthesis in intact chloroplasts. This hypothesis is supported by observations that suggest a carotenoid exchange between envelope and thylakoids (see above). In addition, the envelope has been found in all plastids examined so far[1] and the existence of the envelope structure precedes the thylakoids[1]. Furthermore, Moore and Shepard[39] have indicated that in *Acetabularia* both the pattern and the rates of pigment synthesis are comparable *in vivo* and *in vitro*. They conclude that the complete pigment synthesis pathways and their control mechanisms reside within the chloroplasts. According to Goodwin[40] and Bickel and Schultz[41], intact isolated spinach chloroplasts incorporated ^{14}C from ^{14}CO$_2$ into β-carotene under photosynthetic conditions[42]. Since efficient synthesis of carotenoids takes also place in chromoplasts which lack thylakoïds, it seems reasonable to suppose that the envelope membranes of chloroplasts, in conjunction with soluble enzymes of the stroma and/or the cytoplasm are active in carotenoid biosynthesis. For example, Costes et al[43] have demonstrated that spinach chloroplast envelopes were a site of violaxanthin synthesis.

Polypeptides. Several reports have described SDS polyacrylamide gel electrophoresis of chloroplast envelope proteins (for a review, see ref. 44). Practically all workers reported a serie of high molecular weight bands (above 70,000 daltons) and two predominant polypeptides at approximately 52,000 and 29,000 daltons. The 52,000 band is electrophoretically non-coincident with the large subunit of ribulose 1,5-bisphosphate carboxylase[45] and thus could not be accounted for a contamination of the pure envelope fraction by ribulose 1,5-bisphosphate carboxylase. Flügge and Heldt[46,47] have demonstrated that the 29,000 daltons polypeptide, which represent 20-25 % of the total envelope proteins, play a role in the functioning of the phosphate translocator.

The molecular weight distribution of the envelope polypeptides is markedly different from that of the thylakoïd polypeptides[21,45]. The significance of this fact might be that the synthesis of the thylakoïds is not merely an extention of the inner membrane but involves a major change in the types of proteins inserted into the growing membrane[45,48]. As the envelope proteins are mainly concerned with the transport of many low molecular weight metabolites accross the envelope and with the transfer of chloroplastic proteins synthesized on cytoplasmic ribosomes, the envelope proteins may be expected to differ from those of the thylakoïds which are involved in the light dependent reactions of photosynthesis.

FUNCTIONS OF THE CHLOROPLAST ENVELOPE

The chloroplast envelope maintains the soluble enzymes of the stroma in close contact with the thylakoïd network. Furthermore, the envelope insulate the electron carriers of the thylakoids from those of the other cell membrane systems.

In addition to this simple mechanical role, the three main functions of the chloroplast envelope membranes are :

a) the envelope regulates the movements of low molecular weight molecules between the stroma and the cytoplasm.

b) the envelope regulates the uptake of the many chloroplast proteins that are made by the cytoplasmic ribosomes.

c) the envelope regulates the synthesis of galactolipids.

Metabolites transport in intact chloroplasts. The expanding literature on metabolites transport has been reviewed principaly by Heber[49], Walker[50] and Heldt[51]. Experiments with isolated chloroplasts have demonstrated that the two membranes of the envelope delineate two compartments differing in their accessibility to compounds of different molecular weights : the stromal space and the intermembrane space of the envelope. The membrane space situated between the inner and the outer membrane of the chloroplast envelope is found to be non specifically permeable to sucrose and other molecules charged or uncharged, up to a molecular weight of about 10,000. In contrast, the inner envelope membrane surrounding the stroma space, which is impermeable to sucrose, is selectively permeable to a limited number of anions owing to specific translocators. So far, three specific translocators have been well characterized on the inner membrane of the chloroplast envelope from plants having only the Benson-Calvin pathway of photosynthesis :

a) the <u>phosphate translocator</u> is specific for the transport of inorganic phosphate, 3-phosphoglycerate, dihydroxyacetone phosphate and glyceraldehyde 3-phosphate[52]. The phosphate translocator plays an important role in the control of sucrose and starch synthesis in green leaf cells[53].

b) the <u>dicarboxylate translocator</u> is specific for the transport of malate, oxaloacetate, aspartate and glutamate[52]. This translocator allows the indirect transfer of reducing equivalents accross the envelope, it can operate in both directions and depends on the redox potential of pyridine nucleotides outside and inside the chloroplasts.

c) the <u>ATP translocator</u> has a high specificity for external ATP[54,55].

Consequently, during photosynthesis, the principal imports passing into the chloroplasts are : water, CO_2, orthophosphate; the major exports to the cytoplasm are : O_2, dihydroxyacetone phosphate and/or glyceraldehyde 3-phosphate[44] (Figure 5).

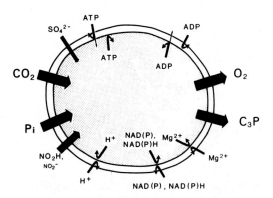

Figure 5 : Summary of the main exchanges between chloroplast and its environnement. Ease of penetration is represented by magnitude of arrows.

In contrast, metabolites transport in intact chloroplasts isolated from plants having the C 4 dicarboxylic acid pathway of photosynthesis (C 4 plants) is practically unknown. The reason for this is that the preparation of chloroplasts from C 4 plants retaining both structural and functionnal integrity has attained only limited success, despite intensive trials.

Protein transport through the chloroplast envelope. Although chloroplasts contain all the compounds for a biological system to be autonomous (DNA, DNA polymerase, RNA, RNA polymerase and protein synthesizing machinery) these compounds neither synthesize nor code for all chloroplast proteins : many of them are synthesized on cytoplasmic ribosomes[56]. Consequently, it is obvious that during plastid development, large amounts of proteins synthesized in cytoplasm must somehow cross the envelope membranes.

So far, several mechanisms have been proposed in order to explain the selection and the passage of several proteins (or proteins subunits) through the chloroplast envelope :

a) direct injection. this mechanism easily explains the passage of the proteins through the outer envelope membrane. It does not explain the passage of the protein through the inner envelope membrane unless the passage occurs at the point where both membranes are in contact. According to this mechanism, nascent polypeptide chains contain a signal peptide at the amino-terminal end of the chain. The signal peptide triggers attachment of the ribosome to special membrane receptor proteins. After transfer across the envelope, the signal peptide is removed by a special enzyme, the "signal peptidase or signalase"[57].

b) specific carriers proteins. this hypothesis implies that a special class of proteins exists in the envelope of chloroplasts which catalyze the unidirectional influx of all those proteins made on cytoplasmic ribosomes but which are destined to function in the chloroplast[58]. In this case, the recognition event seems to be an additional sequence present on the polypeptide which contains the information necessary for binding to a specific envelope protein carrier[59,60].

c) pinocytosis. Electron microscopic studies of developing plastids indicate that the inner membrane of the plastid is rarely completely smooth but possesses frequent discrete folds which invaginate into the stroma[1,44] and form vesicles; it is possible that some proteins interact with the outer surface of the inner membrane and trigger the formation of pinocytotic vesicles, for instance by stimulating galactolipids synthesizing enzymes of the envelope[61].

Role of the chloroplast envelope in lipid synthesis. The major chloroplast membrane acyl lipids consist of three glycolipids rich in polyunsaturated fatty acids (MGDG, DGDG and SL) and two phospholipids (PG and PC), whether small quantities of PE also occur in chloroplasts is a problem which is , as yet, unresolved and is certainly a reflection of the difficulties involved in the isolation of chloroplast membranes free from other cellular constituents.

a) <u>Origin of chloroplast phospholipids</u>. The biosynthesis of PC by the nucleotide pathway has been conclusively demonstrated in spinach leaves[62,63]:

CDP-choline + sn-1,2-diacylglycerol → phosphatidylcholine + CMP

It occurs almost exclusively in the microsomal fraction. Unfortunately, the microsomal fraction is a 100,000 g pellet collected from a cell-free homogenate after removal of the larger cell organelles such as chloroplasts (4,000 g pellet) and mitochondria (10,000 g pellet). In fact, the microsomal fraction contains heterogenous vesicles deriving from all the cell membranes. If the major vesicles present in the microsomal fraction could derive from endoplasmic reticulum, it is likely that a large amount of the vesicles derive from mitochondria, chloroplasts (thylakoids and envelope membranes), dictyosomes and microbodies. If chloroplast envelopes sediment at 30,000 g, the membranes also fragment into smaller vesicles which sediment at 100,000 g or more. Therefore, much of the phosphorylcholine transferase activity previously believed to be associated with the microsomal fraction could, in fact, be derived from the chloroplast envelope vesicles. However, Joyard and Douce[36] have demonstrated that the purified envelope from spinach chloroplasts although containing large amounts of diacylglycerol[37] are devoid of phosphorylcholine transferase activity (Figure 6) indicating probably that PC is synthesized outside of the chloroplasts. These results are essentially in agreement with the finding of Devor and Mudd[62].

It is possible that PC could be synthesized in the envelope by methylation of PE. Indeed this pathway as a route for biosynthesis of PC in spinach leaves has been established *in vivo* as well as direct methylation of phosphatidyl-N-

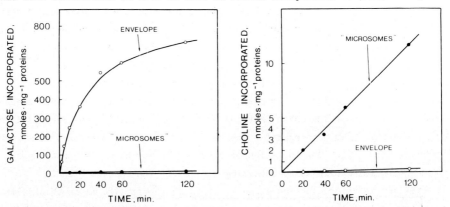

Figure 6 : Galactolipids are synthesized by envelope membranes and not by "microsomes". On the contrary, envelope membranes are unable to catalyse the transfer of choline from CDP-choline to PC, while "microsomes" can do it easily. Experimental conditions, see ref. 36.

methylethanolemine to phosphatidyl-N,N-dimethylethanolamine and of the latter to PC by S-adenosylmethionine[63]. This observation may explain the absence of PE in envelope membranes.

Similarly, Marshall and Kates[64] provided evidence that the biosynthesis of PG which involves the intermediate formation of phosphatidylglycerophosphate from CDP-diacylglycerol and sn-glycerol 3-phosphate occurs in spinach leaves :

sn-glycerol 3-phosphate + CDP-diacylglycerol →
3-sn-phosphatidyl-1'-sn-glycerol 3-phosphate + CMP

3-sn-phosphatidyl-1'-sn-glycerol 3-phosphate →
3-sn-phosphatidyl-1'-sn-glycerol + phosphate

Subcellular localisation studies of PG synthetase revealed that 3 % of the total activity is associated with the chloroplast fraction. Activity is also found in 15,000 g and 90,000 g pellets and in the 90,000 g supernatant. The fraction with the highest specific activity and largest proportion of the total activity is the 40,000 g pellet. Although phosphatidyl-PG-synthetase in green leaves is called a "microsomal" enzyme[64], it could be possible that it originates in the chloroplast envelope, one of the "microsomes" component. Joyard and Douce (unpublished data) have conclusively demonstrated the absence of a CDP-diacylglycerol synthesizing enzyme in the envelope membranes and thylakoids isolated from spinach chloroplasts. Such a result indicates that probably the chloroplast membranes do not contain the enzymes involved in PG synthesis. This problem should be further investigated in order to determine whether or not the envelope is able to catalyze the final steps of PG synthesis.

All these results strongly suggest, but do not prove, that the chloroplast membrane phospholipids may be synthesized by the endoplasmic reticulum system and then incorporated into chloroplast envelope membranes. The same is probably true for the major plant mitochondrial phospholipids. In this case, in spite of many attempts, definitive evidence for the synthesis of PE and PC by intact plant mitochondria $in\ vitro$ has yet to be obtained even though these organelles are rich in these phospholipids[28,65,66].

If the major phospholipids of the chloroplasts are really synthesized at the level of the endoplasmic reticulum, we must imagine a direct transfer of phospholipids between the reticulum and the chloroplast envelope. In animal[67,68] and plant[69] cells exchange of phospholipids has been demonstrated between various intracellular membranes including rough and/or smooth endoplasmic reticulum, whole mitochondria, inner and outer membranes of the mitochondria and plasma membranes. This exchange is catalyzed by one or several specific phospholipid exchange proteins located in the cytoplasm[69] and operating as phospholipid carriers. Unfortunately, evidence for phospholipid exchange between chloroplasts and other cellular fractions is lacking[69].

Finally, the very interesting and attractive proposal concerning possible structural continuity between various cellular membranes[70] and particularly between the endoplasmic reticulum and the outer envelope membrane, if correct, would mean that phospholipids synthesized in the reticulum could diffuse laterally within a continuous network to the outer membrane of the envelope. In other words, the reticulum endoplasmic behaves as a *"generating element"*, the *"end product"* being the envelope membranes and subsequently the thylakoïds.

b) <u>Origin of the chloroplast galactolipids</u>. Early reports suggested that the chloroplast is the main site of galactolipid synthesis in leaves[71,72]. A later report showed that the highest specific activities of UDP-gal incorporation into galactolipids was associated with the 40,000 g and 100,000 g pellet and the conclusion was drawn that the activity is associated with the "microsomes"[73]. It has subsequently been unequivocally demonstrated that the chloroplast envelope is a site of UDP-gal incorporation into both MGDG and DGDG in leaf cells[61]. The reason for the confusion of the earlier reports is the tendency for chloroplast envelopes to lyse during isolation procedure and then contaminate the microsomal fraction. The ability of the isolated chloroplast envelope to synthesize galactolipids was confirmed by other workers using leaves[74,75] and cells of *Euglena gracilis*[76]. Furthermore, Liedvogel and Kleinig[75,77] have clearly demonstrated that the non-photosynthetic chromoplasts inner membranes from the corona of *Narcissus pseudonarcissus* are also found to contain galactolipid synthesizing activities. The same is probably true for the envelope of amyloplasts from etiolated corn coleoptile (Hartman-Bouillon and Benveniste, personal communication). In spinach leaves, the specific activity of the galactolipid synthesizing enzyme is extremely high (45 nmoles/mg protein/min) in carefully prepared envelope[37,78] and exceeds the corresponding figures for total cell proteins by a factor of at least 100. Moreover, the "microsomes" fraction, containing only small amounts of envelope membrane vesicles, is practically unable to synthesize the major galactolipids[36] (Figure 6). All these results demonstrate definitively that in plant cells the plastid envelope, and probably the inner membrane, catalyze specifically the final steps in galactolipid biosynthesis.

Furthermore, we have also demonstrated[80] that the assembly of the three parts of the galactolipids molecules (galactose-glycerol-fatty acids) occurs on the chloroplast envelope although the different substrates required (UDP-galactose, *sn*-glycerol 3-phosphate and fatty acids) originate either from the chloroplast stroma or from the cytoplasm[79]. All the enzymes involved in galactolipid synthesis operate in a multi-enzyme sequence[80] on the envelope. It is clear that the chloroplasts, owing to their envelope membranes, contain the complete array

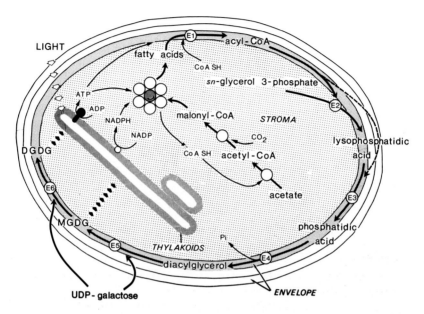

Figure 7 : The chloroplast envelope plays a predominant role in the assembly of the three parts of the galactolipids molecules (galactose, glycerol, fatty acids). Saturated and monounsaturated fatty acids are synthesized in the stroma by a multi-enzyme complexe (fatty acid synthetase). Then, the different steps occur on the envelope, probably at the level of the inner membrane. In these conditions massive transports of galactolipids should occur very rapidly between envelope and thylakoïds. For details, see ref. 80.
E 1 : acyl-CoA synthetase; E 2 : acyl-CoA : sn-glycerol 3-phosphate acyltransferase; E 3 : acyl-CoA : acyl-sn-glycerol 3-phosphate acyltransferase;
E 4 : phosphatidate phosphatase; E 5 : UDP-gal : diacylglycerol galactosyltransferase; E 6 : UDP-gal : MGDG galactosyltransferase and/or galactolipid : galactolipid galactosyltransferase.

of galactolipid biosynthetic machinery (Figure 7). Consequently, chloroplast envelope membranes play a key role in chloroplast biogenesis.

It is well established that the main localization of the polar lipid synthesizing enzymes in plant and animal cells is the endoplasmic reticulum and it is possible that no cellular membrane is independant of the endoplasmic reticulum for its biogenesis[70]. However, the data we present here show, for the first time, that the endoplasmic reticulum is not directly involved in the synthesis of an organelle's major structural lipids.

CONCLUSION

The chloroplast is an unusual cell membrane system. The polar lipid composition of this yellow membrane is strikingly different from that of all other cell membranes (mitochondria, endoplasmic reticulum...).

Furthermore, the envelope is devoid of b cytochrome and of NADH : cytochrome c oxidoreductase activity, contrarily to the other cell membranes. It allows specific transfer of metabolites between chloroplasts and cytoplasm.

Finally, owing to their envelope membranes, the chloroplasts (which are unable to synthesize their own phospholipids) contain the complete array of galactolipids biosynthetic machinery.

ACKNOWLEDGEMENTS

The authors are indebded to Professor Andrew A. Benson for his kind interest. Our grateful thanks are owed to Drs. D. Siefermann-Harms, E. Heinz and J.P. Carde.

This work is supported as a continuing program by the "Commissariat à l'Energie Atomique" and the "Centre National de la Recherche Scientifique".

REFERENCES

1. Gunning, B.E.S. and Steer, M.W. (1975) *Ultrastructure and the Biology of Plant Cells*, Edward Arnold, London.
2. Singer, S.J. and Nicolson, G.L. (1972) *Science 175*, 720-731.
3. Weier, T.E., Engelbrecht, A.H.P., Harrison, A. and Risley, E.B. (1965) *J.Ultrastruct.Res. 13*, 92-111.
4. Bisalputra, T. and Bailey, A. (1973) *Protoplasma 76*, 443-454.
5. Miller, K.R. and Staehelin, L.A. (1973) *Protoplasma 77*, 55-78.
6. Sprey, B. and Laetsch, W.M. (1976) *Z.Pflanzenphysiol. 78*, 360-371.
7. Fineran, B.A. (1975) *Phytomorphology 25*, 398-415.
8. Billecocq, A., Douce, R. and Faure, M. (1972) *C.R.Acad.Sci.Paris 275*, 1135-1137.
9. Billecocq, A. (1974) *Biochim.Biophys.Acta 352*, 245-251.
10. Billecocq, A. (1975) *Ann.Immunol. (Inst.Pasteur) 126 C*, 337-352.
11. Neuburger, M., Joyard, J. and Douce, R. (1977) *Plant Physiol. 59*, 1178-1181.
12. Westrin, H., Albertsson, P. and Johansson, G. (1976) *Biochim.Biophys.Acta 436*, 696-706.
13. Benson, A.A. and Jokela, A.T. (1976) in *Plant Biochemistry*, 3rd edition, Bonner, J. and Varner, J.E. eds., Academic Press, New York, pp. 65-89.
14. Douce, R., Holtz, R.B. and Benson, A.A. (1973) *J.Biol.Chem. 248*, 7215-7222.
15. Mackender, R.O. and Leech, R.M. (1972) in *Proc.2nd Int.Cong. on Photosynthesis*, Forti, G., Avron, M. and Melandri, A. eds., Junk, The Hague, pp. 1431-1440.
16. Mackender, R.O. and Leech, R.M. (1974) *Plant Physiol. 53*, 496-502.
17. Heinz, E., Siebertz, H.P., Linscheid, M., Joyard, J. and Douce, R. *this volume*
18. Poincelot, R.P. (1976) *Plant Physiol. 58*, 595-598.
19. Hashimoto, H. and Murakami, M. (1975) *Plant Cell Physiol. 16*, 895-902.

20. Bahl, J., Franke, B. and Monéger, R. (1976) *Planta 129*, 193-201.
21. Liedvogel, B., Sitte, P. and Falk, H (1976) *Cytobiology 12*, 155-174.
22. Mazliak, P. (1977) in *Lipids and Lipid Polymers in Higher Plants, Tevini, M. and Lichtenthaler, H.K. eds.*, Springer-Verlag, Berlin, pp. 48-74.
23. Kader, J.C. (1972) *Thèse Doctorat d'Etat, University of Paris, France*.
24. Donaldson, R.P., Tolbert, N.E. and Schnarrenberger, C. (1972) *Arch.Biochem. Biophys. 152*, 199-215.
25. Mc. Carty, R.E., Douce, R. and Benson, A.A. (1973) *Biochim.Biophys.Acta 316*, 266-270.
26. Moreau, F., Dupont, J. and Lance, C. (1974) *Biochim.Biophys.Acta 345*, 294-304.
27. Schwertner, H.A. and Biale, J.B. (1973) *J.Lipid Res. 14*, 235-242.
28. Douce, R., Guillot-Salomon, T., Lance, C. and Signol, M. (1968) *Bull.Soc.Fr. Physiol.Vég. 14*, 351-373.
29. Poincelot, R.P. and Day, P. (1974) *Plant Physiol. 54*, 780-783.
30. Poincelot, R.P. (1973) *Arch.Biochem.Biophys. 159*, 134-142.
31. Sprey, B. and Laetsch, W.M. (1975) *Z.Pflanzenphysiol. 75*, 38-52.
32. Jeffrey, S.W., Douce, R. and Benson, A.A. (1974) *Proc.Nat.Acad.Sci.USA 71*, 807-810.
33. Siefermann-Harms, D., Joyard, J. and Douce, R. (1978) *Plant Physiol. 61*, 530-533.
34. Sprey, B. and Laetsch, W.M. (1976) *Z.Pflanzenphysiol. 78*, 146-163.
35. Goodwin, T.W. (1977) in *Lipids and Lipid Polymers in Higher plants, Tevini, M. and Lichtenthaler, H.K. eds.*, Springer-Verlag, Berlin, pp. 29-47.
36. Joyard, J. and Douce, R. (1976) *C.R.Acad.Sci.Paris 282*, 1515-1518.
37. Joyard, J. and Douce, R. (1976) *Biochim.Biophys.Acta 424*, 125-131.
38. Priestley, D.A. (1977) *Ph.D. Thesis, University of Leeds, England*.
39. Moore, F.D. and Shepard, D.C. (1977) *Protoplasma 92*, 167-175.
40. Goodwin, T.W. (1958) *Biochem.J. 68*, 503-511.
41. Bickel, H. and Schultz, G. (1976) *Phytochemistry 15*, 1253-1255.
42. Qureshi, A.A., Kim, M., Qureshi, N. and Porter, J.W. (1974) *Arch.Biochem. Biophys. 162*, 108-116.
43. Costes, C., Burghoffer, C., Joyard, J. and Douce, R. *in preparation*.
44. Douce, R. and Joyard, J. (1979) *Adv.Bot.Res. 7*, in press.
45. Pineau, B. and Douce, B. (1974) *FEBS Lett. 47*, 255-259.
46. Flügge, U.I. and Heldt, H.W. (1976) *FEBS Lett. 68*, 259-262.
47. Flügge, U.I. and Heldt, H.W. (1977) *FEBS Lett. 82*, 29-33.
48. Joy, K.W. and Ellis, R.J. (1975) *Biochim.Biophys.Acta 378*, 143-151.
49. Heber, U. (1974) *Ann.Rev.Plant Physiol. 25*, 393-421.
50. Walker, D.A. (1974) in *MTP Int.Rev. of Biochemistry, serie one, vol. n°11, Plant Biochemistry, Northcote, D.H. ed.*, Butterworths and University Park Press, London and Baltimore, pp. 1-49.

51. Heldt, H.W. (1976) in *The Intact Chloroplast, Barber, J. ed.*, Elsevier, Amsterdam, New York and Oxford, pp. 215-234.
52. Heldt, H.W. and Rapley, L. (1970) *FEBS Lett. 10*, 143-148.
53. Heldt, H.W., Chon, C.J., Maronde, D., Herold, A., Stankovic, Z.S., Walker, D.A., Kraminer, A., Kirk, M.R. and Heber, U. (1977) *Plant Physiol. 59*, 1146-1155.
54. Heldt, H.W. (1969) *FEBS Lett. 5*, 11-14.
55. Strotmann, H. and Berger, S. (1969) *Biochem.Biophys.Res.Comm. 35*, 20-26.
56. Ellis, R.J. (1976) in *The Intact Chloroplast, Barber, J. ed.*, Elsevier, Amsterdam, New York and Oxford, pp. 335-364.
57. Campbell, P.N. and Blobell, G. (1976) *FEBS Lett. 72*, 215-226.
58. Blair, E.G. and Ellis, R.J. (1973) *Biochim.Biophys.Acta 319*, 223-234.
59. Dobberstein, B., Blobell, G. and Chua, N.H. (1977) *Proc.Nat.Acad.Sci.USA 74*, 1082-1085.
60. Ellis, R.J., Highfield, P.E. and Silverthorne, J. (1978) in *Photosynthesis 77 Hall, D.O., Coombs, J. and Goodwin, T.W. eds.*, Plenum, New York, pp. 497-506.
61. Douce, R. (1974) *Science 183*, 852-853.
62. Devor, K.A. and Mudd, J.B. (1971) *J.Lipid Res. 12*, 403-411.
63. Marschall, M.O. and Kates, M. (1974) *Can.J.Bot. 52*, 469-482.
64. Marschall, M.O. and Kates, M. (1972) *Biochim.Biophys.Acta 260*, 558-570.
65. Mazliak, P., Stoll, U. and Abdelkader, A.B. (1968) *Biochim.Biophys.Acta 152*, 414-417.
66. Mazliak, P. (1973) *Ann.Rev.Plant Physiol. 24*, 287-310.
67. Dawson, R.M.C. (1973) *Sub.Cell.Biochem. 2*, 69-89.
68. Wirtz, K.W.A. (1974) *Biochim.Biophys.Acta 344*, 95-117.
69. Kader, J.C. (1977) in *Dynamic Aspects of Cell Surface Organisation, Poste, G. and Nicolson, G.L. eds.*, North-Holland, Amsterdam, pp. 127-204.
70. Morré, D.J. (1975) *Ann.Rev.Plant Physiol. 26*, 441-481.
71. Neufeld, E.F. and Hall, C.W. (1964) *Biochem.Biophys.Res.Comm. 14*, 503-508.
72. Ongun, A. and Mudd, J.B. (1968) *J.Biol.Chem. 243*, 1558-1566.
73. Van Hummel, H.C. (1974) *Z.Pflanzenphysiol. 71*, 228-241.
74. Van Hummel, H.C., Hulsebos, T.J.M. and Wintermans, J.F.G.M. (1975) *Biochim. Biophys.Acta 380*, 219-226.
75. Liedvogel, B. and Kleinig, H. (1976) *Planta 129*, 19-21.
76. Blee, E. and Schantz, R. (1978) *Plant Sci.Lett.*, in press and *this volume*.
77. Liedvogel, B. and Kleinig, H. (1977) *Planta 133*, 249-253.
78. Van Besouw, A. and Wintermans, J.F.G.M. (1978) *Biochim.Biophys.Acta 529*, 44-53 and *this volume*.
79. Heinz, E. (1977) in *Lipids and Lipid Polymers in Higher Plants, Tevini, M. and Lichtenthaler, H.K. eds.*, Springer-Verlag, Berlin, pp. 102-120.
80. Joyard, J., Chuzel, M. and Douce, R. *this volume*.

INVESTIGATIONS ON THE ORIGIN OF DIGLYCERIDE DIVERSITY IN LEAF LIPIDS

E. HEINZ, H. P. SIEBERTZ AND M. LINSCHEID
Institutes for Botany and Organic Chemistry, University of Cologne, Cologne, Germany
J. JOYARD AND R. DOUCE
DRF, Biologie Végétale, CENG and USMG, Grenoble-Cedex, France

CONTENTS

Introduction
Prokaryotic patterns
Diglyceride portions in envelope lipids
Galactolipids synthesized by isolated envelopes
Lipids from in vivo labelled envelopes
Labelling of MGD species
Labelling of DGD species
Attempts to desaturate and ether analogue of MGD

INTRODUCTION

Diglyceride portions of leaf lipids usually contain only a limited selection of different fatty acids. Specificities are observed with respect to occurrence, localization and combination of these fatty acids within different lipids. Restrictions in occurrence are observed for example with unsaturated C_{16}-acids (3-trans-$C_{16:1}$, $C_{16:3}$)[1,2], which are confined to specific chloroplast lipids (PG, MGD). In addition these two fatty acids also show a specific positioning within their lipids, as they are esterified rather ex-

ABBREVIATIONS

Fatty acids are characterized by number of carbon atoms and double bonds. 16:3-plants are those, which contain appreciable proportions of $C_{16:3}$ in MGD in contrast to 18:3-plants, which have nearly exclusively $C_{18:3}$ in this lipid. MGD, DGD, TGD, TeGD and SQD = mono- to tetragalactosyl- and sulfoquinovosyl diacylglycerol. PC, PE, PG, PI = phosphatidyl choline, - ethanolamine, - glycerol, inositol.

clusively at the C-2-position[1,3]. In agreement with the suggestion of its role as possible precursor[4-6] of these unsaturated acids, $C_{16:0}$ is also found in appreciable proportion at C-2 of these and other chloroplast lipids (DGD, SQD), particularly in 16:3-plants[3]. On the other hand phospholipids of extraplastidic origin (PC, PE, PI) carry $C_{16:0}$ at C-1 (ref.7). Probably resulting from positional distribution fatty acids are paired in such a way, that 16/16-combinations are avoided as far as possible in all lipids. An exception is SQD, which always contains some dipalmitoyl species[8]. Another intriguing difference exists between diglyceride moieties of MGD and DGD, since in most cases DGD differs from MGD by the presence of a 18:3/16:0-species[9]. According to presently known schemes of biosynthesis[10,11], both lipids should differ only in number of galactoses, but not in nature of diglyceride moieties. As discussed below, it is this unique species, which may say something about the physiological significance of the in vitro incorporation of UDP-Gal by chloroplast envelopes after isolation.

Analysis of fatty acid labelling in various lipids including their positional distribution led to the recognition of several pools of lipids and fatty acids, which are characterized by differences in labelling kinetics[6,12,13]. In 16:3-plants there are at least three pools of $C_{16:0}$, from which only those related to C-2

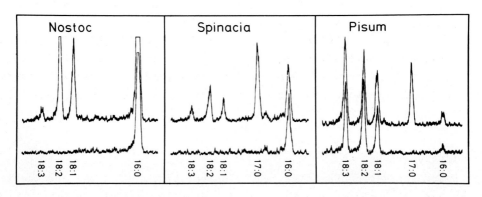

Fig. 1. GLC of $^{14}CO_2$-labelled fatty acids from total MGD (upper traces, representing fatty acids from C1 + C2) and from enzymatically produced lyso-MGD (lower traces, representing fatty acids at C2). Nostoc represents a prokaryotic, Spinacia a 16:3 and Pisum a 18:3 pattern. $C_{17:0}$ was included as internal mass and radioactivity standard.

of MGD and PG may be desaturated. There are at least two pools of
MGD, from which only one free of C_{16}-unsaturated acids may be rapidly galactosylated to DGD. There are at least two pools of $C_{18:3}$.
Reflecting the prokaryotic situation (see below) only one is feeding rapidly into the C-1-position of MGD and DGD, whereas $C_{18:3}$
at C-2 of these and in other lipids is labelled at a definitely
slower rate (fig.1). This may be true only for 16:3-plants, since
in similar experiments with 18:3-plants both positions of MGD were
labelled symmetrically thus reducing the number of pools in these
plants. (fig.1).

Differences and pools just mentioned may be caused by different
factors. Differences in diglyceride moieties may be due to specificities of enzymes catalyzing the primary, stepwise acylation of
glycerol phosphate. It is well known that such enzymes are present
in several compartments (see for example ref.14). Therefore we try
to characterize the acyl-CoA:glycerol phosphate acyltransferase
from chloroplast stroma (see contribution by M.Bertrams and E.Heinz
in this volume), which may play a crucial role in establishing primary asymmetries. The different diglyceride portions of MGD and DGD
could be caused by a selectivity of the galactosyltransferase forming DGD by galactosylation of MGD. But our previous experiments
have shown, that the specificity observed to operate at this level
in vitro most likely is not responsible for these differences[15].
This conclusion is supported by the labelling patterns to be shown
below. Finally it is obvious to assume that pools observed in experiments with whole leaves may partly be due to subcellular compartimentation of lipid localization. Therefore it is important to
localize these differences and their origin within subcellular compartments and membranes. We now investigated the lipids from chloroplast envelopes, since it is in this lipogenic membrane system,
where diglyceride specificities may be introduced. In the following
a detailed analysis of envelope lipids with special emphasis on diglyceride moieties is presented. In addition we separated molecular
species of lipids from envelopes and thylakoids, which were isolated from $^{14}CO_2$-labelled leaves, and compared them with species
synthesized by isolated envelopes from radioactive UDP-Gal. More
and also experimental details will be included in original publications which are in preparation.

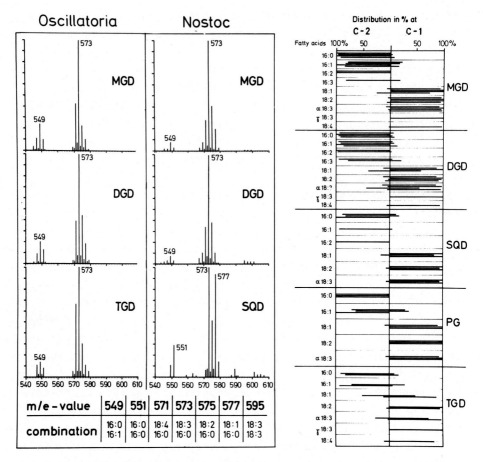

Fig. 2 (right). Positional distribution of fatty acids in glycerolipids from blue-green algae. The uppermost bar in each fatty acid block represents Anabaena, the next one Nostoc, the third one Tolypothrix and the lowest one Oscillatoria(from ref. 16).

Fig. 3 (left). Diglyceride profiles in prokaryotic glycolipids as analyzed by mass spectrometry of intact, acetylated glycolipids. Major species are indicated by m/e values and characterized in terms of predominant fatty acid combinations(from ref. 16).

PROKARYOTIC PATTERNS

With respect to the role played by compartmentation it should be interesting to look at diglyceride moieties from lipids of such photosynthetically active cells, which do not contain membrane-enclosed organelles. Therefore we investigated the lipids from several prokaryotic blue-green algae, which show what kind of differences may be expected in case of largely reduced compartmentation[16].

Positional distribution of fatty acids in four predominant acyl lipids is shown in fig.2, and chain length specificity is clearly evident. This specificity is observed in all lipids and directs C_{16}-fatty acids irrespective of degree of unsaturation into the C-2-position, whereas C_{18}-acids are found at C-1. Such a simple and generally valid scheme is missing in eukaryotic cells. But a similar distribution is found in MGD from 16:3-plants, the distribution of which therefore may be considered to be prokaryotic.

Mass spectrometry of acetylated glycolipids (fig.3) shows that not only identical distribution but also identical fatty acid pairings and therefore identical diglyceride moieties are found in the galactolipid series from the prokaryotic organisms investigated (Anabaena, Microcystis, Tolypothrix, Nostoc and Oscillatoria, from which only the last two are included in fig. 3). The small proportion of 16/16-combinations shows the previously mentioned discrimination against this pairing and demonstrates on the other hand, that localization of C_{16}-acids at C-2 cannot be absolute. When looking at SQD and PG (data not presented here) it is clear, that differences in diglyceride portions do exist in prokaryotic cells too. They are due to the presence of different fatty acids (γ-$C_{18:3}$ and $C_{18:4}$ in galactolipids, but not in PG or SQD of Tolypothrix) or to a shift in proportion of fatty acids (reduced levels of $C_{18:3}$ in PG and SQD from Anabaena, Nostoc and Oscillatoria). Therefore the diversity in diglyceride portions of lipids from prokaryotic cells is reduced as compared to eukaryotic organisms, but nevertheless does exist. Accordingly compartimentation may not be the only reason for such differences.

DIGLYCERIDE PORTIONS IN ENVELOPE LIPIDS

In eukaryotic leaf cells galactolipids are localized in at least two completely different membrane systems, in thylakoids and envelopes. In recent years increasing evidence points to a predominant if not exclusive role of envelopes in galactolipid assembly (see contribution by J.Joyard and R.Douce in this volume and ref. 10). Therefore we analyzed diglyceride profiles of lipids from chloroplast envelopes to see whether or not the above mentioned differences in galactolipids are already present at the site of their biosynthesis.

Before discussing these results it should be mentioned that envelope preparations may contain minor proportions of acylated gly-

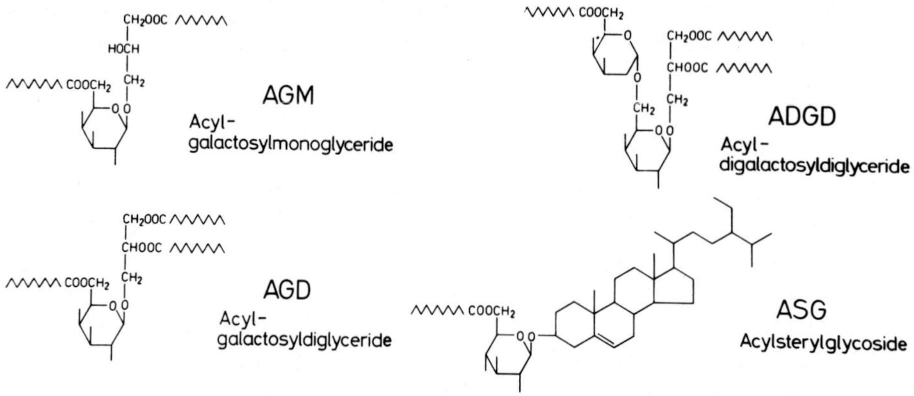

Fig. 4. Acylated glycolipids which may be formed by the acyltransferase present in chloroplast envelopes. Literature is summarized in ref. 17.

colipids (structures shown in fig. 4), which most likely are due to the activity of a latent acyltransferase present in these membranes and only becoming active after shifting the pH into the acid range[17]. We found AGD and ADGD, which runs very close to MGD in the usual solvent systems and accounted for about 8% of envelope acyl lipids.

Diglyceride profiles from major envelope lipids as analyzed by mass spectrometry of acetyl derivatives of intact glycolipids or of acetylated diglycerides prepared enzymatically from phospholipids are shown in fig. 5. It is evident that with the exception of DGD diglycerides, MGD, TGD and TeGD share mass-spectrometrically similar diglyceride portions, in which 18:3/16:3- and 18:3/18:3-combinations predominate. Even in the small amount of triglycerides found in envelopes, combinations of these two fatty acids (18:3/18:3/18:3 and 18:3/18:3/16:3 predominate. It should be pointed out that free diglycerides are highly unsaturated (only 5.7 % of $C_{16:0}$,

Fig. 5. Fatty acid pairings in lipids from spinach chloroplast envelopes as analyzed by mass spectrometry. Portions of spectra cover molecular ions from triglycerides = TG and acetylated diglycerides (DG from free diglycerides, PG and PC from diglycerides released by phospholipase C hydrolysis from parent lipids) or fragment ions of diglyceride structure from glycolipid derivatives. Ordinates show relative abundance in %, abscissae m/e values. Major combinations are identified by constituent fatty acids.

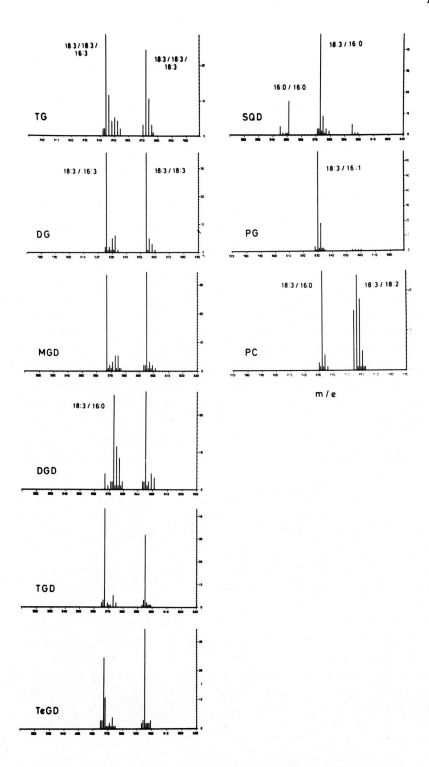

although 16.8 % were found before in chloroplast diglycerides[18]).
At first glance this does not fit into the hypothesis that fatty
acids may be desaturated only after incorporation into polar lipids.
Furthermore a 18:3/16:0-pairing is nearly absent from the pool of
free diglycerides, which is important for experiments on DGD formation by isolated envelopes presented below. In contrast, the diglyceride mixture extractable from whole spinach leaves may contain
this species, because a high proportion (42.7 %) of $C_{16:0}$ was found
in this fraction[2]. The large pool of free diglycerides and the rapid use of UDP-Gal for galactolipid synthesis by isolated envelopes have been interpreted as indicating that envelopes suffer in
vivo from a UDP-Gal deficiency and therefore pile up large quantities of diglycerides waiting for galactosylation. As discussed below this diglyceride pool may be looked at in a different way.

The mass-spectrometrically identical diglyceride portions of the
above mentioned lipids are in agreement with a consecutive galactosylation sequence operating in galactolipid biosynthesis with the
exception of DGD, which interrupts this homogenous series. The 18:3/
16:3-pairing of DGD represents only a very small proportion and the
18:3/16:0-species is already abundant in envelopes, although MGD
and diglyceride precursor pools do not provide sufficient quantities of this species. Therefore we conclude that the two major species of DGD observed in extracts from whole leaves (18:3/18:3 and
18:3/16:0) are not due to extraction from different subcellular
membranes, since they are also present at the site of galactolipid
biosynthesis.

The positional distribution of fatty acids in the lipids discussed so far (fig. 6) is in accordance with previously established
patterns from 16:3-plants, for which enrichment of $C_{16:3}$ and a high
proportion of $C_{16:0}$ at C-2 are most characteristic[3]. This has now
been extended to MGD, DGD, TGD and diglycerides from envelopes. A
close biogenetic relation between MGD and diglycerides seems to
be obvious in view of almost identical pairings and positions of
fatty acids. Therefore the preferred galactosylation of other diglycerides as observed in vivo is rather surprising (see below).

When looking at SQD and phospholipids from the envelope, it is
evident that each one displays a unique diglyceride profile pointing to possibly divergent origins of these lipids. Remarkable are
the 16/16-combination in SQD (fig.5) as found before in whole leaf

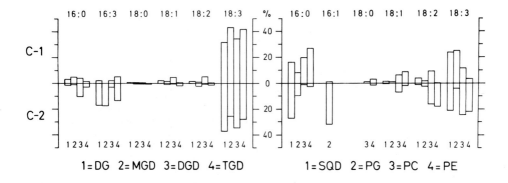

Fig. 6. Proportion and positional distribution of fatty acids in glyco- and phospholipids from spinach chloroplast envelopes. DG = diglycerides.

extracts[8] and the absence of 18/18- and 16/16-pairings in PG, which consists mostly of 18:3/16:1-species reflecting the prokaryotic type of pairing. PC is composed of many species, among which 18:3/16:0 and 18:2/18:3 predominate. Therefore diglyceride profiles at the level of envelopes are not in particular support of the original suggestion that PC is the direct source of diglycerides or fatty acids found in galactolipids[19]. Accordingly it has been postulated that PC-derived precursors are desaturated on their way from PC into galactolipids[13,20]. The positional distribution of fatty acids in PC (fig. 6) differs from that found in envelope glycolipids and especially from the 18:3/16:0-species of DGD. In PC $C_{16:0}$ is exclusively at C-1 in agreement with previous studies[21], whereas in DGD and SQD more than half is at C-2.

GALACTOLIPIDS SYNTHESIZED BY ISOLATED ENVELOPES

Since the 18:3/16:0-species of DGD is already present in envelopes, we wanted to know, whether or not this combination is synthesized by isolated envelopes. For this purpose we incubated isolated envelopes with radioactive UDP-Gal and subjected the radioactive galactolipids to $AgNO_3$-TLC for species separation according to degree of unsaturation. In the case of MGD these were tentatively identified by comparison with a selection of authentic species prepared synthetically[22] or isolated from various sources including blue-green algae. It should be noted that 18:3/18:2- and 18:3/18:3-

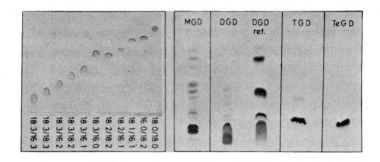

Fig. 7 (left). Argentation chromatography of authentic MGD species.
Fig. 8 (right) Argentation chromatography of galactolipid species synthesized by isolated envelopes from radioactive UDP-Gal. Labelled DGD references of decreasing mobility represent 18:1/18:1, 16:0/18:3, 18:2/18:2 and 18:3/18:3 species, which were obtained enzymatically by in vitro galactosylation of corresponding MGD species with radioactive UDP-Gal.

species run just ahead of the corresponding 18:3/16:2- and 18:3/16:3-combinations, respectively (fig. 7). Radioactive DGD species, which are included as references in fig. 8, were obtained by enzymatic galactosylation[15] of appropriate MGD species with radioactive UDP-Gal. As seen in fig. 8, in all four galactolipids only such species are labelled, which are derivable from envelope diglycerides, since all four galactolipids are mainly composed of hexaene species. These stay closest to the start with 18:3/18:3- and 18:3/16:3-species predominating in MGD, whereas both species, if present in the higher homologues, were not separated by TLC. These patterns did not change over an incubation time of 3 h in contrast to previous experiments with leaves[6] and also in contrast to patterns from in vivo labelled envelopes shown below. With the exception of DGD the above results agree with the mass spectra. But it has to be emphasized that isolated envelopes, although they do contain the 18:3/16:0-species of DGD, cannot synthesize this pairing in vitro from UDP-Gal in the absence of stroma and other components. Apparently the abundance of the appropriate precursor species in digly-

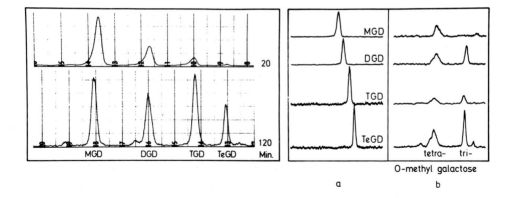

Fig. 9 (left). Distribution of radioactivity in galactolipid homologues (separation by TLC) after incubating isolated envelopes for different times with radioactive UDP-Gal. In all chromatograms, also those from fig. 10, solvent fronts are at the left.

Fig. 10 (right) a: TLC of permethylated cores of galactolipids obtained in vitro by incubating envelopes with radioactive UDP-Gal. b: TLC separation of partially methylated galactoses released by acid hydrolysis from corresponding permethylated compounds shown in a.

ceride or MGD precursor pools is too low (but compare the contribution by J. Joyard, M. Chuzel and R. Douce in this volume).

Another unusual feature of isolated envelopes is the completely unnatural proportion of radioactivity incorporated from UDP-Gal into higher homologues of galactolipids such as TGD and TeGD (fig.9). This is in contrast to their proportion on a weight basis[23] and also differs from their labelling in envelopes isolated from $^{14}CO_2$-labelled leaves as will be shown below. On the other hand the short time pattern (fig. 9, 20 min) is dominated by radioactive MGD. This may show that pools of MGD and DGD cannot efficiently compete with diglycerides for UDP-Gal, although these three lipids represent roughly equally sized pools[24]. If they are not used as galactose acceptors, then higher homologues may be formed by de novo synthesis starting from diglycerides and by-passing existing pools.

To find out which of the pre-existing pools of envelope lipids are used for galactosylation to higher homologues, we applied the methylation technique used before by Williams et al[25]. By this method external and internal galactoses from galactolipids are converted to tetra- and trimethylgalactose, respectively, which are easily separated by TLC. This procedure gives unequivocal results only for DGD, but we also subjected the other galactolipids to this treatment with the results shown in fig. 10. From MGD as expected only radioactive tetramethylgalactose was obtained, whereas DGD, TGD and TeGD gave roughly equal proportions of radioactivity in tetra- and trimethylgalactose. This means that most of the DGD is made de novo from diglycerides via MGD. This in turn shows that the pool of cold MGD present in isolated envelopes is not used and therefore apparently does not equilibrate with the small proportion of newly made MGD which is the precursor for DGD. This would indicate restrictions in pool mixing. We shall find this phenomenon in another group of envelope lipids as discussed below.

In summary we can say that all diglyceride specificities known before from whole leaves are already present in envelopes, particularly the 18:3/16:0-species of DGD. Therefore diglyceride diversities of galactolipids most likely are not due to different subcellular localization of different molecular species. In vitro incorporation of UDP-Gal by envelopes produces galactolipids in unnatural proportions, but not the 18:3/16:0-pairing of DGD. Methylation and argentation chromatography show that DGD is mainly derived from newly made MGD by by-passing the existing pool of MGD.

LIPIDS FROM IN VIVO LABELLED ENVELOPES

Next we studied the in vivo labelling of envelope lipids and their species to find out, how the various pools are built up and to what extent they are subject to turnover. Leaves were labelled with $^{14}CO_2$ for subsequent isolation of radioactive chloroplasts, from which thylakoids and envelopes were released. Lipids were extracted from these in vivo labelled membranes and subjected to the previously used procedures. The first question we wanted to be answered was whether envelopes are sites of highest galactolipid synthesizing capacity in vivo as well. In this case one would expect higher total activity at short labelling times in the envelope than in the corresponding thylakoid system. Fig. 11 shows patterns of labelled lipids from thylakoids, envelopes and whole spinach leaves obtained after various labelling times. Surprising is the low total radioactivity in the envelope (6 - 8 % of that present

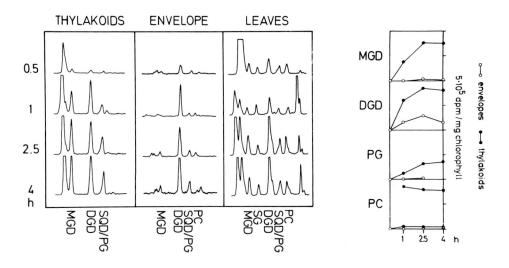

Fig. 11 (left). Radioactive lipids present in whole spinach leaves, thylakoids and chloroplast envelopes after CO_2-labelling for the times indicated at the left (radioscans of TLC separations, solvent fronts at the left).

Fig. 12 (right). Total radioactivity in individual lipids recovered from gradient-separated envelope and total corresponding thylakoid system normalized to 1 mg of chlorophyll. Total radioactivity in PC from whole leaves per mg chlorophyll is marked by asteriks.

in the corresponding thylakoid system) and the simple pattern of radioactive lipids, since only DGD is heavily labelled accounting for about 80 % at all time points investigated. Similar patterns with a slightly reduced proportion of label in DGD were reproduced in a second series of experiments with labelling times as short as 5 min. Quantitative evaluation shows (fig. 12) that at all time points investigated each one of the envelope lipids contained less total activity than the corresponding lipid from thylakoids. The activity levels differed by a factor of about 20 in MGD, 15 in PG, 10 in SQD, 3 in DGD and 2 in PC, but specific radioactivities of glycolipids were higher in envelopes than in thylakoids. In addition the total radioactivity in envelope lipids is increasing up to 2.5 h showing that pools are still filled up and steady state is not yet reached. At present and at first sight the distribution of total radioactivity between envelopes and corresponding thylakoid system obviously does not satisfy expectations originating in the picture of exclusive formation of galactolipids in envelopes.

But it has to be pointed out that some data of prime importance in this context are not available at present. First of all, the figures given for total radioactivity in envelope lipids are minimal ones, since recovery of the envelope is not complete (only about 40 %), and since part of it may be present in the thylakoid fraction. Another important factor is the time elapsed between harvest and final separation of thylakoids from envelopes. During this time, which is in the order of 2 h, a large proportion of lipids and in particular the newly made MGD may have been exported from the envelope. In relation to the picture presented below it may be speculated that the small but steadily increasing proportion of labelled MGD in the envelope represents MGD released into the bilayer portion of envelope lipids, which escape the massive and rapid export transferring newly assembled lipids from synthetic units in the envelope into thylakoids.

Furthermore it should be pointed out that neither in envelopes nor in thylakoids the proportion of radioactivity in PC was ever as dramatic as expected from its role in the lipid and acyl shuttle proposed to operate between endoplasmic reticulum and chloroplasts [20]. Finally, the alkali-stable compound migrating similar to steryl glycoside between MGD and DGD is not found in chloroplasts.

LABELLING OF MGD SPECIES

Another interesting fact is the very low labelling of free diglycerides from envelopes, although labelled galactolipids contain a small proportion of label in fatty acids. This confirms earlier work with whole leaves, where also no activity was detected in free diglycerides[26]. Both observations taken together could mean that newly formed diglycerides are immediately galactosylated and therefore cannot equilibrate with the large pool of pre-existing diglycerides in envelopes. Support for this idea comes from $AgNO_3$ separation of labelled envelope galactolipids (fig. 13). For comparison we have included species synthesized in vitro by isolated envelopes from UDP-Gal, where the predominance of hexaene species is seen again. In vivo labelled envelope MGD shows a completely different pattern. We see many species, among which the more saturated ones and not the hexaenes predominate. This demonstrates unspecificity

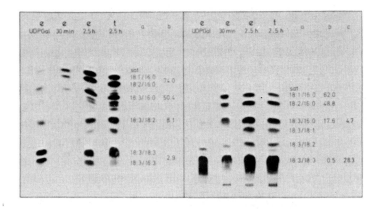

Fig. 13 (left). Separation of MGD species by argentation chromatography. MGD samples were obtained by incubating isolated envelopes with radioactive UDP-Gal (e UDP-Gal) or by extracting envelopes(e) and thylakoids (t) prepared from spinach leaves, which were labelled for the times indicated with CO_2. Individual species are characterized in terms of fatty acid composition (column a). Proportions of radioactivity in fatty acids after hydrolysis of separated species from the 2.5 h thylakoid sample are given as % in column b.

Fig. 14 (right). Separation of DGD species by argentation chromatography. Identification of samples and of columns a and b as in fig. 13. Column c represents ratios of radioactivity in tetra-/trimethylgalactose obtained by permethylation and hydrolysis of separated species from the 2.5 h thylakoid sample.

of the first galactosyltransferase with respect to diglycerides (compare ref. 18 and 27). Furthermore we conclude that the large pool of free hexaene diglycerides is galactosylated to MGD at a comparatively slow rate as evident from the proportion of radioactivity in hexaene species of MGD, which does not reflect their predominance on a weight basis. Very similar patterns were observed in the corresponding thylakoid MGDs (see below). Finally, the diglyceride pool in vivo available for galactosylation in the envelope differs significantly from and does not equilibrate with the large pool of mainly hexaene diglycerides. This is the next restriction in pool mixing of envelope lipids.

That the labelled oligoene species of envelope MGD may in fact be derived from newly made diglycerides is suggested by the data included in column b of fig. 13, which were obtained by hydrolyzing thylakoid MGD after a labelling period of 2.5 h. They show the percentage of radioactivity in fatty acids from separate MGD species differing in degree of unsaturation. When a 18/16-diglyceride and a galactose, both having the same specific radioactivity per C-atom, are linked to MGD, 79 % of the radioactivity should be present in fatty acids. This percentage is approached by the uppermost zone analyzed, which according to fatty acid analysis contained 18:1/16:0- and 18:2/16:0-pairings. In short time envelope patterns these species contained the highest proportion of radioactivity, which may be interpreted as showing that the more saturated a species, the higher the rate of its synthesis. Despite their rapid appearance these oligoene species do not persist and cannot accumulate similar to oligoene MGD species labelled with ^{3}H-glycerol[20]. One possible way of turnover already discussed before[6,7] is desaturation of lipid linked acyl chains, which would result in accumulation of hexaene species as required.

LABELLING OF DGD SPECIES

Similar considerations have to be applied to the interpretation of $AgNO_3$ patterns of DGD from in vivo labelled envelopes and thylakoids as shown in fig. 14. For comparison we have again included hexaene species made in vitro by envelopes from UDP-Gal. In vivo labelled envelope DGD contains far more and also more saturated species, among which we finally find the desired 18:3/16:0-pairing. In view of the in vitro experiments we assume that this species most likely cannot be derived from the large pool of pre-existing cold MGD. But it may rather originate from newly made MGD, which in turn is derived from newly made diglycerides. This is supported by the proportion of radioactivity in fatty acids, which is included in column b of fig. 14 (data obtained from thylakoid DGD after 2.5 h labelling) and may serve again as an indicator for assembly of de novo made precursors. A 18/16-DGD made de novo from precursors of identical specific radioactivity per C-atom would have 70 % of its radioactivity in fatty acids. This percentage is approached by the faster moving species. The fact that part of the fraction of newly made MGD, which is very small compared to the large pool of cold hexaene species, is carried through to DGD indicates that also in vivo newly made MGD does not disappear in the cold pool of preexisting MGD. This points again to the restriction in mixing of MGD pools from envelopes.

As described above DGD made in vitro by isolated envelopes contained about equal proportions of radioactivity in terminal and inner galactoses. For DGD from in vivo labelled membranes methylation analysis gave completely different proportions of radioactivity in tetra- and trimethylgalactose. After 2.5 h labelling the ratio of tetra-/trimethylgalactose-labelling was 35, 22 and 13, respectively, for DGD from envelopes, thylakoids and whole chloroplasts, respectively. At present we have no explanation for these differences, but the large ratios demonstrate that after 2.5 h labelling in addition to newly made DGD a large proportion is derived from cold MGD by galactosylation with radioactive galactose. This is in sharp contrast to the in vitro results. When methylation analyses were carried out with individual DGD species separated according to degree of unsaturation a dramatic increase in the ratio of tetra-/trimethylgalactose-labelling was observed with increasing unsaturation (data obtained from the 2.5 h thylakoid sample are included in column c of fig. 14). This ratio may be used as an additional indicator for de novo synthesis of DGD complementing the data of fatty acid labelling. From both series of analyses we conclude that oligoene species of DGD are derived rather directly from newly made oligoene diglycerides, whereas hexaene DGD is synthesized by galactosylation of pre-existing cold hexaene MGD. Previous experiments on fatty acid labelling have shown that 18:3/16:3-species of MGD are excluded from galactosylation[6]. Therefore hexaene species of MGD available for rapid galactosylation to DGD can only be 18:3/18:3-pairings.

Taking into account this additional separation, envelope MGD may be divided into three pools: newly made oligoene species, which are rapidly galactosylated to DGD and therefore do not equilibrate with the large pool of pre-existing hexaene species. In turn these hexaenes are separated into 18:3/16:3- and 18:3/18:3-combinations, from which only the last ones are available for galactosylation to DGD. Fig. 15 summarizes the various data and conclusions drawn from labelling experiments with envelopes. The most significant differences between in vitro and in vivo experiments are related to the appearance of oligoene species in various lipids during in vivo labelling. This may indicate that isolated envelopes, though able to assemble various precursor molecules to galactolipids, cannot desaturate these compounds, since such reactions need additional, yet unknown cofactors or enzymes not present in envelopes.

A next step would be to correlate the different pools and restrictions in their mixing with cytological structures. Although envelopes are composed of four bilayer halves providing ample possibilities for location of separate pools, we do not know at present how to make use of these different lipid layers. Our efforts to press experimental details into a unifying picture resulted in a speculative separation of lipids into two groups. The minor one is labelled rapidly and represents intermediates and unfinished molecules in close contact or bound to active sites of membrane-embedded enzyme complexes. These compounds are not able to diffuse away into the membrane lipid bilayer, in which case they would escape further alteration and therefore could accumulate. The other group of lipids represents completely finished and desaturated molecules, which have left enzyme complexes to diffuse away and build up the large pool of bilayer lipids. The speculative complexes comprise a multiplicity of enzymes responsible for many reactions such as activation of de novo made $C_{18:1}$ and $C_{16:0}$ fatty acids, assembly of an accordingly oligoene diglyceride and subsequent galactosylation to oligoene MGD and DGD. $C_{18:1}$ acyl chains from both galactolipids may be subject to desaturation reactions which are also performed in these complexes. But we do not know how to fit in the observations that desaturation of $C_{16:0}$ is possible only in relation to MGD and that this particular MGD with its C_{16}-chains being desaturated cannot be galactosylated to DGD. Another difficulty shows up when comparing molecular species of MGD from envelopes with those from thylakoids. Both membranes display very similar patterns, and in both cases the early labelled oligoene species do not accumulate. This could mean that after import of oligoene or more desaturated intermediates further desaturation reactions have to take place in thylakoids as well. On the other hand a contamination of thylakoids by labelled envelopes may also explain the similarity in

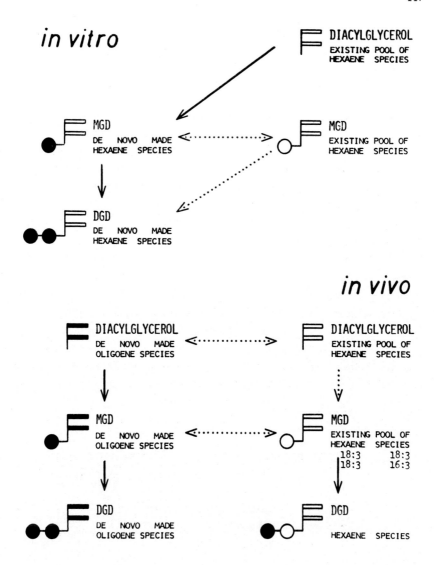

Fig. 15. Separation of lipid pools in envelopes based on differences in labelling kinetics of molecular species and distribution of radioactivity within labelled lipids as analyzed by hydrolytic release of fatty acids or methylation analysis. Fatty acids are symbolized by elongated rectangles, galactoses by circles, open symbols = unlabelled, closed symbols = labelled molecules. Solid arrows = rapid galactosylation, dashed arrows = slow galactosylation or slow pool mixing.

patterns and would not require desaturation in thylakoids.

ATTEMPTS TO DESATURATE AN ETHER ANALOGUE OF MGD

The above shown $AgNO_3$ patterns confirm our previous studies with leaves[6] and demonstrate again the predominance of MGD and DGD oligoene species after short labelling times. We have now shown that these oligoene species synthesized at highest rates are assembled from newly made diglyceride and galactose precursors, whereas at the same time far less radioactive galactose is reacted with unlabelled hexaene diglycerides (fig. 13, e 30 min). Since the oligoene species do not accumulate, it is still and again tempting to interpret such patterns and their change with time as showing a lipid linked desaturation of acyl chains, which also would contribute to finally observed diglyceride specificities in lipids. To provide unequivocal proof for direct lipid linked desaturation in galactolipids, we synthesized an ether analogue of MGD[28]. Instead of ester it had ether linked monoene side chains, which cannot be shuttled between oxygen ester linkage in lipids and thioester bonds in CoA, ACP or enzymes[7]. The synthesis started with D-mannitol and resulted in the ether analogue having ß-D-galactopyranose linked to C-3 of glycerol and two alkyl groups at the sn-1,2-positions. TLC did not

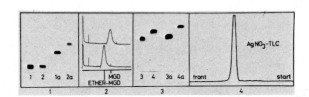

Fig. 16. Experiments with an ether analogue of MGD. 1: TLC before (1, 2) and after (1a, 2a) acetylation of ester (1, 1a) and ether (2, 2a) MGD.
2: GLC of trimethylsilyl derivatives of 16/16-ester and ether MGD.
3. TLC before (3, 4) and after (3a, 4a) acetylation of enzymatically prepared 6-O-acyl derivatives of ester (3, 3a) and ether (4, 4a) MGD.
4: argentation chromatography of tritiated ether MGD (1,2-di-O-octadec-9'-enyl-3-O-ß-D-galactopyranosyl-sn-glycerol) used in desaturation experiments.

differentiate between ether and ester MGD in underivatized form, but acetates had different Rf-values (fig. 16, 1). Ester and ether MGD were also differentiated by GLC as trimethylsilyl derivatives due to differences in molecular weights (fig. 16, 2). Acceptance of ether galactolipids by plant enzymes was tested in two reactions. Galactosylation with radioactive UDP-Gal to the DGD analogue could not be demonstrated with an assay system, which galactosylates the ester MGD[15]. Acylation by the envelope acyltransferase to the acylated derivative with fatty acids originating from DGD as acyl donor[29] proceeded very well with the ether MGD. The 6-O-acyl derivative was isolated by preparative TLC (fig. 16, 3) and characterized by mass spectrometry after acetylation.

After this positive result we synthesized an ether MGD having two octadec-9-ynyl residues. This compound was reduced with tritium gas to the tritiated di-octadec-9-enyl compound (fig. 16, 4) having the necessary high specific activity of about 20 mCi/μmole. This substance was added dissolved in organic solvents or dispersed in water onto the surface of pea and bean leaves or infiltrated through the petioles. Desaturation was checked after several days by lipid extraction and argentation chromatography of the recovered MGD band. So far this procedure did not result in the production of a more unsaturated ether. Therefore this straight approach to demonstrate the existence of hypothetical galactolipid desaturases failed so far, but experiments will be continued.

ACKNOWLEDGEMENT

Support by the Deutsche Forschungsgemeinschaft, Commissariat à l'Energie Atomique and Centre National de la Recherche Scientifique is gratefully acknowledged.

REFERENCES

1. Haverkate, F. and van Deenen, L.L.M. (1965) Biochim. Biophys. Acta, 106, 78-92.
2. Jamieson, G.R. and Reid, E.H. (1971) Phytochemistry, 10, 1837-1843.
3. Heinz, E. (1977) in Lipids and Lipid Polymers in Higher Plants, Tevini, M. and Lichtenthaler, H.K. eds., Springer-Verlag, Berlin-Heidelberg-New York, pp. 102-120.
4. Nichols, B.W., Harris, P. and James, A.T. (1965) Biochem. Biophys. Res. Commun., 21, 473-479.

5. Bartels, C.T., James, A.T. and Nichols, B.W. (1967) Eur. J. Biochem., 3, 7-10.
6. Siebertz, H.P. and Heinz, E. (1977) Z. Naturforsch., 32 c, 193-205.
7. Hitchcock, C. and Nichols, B.W. (1971) Plant Lipid Biochemistry, Academic Press, London and New York.
8. Tulloch, A.P., Heinz, E. and Fischer, W. (1973) Hoppe-Seyler's Z. Physiol. Chem., 354, 879-889.
9. Rullkötter, J., Heinz, E. and Tulloch, A.P. (1975) Z. Pflanzenphysiol., 76, 163-175.
10. Douce, R. and Joyard, J. (1979) Adv. Bot. Res., 7, in press.
11. van Besouw, A. and Wintermans, J.F.G.M. (1978) Biochim. Biophys. Acta, 529, 44-53.
12. Williams, J.P., Watson, G.R., Khan, M.U. and Leung, S. (1975) Plant Physiol., 55, 1038-1042.
13. Williams, J.P., Watson, G.R. and Leung, S.P.K (1976) Plant Physiol., 57, 179-184.
14. Boehler, B.A. and Ernst-Fonberg, M.L. (1976) Arch. Biochem. Biophys., 175, 229-235.
15. Siebertz, M. and Heinz, E. (1977) Hoppe-Seyler's Z. Physiol. Chem., 358, 27-34.
16. Zepke, H.D., Heinz, E., Radunz, A., Linscheid, M. and Pesch, R. (1978) Arch. Microbiol., 119, 157-162.
17. Heinz, E., Bertrams, M., Joyard, J. and Douce, R. (1978) Z. Pflanzenphysiol., 87, 325-331.
18. Eccleshall, T.R. and Hawke, J.C. (1971) Phytochemistry, 10, 3035-3045.
19. Roughan, P.G. (1975) Lipids, 10, 609-614.
20. Slack, C.R., Roughan, P.G. and Balasingham, N. (1977) Biochem. J., 162, 289-296.
21. Devor, K.A. and Mudd, J.B. (1971) J. Lipid Res., 12, 396-402.
22. Heinz, E. (1971) Biochim. Biophys. Acta, 231, 537-544.
23. Douce, R. and Benson, A.A. (1974) Portug. Acta Biol., 14, 45-64.
24. Joyard, J. and Douce, R. (1976) Biochim. Biophys. Acta, 424, 125-131.
25. Williams, J.P., Khan, M. and Leung, S. (1975) J. Lipid Res., 16, 61-66.
26. Roughan, P.G. (1970) Biochem. J., 117, 1-8.
27. Mudd, J.B., van Vliet, H.H.D.M. and van Deenen, L.L.M. (1969) J. Lipid Res., 10, 623-630.
28. Heinz, E., Siebertz, H.P. and Linscheid, M. (1979) in preparation
29. Heinz, E. (1973) Z. Pflanzenphysiol., 69, 359-376.

THE ENZYMIC DEGRADATION OF MEMBRANE LIPIDS IN HIGHER PLANTS

TERRY GALLIARD

RHM Research Ltd. Lord Rank Research Centre, High Wycombe. U.K.

INTRODUCTION

In comparison with lipid biosynthesis, the breakdown of membrane lipids is a largely neglected area of research in plant biochemistry. Despite an accumulation of evidence for 'turnover' of lipids and their component parts in membranes of plant tissues, only the biosynthetic and exchange reactions have been studied with respect to the enzymes involved in these processes.

Most of our meagre knowledge on lipid-degrading enzymes in plants has come from studies on tissues where a *net loss* of membrane lipid can be clearly demonstrated. This review will deal firstly with the enzyme activities that are known with some certainty to be present in various tissues. These include both hydrolytic and oxidative enzymes. Some examples of sequential enzyme reactions will be presented, illustrated by recent work from the author's former laboratory in Norwich. Finally, in the absence of hard facts, some speculations on the involvement of lipid degrading enzymes in membrane turnover, will be presented to stimulate discussion and, hopefully, to initiate productive research.

MEMBRANE DEGRADATION

Net loss of membrane lipids occurs under certain physiological conditions and in disrupted plant tissues. Some examples include :-
1) lipids in endosperm or cotyledons of germinating seeds;
2) senescent tissues;
3) ripening fruits (loss of chloroplast lipids);
4) tissues disrupted either by damage or infection, resulting in organelle rupture and autolysis.

Initial events in all these processes are probably due to the action of hydrolases (transferases) acting on endogenous lipids. Since this review is limited to membrane lipids, lipolysis of storage lipids by lipases will not be discussed here. The acyl lipids of interest in membranes are predominantly the phospholipids and glycolipids and therefore, discussion is limited to these classes.

The known pathways of phospholipid hydrolysis in plants are indicated in Fig. 1 by the solid arrows.

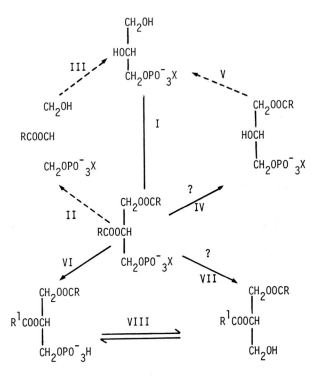

Fig. 1 Possible routes for the enzymic degradation of phospholipids

Enzyme reactions which have been characterized in plants are shown by solid arrows; reactions known in animals or microbial organisms but not yet confirmed in plants are shown by broken arrows; reactions for which some evidence in plants exists, but are not fully characterized, are shown with queries.

Lipid acyl hydrolases

Deacylation of phospholipids is catalysed by a lipid acyl hydrolase of wide substrate specificity (I in Fig. 1). This enzyme, first demonstrated in potato tubers [1], has been found in a variety of plants. Its properties have been reviewed [2] and three points only will be emphasised here. Firstly, a wide range of lipids is hydrolysed; lysophospholipids and monoacylglycerols are deacylated most rapidly; phospholipids, galactolipids, sulpholipids and some synthetic esters of long chain acids are also hydrolysed but triacylglycerols ("lipase" substrates) are not attacked. Table 1 compares results on substrate specificity of the enzyme from potato and from *Phaseolus vulgaris* leaves. It is probable that the enzyme from *Phaseolus* leaves is identical to that, previously described as a galactolipase [3]. The second point to note is that, like most hydrolases, the enzyme acts as an *acyl transferase* in the presence of a hydroxyl acceptor; e.g. it has been shown that the affinity of the enzyme for methanol (producing fatty acid methyl esters) is 10-fold that for water [4]. Thirdly, all the evidence so far indicates that the enzyme removes both fatty acids from diacyl lipids and thus differs from phospholipases -1 and -2 of microbial or animal origin.

TABLE 1

COMPARISON OF LIPID ACYL HYDROLASES FROM POTATO TUBER AND *PHASEOLUS VULGARIS* LEAVES

Substrate	Specific activity (relative to Monoleoylglycerol)	
	Potato tuber (Ref 1)	*Phaseolus* leaves (Ref 5)
monoleoylglycerol	100	100
lysophosphatidylcholine	71	88
monogalactosyldiacylglycerol	31	62
dioleoylglycerol	21	-
digalactosyldiacylglycerol	17	38
phosphatidylcholine	13	15
phosphatidylglycerol	12	-
sulphoquinovosyldiacylglycerol	-	9
phosphatidylethanolamine	11	-
trioleoylglycerol	1	-

Although the acyl hydrolases are normally extracted from particle-free supernatant fractions, they are compartmentalized within healthy cells. Activity readily released is probably 'lysosomal' or contained in vacuoles. However Bligny and Douce [6] have recently demonstrated a Ca^{++} - dependent lipid acyl hydrolase activity associated with mitochondical membranes of potato tubers. A related acyl transferase system catalysing the acylation of galactose residues of galactolipids has been demonstrated in leaf tissues [7] and recent results from Heinz et al. [8] show clearly that this activity is contained in chloroplast envelope membranes.

Of the other possible acyl hydrolyase activities indicated in Fig. 1, there is no direct evidence for specific phospholipases -1 (II) or -2 (IV) or for lysophospholipases (III and V). An acyl hydrolase in cereal seeds [9] forms lyso-intermediates from phospholipids during complete deacylation and it is not yet clear whether such an enzyme differs fundamentally from the type catalysing reaction I in Fig. 1.

Other polar lipid hydrolyases

Phospholipase D (VI) is well known and exists in a wide range of plants. Like the acyl hydrolase, this enzyme is manifested in disrupted tissues causing accumulation of phosphatic acid and it also acts as a transferase. E.g. in the presence of some alcohols (eg. methanol), natural phospholipids may be converted to artifacts such as phosphatidylmethanol. Because of the possibility of hydrolytic or transferase reactions acting rapidly when plant cells are disrupted, lipid extractions and analyses are only meaningful if such enzymes are inactivated prior to cell disruption. The physiological role of phospholipase D is not known; in fact, Roughan and Slack [10] recently published a paper with the provoactive title :- "Phospholipase D - is it really an enzyme ?", in which they suggested that the hydrolytic activity was an accidental property of a structural protein. Some early evidence for the presence of phospholipase C (VII) in plants [11] has not been further substantiated. Phosphatidic acid hydrolysis to diacylglycerol (VIII) is an essential step in glycerolipid biosynthesis but its possible role in catabolism has not been studied.

The above discussion has concentrated on hydrolysis of membrane phospholipids. Glycolipid hydrolysis in higher plants has received some attention.

'Galactolipase' activity has been demonstrated in leaves and the enzyme purified [12]. However, as discussed previously, it is probable that 'galactolipase' activity is but one facet of a more general lipid acyl hydrolase activity as indeed is sulpholipid hydrolysis in higher plants [13].

Degradative reactions on hydrolysis products.

In systems undergoing a net loss of membrane lipid, the initial reaction is normally the hydrolytic removal of fatty acids. Free fatty acids are themselves toxic to biological materials, eg. they disrupt membrane systems, cause lysis of organelles, bind to, and inactivate, proteins. Furthermore, fatty acids are substrates for further enzyme reactions and, as such, are links in a sequence of enzyme reactions, initiated by the initial hydrolysis of lipid.

Some reactions of fatty acids involve activated forms (eg. Coenzyme A or ACP-thioesters) wherease others occur with the free fatty acid form. The β-oxidation of fatty acids in germinating seeds requires CoA derivatives, but in systems undergoing net loss of membrane lipids - and certainly in disrupted tissues - activation of fatty acids is less likely than their transformation by enzymes requiring only the free carboxylate form. Of these, two processes appear to be of particular relevance :-

a) _α-oxidation_ attacks long-chain, C_n-fatty acids to produce either C_n-2-hydroxyacids (e.g. as found in cerebrosides) or C_{n-1}-aldehydes which, after oxidation to the corresponding C_{n-1} fatty acids, may recycle through the α-oxidation system. Thus odd chain-length derivatives can result from this process (cf. β-oxidation). The following mechanism, involving 2-_hydroperoxyacid_ intermediates, was proposed by Shine and Stumpf [14] :-

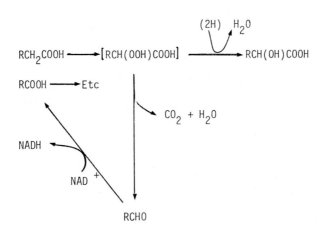

The involvement of α-oxidation in the formation of volatile long-chain aldehydes in disrupted cucumber fruit was recently demonstrated in our laboratory [15]. Laties' group [16] have shown that respiration, induced on cutting potato tuber tissue, is based on α-oxidation of fatty acids (released by acyl hydrolases on cell disruption) rather than carbohydrate oxidation.

b) *Lipoxygenase-mediated reactions*. Linoleic acid and linolenic acids are predominant in most plant tissues and these fatty acids are substrates for the enzyme, lipoxygenase, which produces hydroperoxide derivatives. The hydroperoxides do not normally accumulate, but are converted, by one of several routes, to a range of products. These reactions, summarized in Fig. 2, have been the subject of several recent reviews [17-20] and will not be covered in detail in this review.

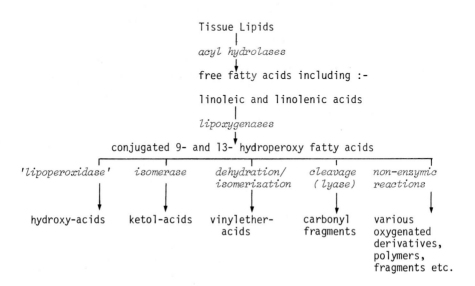

Fig. 2 Lipoxygenase-mediated conversions of polyunsaturated fatty acids

Here, I would like to mention briefly some recent work from our laboratory.

For several years, we had been interested in the sequential reactions involved in the breakdown of membrane lipids, via free fatty acids and their hydroperoxides, to a range of oxygenated fatty acids and fragmentation products. Much of our earlier work was on potato tubers in which, for reasons we still don't understand, activities of many of the enzymes involved are very high. These studies have been reviewed [18-20].

Biogenesis of volatile aldehydes

More recently, we became interested in the mechanism by which volatile aldehydes with characteristic flavour properties, were formed from endogenous lipids when fresh tissues of many plants were disrupted. We selected cucumber fruits because it was already known that (a) the characteristic cucumber flavour was due to C_9- unsaturated aldehydes derived from fatty acids and (b) these aldehydes were not present in the intact fruit but were formed only on disruption by cutting etc. Subsequently, we also worked with tomato fruits and bean leaves. (In both these latter tissues C_6- volatile aldehydes released on tissue disruption, contribute to the aroma).

As a result of this work, we think that we are able to propose an explanation for the biogenesis of these volatile compounds from endogenous lipids and this will be briefly summarized below (see also refs. 20 - 22).

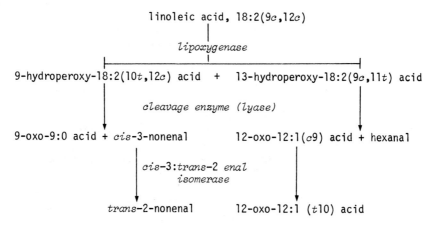

Fig. 3. Enzymic formation of fatty acid cleavage products in cucumber fruits

Fig. 3 lists the reactions involved in cucumber fruit for the conversion of linoleic acid (released from endogenous lipids by acyl hydrolases) via its hydroperoxides to cleavage products. Analogous reactions take place with linolenic acid. The 9- and 13-hydroperoxides, formed by lipoxygenase action, are cleaved by a specific cleavage enzyme to C_9 and C_6 volatile aldehydes respectively and the C_9 or C_{12} oxoacid, non-volatile fragments. The *cis* -3 enals are rapidly isomerized by an enzyme also present in cucumber [23] to the more stable trans-2-forms. A similar process has been shown in water melon seedlings by Zimmerman's group [24] who have also recently shown that the C_{12} oxoacid derived from linoleic or linolenic acids is the active principle in the 'wound hormone', traumatin [25].

The sequence observed in cucumber could not be translated directly to other plants because the C_9 volatile aldehydes, with their characteristic aromas, are not found in significant amounts in most other plant extracts. When we investigated tomato fruits we were puzzled by the fact that the lipoxygenase produced predominantly (ca. 95%) the 9-hydroperoxide isomer and yet we found no evidence of C_9 cleavage products. Subsequent studies[26,27] explained this when we found that, unlike the cucumber enzyme, the hydroperoxide cleavage enzyme in tomato was specific for the 13-hydroperoxide isomer. Thus, although this isomer was a minor (ca. 5%) product of tomato lipoxygenase, it alone was cleaved by the enzyme to produce C_6 volatile products. This is indicated diagrammatically for linoleic acid breakdown in Fig. 4.

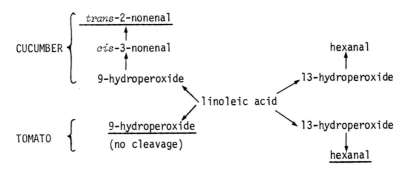

Fig. 4. Comparison of specificities of fatty acid hydroperoxide cleavage systems in cucumber and tomato fruits

In *Phaseolus* sp. leaves, large amounts of C_6 aldehydes are released on maceration and our limited studies [28] with this tissue indicate that these aldehydes are produced from linoleic and linolenic acids of endogenous lipids by the same process as outlined above. However, Hatanaka's group in Japan have proposed [29] that a direct attack on the fatty acids (not involving lipoxygenase) takes place in leaf chloroplasts to produce the volatile aldehydes. Further research will no doubt clarify these points.

Autoxidation of membrane lipids

The above discussion has concentrated on enzymic processes of membrane lipid degradation. However, it should be noted that plant membranes, with a high content of polyunsaturated fatty acids, are potentially liable to autoxidation. In healthy cells, the presence of natural antioxidants, physical constraints, and mechanisms for destruction of activated oxygen species (eg. superoxide dismutase) no doubt inhibit autoxidation. Nevertheless, these restrictions apply also in mammalian cells where membrane peroxidation still occurs and this is an active field of research in relation to toxicity, carcenogenicity and ageing processes. This area has not been researched in detail in plants but two recent observations illustrate the possible importance of non-enzymic oxidation of membrane lipids. de Kok and Kuiper[30] have recently demonstrated a light - and low temperature - catalysed degradation of lipids in cucumber leaf disks; this was not an enzymic process and probably resulted from lack of protection of the membrane lipids to photo-oxidation at low temperatures where membrane structure in cold-sensitive plants would be disturbed. The second example is given by the appearance of 'lipofuscin' - like fluorescent products in ripening fruit;[31] these pigments are characteristic of peroxidized membranes in ageing mammalian membranes. In ripening fruits, onset of senescence and loss of cellular control processes possibly lead to enhanced membrane oxidation.

MEMBRANE TURNOVER

Here we progress even further into realms of speculation. As stated above, most of our meagre knowledge of lipid catabolism in plants is based on cases where net losses of membrane lipid occur. However, there is no doubt that plant membrane lipids are dynamic systems in which breakdown and resynthesis of all or part of individual lipid molecules take place. For example, results from developing soybean[32] have shown that the rate of synthesis of phospholipids is much greater than the net accumulation rate and hence a high catabolic activity is indicated.

At this point, the difficulties in interpreting data on turnover in terms of specific enzyme activities should be emphasised. Further complications are introduced by the existence of exchange reactions. In brief, one is faced with a dynamic system of several inter-dependent processes:

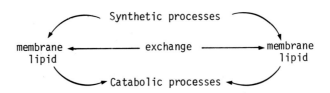

The exchange proteins, known (from the work of Mazliak, Kader and colleagues in Paris [33]) to exist in plant tissues, raise some interesting questions. For example, if membrane lipids are subject to turnover, do the catabolic reactions take place *in situ* in the membrane, or are the lipids removed from the membrane by the exchange protein and metabolised elsewhere? What are the relative turnover rates of lipids in the two layers of a double membrane and how do catabolic enzymes attack both inner and outer layers of a bilayer?

Other questions arise when one considers the possible routes by which acyl lipids can be catabolized. For example, there is good evidence for the involvement of diacylglycerols in phospholipid turnover. The most direct process is the hydrolysis of phospholipid by a phospholipase C (see Fig. 1) but evidence for such an enzyme in plants is scant. A second possibility involves intermediacy of phosphatidic acid via phospholipase D and phosphatidate phosphohydrolase enzymes (Fig. 1). A third interesting possibility, recently suggested by Slack et al [34] is that the cytidine nucleotide pathway of biosynthesis could be reversed :-

Evidence for such a process exists in rat liver [35].

The fatty acid composition of membrane lipids is known to change in response to environmental factors. Does this indicate complete *de novo* synthesis of new lipid species or does a turnover of fatty acyl groups take place on the glycerol backbone? Some difficult problems arise here. Since fatty acid species tend to be localized on specific glycerol hydroxyl groups (the *sn*-1-position generally contains a higher proportion of unsaturated acids and the sn-2 position the more unsaturated acids), does this mean that positionally-specific acyl hydrolases (eg. phospholipases -1 and -2) selectively remove the fatty acids? We do not yet have much evidence for such enzymes in plants. Furthermore, deacylation and reacylation represents an energy-intensive process; a more efficient process, perhaps, would be specific replacement of fatty acids by acyl transferase reactions. Again the problem of positional specificity arises. A heretical suggestion is that, maybe, specific enzymes are not required but that the specificity may be built into the membrane. That is, the physical environment in a membrane may determine the type of fatty acid allowed.

The above discussion has raised many questions and answered few. Perhaps, subsequent meetings of plant lipid biochemists will be enlightened by answers to at least some of these problems.

Acknowledgements

I would like to thank the organisers of this symposium for inviting me to participate. Our own work referred to in this presentation was performed in the author's previous laboratory at the A.R.C. Food Research Institute, Norwich and I acknowledge the contributions of my former colleagues, in particular, David Phillips, Jenny Matthew, Dennis Wardale and Henry Chan.

REFERENCES

1. Galliard, T. (1970) Biochem. J. 121, 379-390
2. Galliard, T. (1975) In "Recent Advances in the Chemistry and Biochemistry of Plant Lipids" (T. Galliard and E.I. Mercer, eds) Academic Press, London pp 319-357
3. Sastry, P.S. and Kates M. (1964) Biochemistry 3, 1280-1287
4. Galliard, T. and Dennis, S. Phytochemistry 13, 1731-1735
5. Burns, D.D., Galliard, T. and Harwood, J.L. (1977) Biochem. Soc. Trans. 5, 1302-1304
6. Bligny, R and Douce, R. (1978) Biochim. Biophys. Acta 529, 419-428
7. Heinz, E.Z. Pflanzenphysiol. (1973) 69, 359-376
8. Heinz, E., Bertrams, M., Joyard, J. and Douce, R. (1978) Z. Pflanzenphysiol. 87, 325
9. Acker, L. and Schmitz, H.J. (1967) Die Stärke 19, 233-239.
10. Roughan, P.G. and Slack, C.R. (1976) Biochim. Biophys. Acta, 431, 86-95
11. Kates, M. (1955) Can. J. Biochem. Physiol. 33, 575-589
12. Helmsings, P.J. (1969) Biochim. Biophys. Acta 178, 519-533

13. Burns, D.D. Galliard, T. and Harwood, J.L. (1977) Phytochemistry 16, 651-654
14. Shine, W.E. and Stumpf, P.K. (1974) Archs. Biochem. Biophys. 162, 147-157
15. Galliard, T. and Matthew, J.A. (1976) Biochim. Biophys. Acta 424, 26-35
16. Laties, G.G. Hoelle, C. and Jacobson, B.S. (1972) Phytochemistry, 11, 3403-3411
17. Gardner, H.W. (1975) J. Ag. Fd. Chem. 29, 129-135
18. Eriksson, C. (1975) J. Ag. Fd. Chem. 23, 126-128
19. Veldink, G.A., Vliegenthart, J.F.G. and Boldingh, J. (1977) Prog. Chem. Fats, Lipids 15, 131-166
20. Galliard, T. (1978) In "Biochemistry of Wounded Plant Tissues" (G. Kahl. ed) Springer, Berlin. pp 155-201
21. Galliard, T., Phillipds, D.R. and Reynolds, J. (1976) Biochim. Biophys Acta 441, 1731-4
22. Phillips, D.R. and Galliard, T. (1978) Phytochemistry 17, 355-358
23. Phillips, D.R., Matthew, J.A., Reynolds, J. and Fenwick, G.R. (1979) Phytochemistry (in press)
24. Vick, B.A. and Zimmerman, D.C. (1976) Plant Physiol. 57, 780-788
25. Zimmerman D.C. and Olson, C.A. (1978) Plant Physiol (In press)
26. Galliard, T., and Matthew, J.A. (1977) Phytochemistry 16, 1731-4
27. Galliard, T., Matthew, J.A., Wright, A.J., and Fishwick, M.J. (1977) J. Sci. Fd. Agric. 28, 863-868
28. Matthew, J.A. and Galliard, T (1978) Phytochemistry 17, 1043-4
29. Hatanaka, A., Sekiya, J. and Kajiwara T. (1977) Plant Cell Physiol. 18, 107-116
30. de Kok, L.J. and Kuiper, P.J.C. (1977) Physiol. Plant 39, 123-128
31. Maguire, Y.P. and Haard, N.F. (1976) J. Food Sci. 41, 551-4
32. Wilson, R.F. and Rinne, R.W. (1976) Plant Physiol. 57, 375-381
33. Douady, D., Kader, J.C. and Mazliak, P. (1978) Phytochemistry, 17, 793-4
34. Slack, C.R., Roughan, P.G. and Balsingham, N. (1978) Biochem. J. 170, 421-433
35. Kanoh, H. and Ohno, K. (1973) Biochim. Biophys. Acta 326, 17-25

PLASTID DEVELOPMENT AND TOCOCHROMANOL ACCUMULATION IN OIL SEEDS

H. BERINGER and F. NOTHDURFT
Department of Plant Nutrition, University Hohenheim, POB 106, D-7000 Stuttgart 70

Although the pathways of tocochromanol - biosynthesis are established[1], the processes which lead to tocochromanol accumulation and specific tocochromanol patterns in developing seeds are poorly understood. The functions of tocochromanols in plant cells are also not clear. α-tocopherol might be involved in the photosynthetic electrontransport[2,3]. On the other hand it has often been assumed that the biochemical significance of tocochromanols in seeds is related to their antioxidative properties and thus to the stabilization of unsaturated reserve lipids[4,5,6]. This assumption was based on the observed concomitant accumulation of reserve lipids and tocochromanols during seed development and on the observation that seeds rich in unsaturated lipids also contain considerable amounts of tocochromanols.

Therefore the biochemical and ultrastructural principles underlying tocochromanol accumulation were studied in an attempt to define the relationship between tocochromanols and storage lipids of seed tissues.

In vegetative plant material the dominant tocochromanol is α-tocopherol. It is located in the chloroplasts and increases in amount during leaf development up to senescence and also during the ripening of green fleshy fruits such as tomatoes and Capsicum annuum[7,8,9,10]. By contrast seeds frequently contain tocochromanols other than α-tocopherol[11].

a) *Avocado mesocarp*. The dominant tocochromanol of the green and pale coloured mesocarp of the avocado fruit is α-tocopherol. In addition variable amounts of γ-tocopherol and γ-tocotrienol can also be found. However, the significance of this variation is not clear since the origin and ripening conditions of the fruits were not known.

Table 1 shows the tocochromanol distribution within different mesocarp zones of a single avocado.

TABLE 1

Lipid-, chlorophyll- and tocochromanol-content of avocado mesocarp

Zone*	Lipid mg/g FW	Chloroph. ug/g FW	Tocochr. ug/g FW	% of Tocochr. γ-T-3	% of Tocochr. α-T
outer	152	125	75	25	75
middle	178	27	45	38	62
inner	186	13	43	40	60

* Mesocarp was dissected into three zones of equal thickness. The notation of the zones as 'outer' etc. refers to their increasing distance from the exocarp.

The lipid content of the pale inner mesocarp was slightly higher than that of the green outer zone, whereas the chlorophyll content was ten times higher in the outer than in the inner zone. The highest chlorophyll content was associated with the highest tocochromanol content. There was also a marked difference in the relative proportions of tocochromanols, α-tocopherol decreased whereas γ-tocotrienol increased from outside to inside. Within each mesocarp zone the tocochromanol patterns of whole tissue and of isolated plastids were identical.

Electronmicroscopical examination of mesocarp revealed that the cells of the outer green coloured mesocarp zone contained mainly chloroplasts (Fig. 1) whereas etiochloroplasts were characteristic of the middle, yellow-green zone (Fig. 2). The cells of the inner yellow zone contained mainly etioplasts. In all types of plastids numerous plastoglobuli were present. The occurrence of various plastid types in avocado mesocarp has also been described by others[12]. All these observations suggest that the tocochromanol patterns within avocado mesocarp may be related to plastid distribution and plastid type.

The presence of tocochromanols in preparations of oil bodies, microscopically free of plastids, appeared to contradict the evidence of tocochromanols being located exclusively in plastids. An explanation could be that the oil body preparations were contaminated with plastoglobuli released from broken plastids. Plastoglobuli, because of their low density, behave like oil bodies during centrifugation[13] and are known to contain α-tocopherol and plastid-quinones[14].

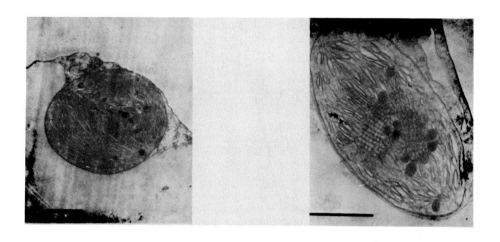

Fig. 1 and 2   Plastid types of avocado mesocarp
left : chloroplast from the green outer zone
*   bar represents 1 /u*
right: etiochloroplast from the yellow middle zone
*   bar represents 1 /u*

b) <u>Sunflower seeds</u>. In sunflower seeds the dominant tocochromanol is also α-tocopherol. It accumulated between 15 and 30 days after pollination (Fig. 3). During this period the amount of tocopherol per embryo increased by a factor of 30 but the ratio of α-tocopherol:lipid decreased from 2.5 to 1.4 /ug/mg lipid.

Electronmicroscopical studies of seeds during this period of seed development again revealed changes in the shape of plastids. 15 days after pollination sunflower seeds contained amoeboid proplastids whereas mature seeds were characterized by amyloplasts containing prominent starch granules.

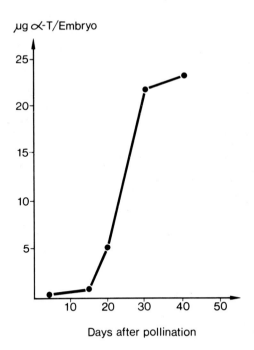

Fig. 3 α-tocopherol accumulation in developing sunflower seeds

c) *Vicia faba seeds*. A remarkable tocochromanol accumulation was also found in seeds of Vicia faba which do not accumulate storage lipids (Table 2).

TABLE 2

Fresh weight, tocopherol and plastoquinone content of developing Vicia faba seeds

Days after pollination	F.W. g/seed	Tocopherol ug/seed α-T	γ-T	PQ_{45} ug/seed
20	0.63	tr.	2.9	n.d.
30	0.73	-	3.5	1.5
40	1.04	-	18.9	n.d.
50	1.10	-	22.7	2.1

During the period of tocochromanol accumulation between 30 and 50 days after pollination plastoquinone also accumulated. Like in sunflowers, the development of mature Vicia faba cotyledons was associated with the appearance of amyloplasts.

In summary the chemical and structural findings on avocado mesocarp and seeds of sunflower and of Vicia faba do not suggest a causal relationship between the content of tocochromanols and reserve lipids in seeds. Tocochromanol accumulation is rather related to plastid differentiation as has been previously suggested from studies of leaves and fleshy fruits[10]. Thus tocochromanols do not seem to contribute to the stabilisation of unsaturated storage lipids in oil bodies of oil seed cells.

REFERENCES

1. Threlfall, D.R. and Whistance, G.R. (1971) in Aspects of Terpenoid Chemistry and Biochemistry, T.W. Goodwin ed., Academic Press, pp. 357-404.
2. Bishop, N.I. and Wong, J. (1974) Ber. Deutsch. Bot. Ges., 87, 354-371.
3. Barr, R. and Crane, F.L. (1977) Plant Physiol., 59, 433-436.
4. Green, J. (1958) J. Sci. Food Agric., 9, 801-812.
5. Janiszowska, W. and Pennock, J.F. (1976) Vitamines and Hormones, 34, 77-105.
6. Gutfinger, T. and Letan, A. (1974) Lipids, 9, 659-663.
7. Booth, V.H. (1963) Phytochem., 2, 421-427.
8. Bucke, C. (1968) Phytochem., 7, 693-700.
9. Tendille, C., Gervais, C. and Gaborit, Th. (1966) Ann. Physiol. veg., 8, 271-283.
10. Lichtenthaler, H.K. (1969) Ber. Deutsch. Bot. Ges., 82, 483-497.
11. Beringer, H. and Dompert, W.U. (1976) Fette, Seifen, Anstrichmittel, 78, 228-231.
12. Cran, D.G. and Possingham, J.V. (1973) Ann. Bot., 37, 993-997.
13. Appelquist, L.Å. (1976) in Recent Advances in the Chemistry and Biochemistry of Plant Lipids, T. Galliard and E.I. Mercer eds., Academic Press, pp. 247-286.
14. Lichtenthaler, H.K. and Sprey, B. (1966) Z. Naturforsch., 21b, 690-697.

SOLUBLE, ISOMERIC FORMS OF GLYCEROLPHOSPHATE ACYLTRANSFERASE IN CHLOROPLASTS

M. BERTRAMS and E. HEINZ
Institute of Botany, University of Cologne, Cologne, Germany

Acyl-CoA: sn-glycerol 3-phosphate acyltransferase, catalysing the initial step in glycerolipid biosynthesis, may be of importance in establishing specific distribution of fatty acids at the glycerol backbone of lipids. The enzyme is localized in several subcellular compartments [1-7]. We investigated the acyltransferase from pea chloroplasts, the only soluble activity in leaves[1], with respect to the following questions:
1) which of the possible C_3-precursors is acylated by the acyltransferase ?
2) which position of the glycerol backbone is acylated, and are specific fatty acids preferred depending on chain length or degree of unsaturation ?
3) does the enzyme also accept acyl-ACP as substrate ?

In order to answer these questions, the enzyme should be free from interfering activities. Therefore, several steps of purification were carried out including acid and ammonium sulphate precipitation, column chromatography on DEAE cellulose and Sephadex G 100, and finally isoelectric focussing. During gel filtration the activity eluted as a single peak. The molecular weight estimated with the aid of standard proteins was about 42,000. Isoelectric focussing demonstrated the existence of two forms of acyltransferase (Fig. 1 A). Apparent isoelectric points differed by 0.3 - 0.35 pH units. This separation is no artifact, as was shown by refocussing isolated peaks (Fig. 1 B and C). In each case we only found one peak of activity and no separation into two forms.

Examination of subcellular localization of both isoenzymes by focussing total soluble leaf and chloroplast stroma proteins, respectively, gave the activity patterns shown in figures 1 D and E. In both cases we observed the same splitting into two forms as before in the enriched enzyme fraction. We, therefore, conclude

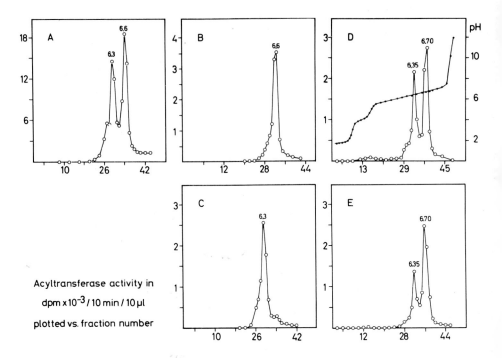

Fig. 1. Acyltransferase activity in different protein fractions after isoelectric focussing in the pH range from 5-7: (A) enriched enzyme fraction, (B) and (C) isolated peaks of the enriched enzyme fraction, (D) total soluble leaf proteins, (E) chloroplast stroma proteins. Apparent pI values are given above each peak.

that the acyltransferase occurs in two soluble, isomeric forms, and that both forms are localized in chloroplasts. The lower total activity of the first isoenzyme after focussing stroma proteins as compared to the total leaf proteins may be due to differences in leaf material used in the different experiments, since total acyltransferase activity strongly depends on the age of leaves[1]. At the present stage, however, we cannot exclude the possibility, that a small proportion of soluble acyltransferase is localized in the cytoplasm as well.

The purification procedure outlined above raised the specific activity by a factor of about one thousand to 1-2 μmole/min/mg protein. The isoenzyme fractions were not pure electrophoretically, but there were no detectable enzymatic activities which could interfere with the characterization of the acyltransferase.

The only reaction product of the enzymatic assay is lysophosphatidic acid: No radioactivity was incorporated into phosphatidic acid or monoacylglycerol. Therefore, the enzyme fractions contained neither monoacylglycerol 3-phosphate acyltransferase nor lysophosphatidate phosphatase. In addition glycerol 3-phosphate phosphatase activity was absent, since no glycerol could be detected after separation of water soluble reaction products on Dowex columns. Furthermore, photometrical tests of acyltransferase activity using Ellman's reagent 5,5'-dithiobis-(2-nitrobenzoic acid) (DTNB) demonstrated that acyl-CoA hydrolase activity was not detectable, since the incubations of enzyme fraction plus acyl-CoA plus DTNB did not show any increase in extinction.

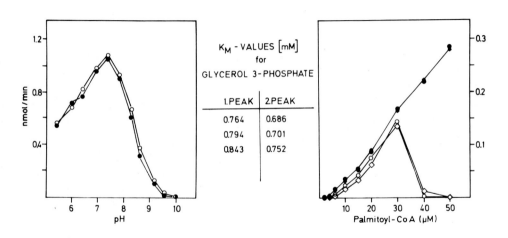

Fig. 2. Kinetic properties of the two acyltransferase forms, peak 1 with pI'=6.3 and peak 2 with pI'=6.6. Left figure: pH-dependence of peak 1 o—o and peak 2 •—• activity, right figure: acylation rate (nmole/min) as a function of palmitoyl-CoA concentration in the absence (o—o peak 1, □—□ peak 2) and presence (•—• peak 1, ■—■ peak 2) of serum albumin.

Figure 2 depicts some of the kinetic properties of both isoenzymes. Both forms showed identical pH optima close to pH 7.4. K_M-values for glycerol 3-phosphate were also very similar. Our efforts to raise the affinity of the enzymes for glycerol 3-phosphate by adding different salts were not successful. Furthermore, the addition of lipids including phosphatidylglycerol[8] had no effect on acyltransferase activity. Acylation rates varied with palmitoyl-CoA concentrations in the same way for both isoenzymes. Both forms seem to use acyl-CoA dispersed in micellar forms, and the inhibition by higher acyl-CoA concentrations was relieved by adding optimal concentrations of serum albumin.

The specificity of the acyltransferases for possible C_3-precursors is of interest in connection with the subcellular origin of these substrates. Glycerol 3-phosphate cannot be formed directly from dihydroxyacetone phosphate in chloroplasts, since glycerol 3-phosphate dehydrogenase was not detectable in leaves[9]. It may be formed by a sequence of reactions localized in the cytoplasm in which glycerol kinase takes part[9]. Therefore, we wanted to see, if also dihydroxyacetone phosphate, which is formed directly in chloroplasts, can serve as acyl acceptor. In all experiments, applying radioactive and photometric assays and using different dihydroxyacetone phosphate concentrations, both isoenzymes did not accept dihydroxyacetone phosphate as substrate. Figure 3 shows a photometric assay in which acylation, as evidenced from release of CoA, does not commence before addition of glycerol 3-phosphate, whereas dihydroxyacetone phosphate is without effect.

The question, whether or not this enzyme is involved in establishing asymmetric distribution of fatty acids in lipids was approached by investigating the positional specificity of both isoenzymes. The reaction product lysophosphatidic acid was dephosphorylated by alkaline phosphatase, and the resulting monoglycerides were separated by thin-layer chromatography. Isomerisation due to acyl migration was excluded by appropriate tests with authentical 2-acyl glycerol. Isomeric monoglycerides were analysed after incubating each isoenzyme with different acyl-CoAs individually at different concentrations and with different acyl-CoA mixtures, including or omitting serum albumin. All assays gave the same results, almost all radioactivity (96.4 $\pm$ 1%) was found in the 1-acyl isomer, and hardly any activity (3.6 $\pm$ 1%) was present in the 2-acyl isomer. Therefore, in vitro both isoenzymes possess a high positional specificity for acylation of C-1 of the glycerol backbone regardless of the acyl-CoAs offered.

First results with regard to the fatty acid specificity of the acyltransferase suggest that the enzyme did not display a preference for specific fatty acids. However, this question is subject to further experiments, using varying acyl-CoA concentrations and combinations.

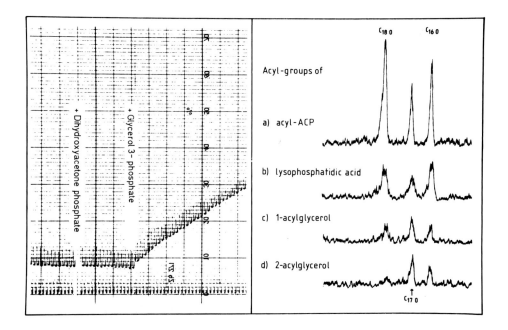

Fig. 3. Photometrical acyltransferase test with Ellman's reagent DTNB. Possible acyl acceptors were added as indicated. Extinction was recorded at 412 nm.

Fig. 4. Radio gaschromatograms showing the fatty acid composition of (a) the acyl-ACP used in the assay,
  (b) the reaction product lysophosphatidic acid obtained from an assay with peak 2 (an assay with peak 1 gave the same result),
  (c) and (d) the 1-acyl and 2-acyl glycerol, released from enzymatically produced lysophosphatidic acid.
Radioactive $C_{17:0}$ was used as internal standard.

Finally we wanted to know, if apart from acyl-CoA also acyl-ACP could act as acyl donor. We incubated both isoenzymes with ^{14}C-acyl-ACP, and obtained the following results (Fig. 4). Both isoenzymes use acyl-ACP as substrate, and both C_{16}- and C_{18}-acyl groups were incorporated, $C_{16:0}$ being slightly preferred. The positional specificity of the acyltransferase with acyl-ACP as substrate is different from the one with acyl-CoA as substrate.

With acyl-ACP both positions of the glycerol backbone were acylated. About two thirds of the radioactivity were found in the 1-acyl isomer and about one third in the 2-acyl isomer. A comparison of the two chromatograms (c and d) may suggest a specificity of the acyltransferase in favour of $C_{16:0}$ when directing fatty acids into the C-2 position.

In summary, two soluble, isomeric forms of glycerol 3-phosphate acyltransferase are localized in chloroplasts. Both forms use glycerol 3-phosphate but not dihydroxyacetone phosphate as acyl acceptor. With acyl-CoA as substrate both isoenzymes exclusively catalyze the acylation of the sn-1-position of the glycerol backbone irrespective of the acyl-CoAs offered, whereas with acyl-ACP as acyl donor the situation may be quite different as shown by the last experiments, which are still in progress and should be considered as preliminary. Therefore, it is at present premature to draw any conclusions about the role played by glycerol 3-phosphate acyltransferase in introducing fatty acid specificities into lipids.

ACKNOWLEDGEMENTS

The authors are indebted to Prof. Dr. Stumpf and Dr. Radunz for the generous supply of radioactive acyl-ACP and hexadecatrienoic acid, respectively. This investigation was supported by the Deutsche Forschungsgemeinschaft.

REFERENCES

1. Bertrams, M. and Heinz, E. (1976) Planta, 132, 161-168.
2. Boehler, B.A. and Ernst-Fonberg, M.L. (1976) Arch. Biochem. Biophys., 175, 229-235.
3. Cheniae, G.M. (1965) Plant. Physiol., 40, 235-243.
4. Gurr, M.J., Blades, J., Appleby, R.S., Smith, C.G., Robinson, M.P. and Nichols, B.W. (1974) J. Biochem., 43, 281-290.
5. Joyard, J. and Douce, R. (1977) Biochim. Biophys. Acta, 486, 273-285.
6. Sastry, P.S. and Kates, M. (1966) Can. J. Biochem., 44, 459-467.
7. Vick, B. and Beevers, H. (1977) Plant Physiol., 59, 459-463.
8. Kito, M., Ishinaga, M. and Nishihara, M. (1978) Biochim. Biophys. Acta, 529, 237-249.
9. Hippmann, H. and Heinz, E. (1976) Z. Pflanzenphysiol., 79, 408-418.

THE INFLUENCE OF PHYTOHORMONES ON PRENYLLIPID COMPOSITION AND PHOTOSYNTHETIC ACTIVITY OF THYLAKOIDS

CLAUS BUSCHMANN and HARTMUT K. LICHTENTHALER
Botanical Institute II (Plant Physiology), University of Karlsruhe,
Kaiserstrasse 12, D-7500 Karlsruhe (FRG)

ABSTRACT

Seedlings of Raphanus sativus L. were grown on a nutrient solution without (control) or with addition of the phytohormones β-indoleacetic acid (IAA) and kinetin. During greening in white light both phytohormones lead to an increased formation of chlorophyll a (higher chlorophyll a/b ratios), plastoquinone-9, phylloquinone K_1, and P 700. Carotenoids and α-tocopherol, in turn, are present in lower amounts after IAA or kinetin treatment. In etiolated plants kinetin leads to a higher accumulation of phototransformable protochlorophyll(ide).

Together with the rise in prenylquinone and P 700 content, chloroplasts isolated from seedlings grown in the presence of phytohormones also exhibit a higher photosynthetic activity (Hill activity). Higher Hill activity rates are also found in chloroplasts from seedlings grown in the presence of kinetin in blue (+ 20 %) and red light (+ 70 %).

The data indicate that the phytohormones IAA and kinetin modify the prenyllipid composition and activity of the photosynthetic apparatus towards a higher efficiency for photosynthetic light quanta conversion. The hypothesis is discussed that blue and high intensity light, leading to similar changes in prenyllipid composition and photosynthetic activity as phytohormone treatment, may act via an increase of the levels of the endogenous phytohormones IAA and cytokinin.

INTRODUCTION

Phytohormones, like β-indoleacetic acid (IAA) and kinetin (6-furfurylaminopurine) are known to influence many developmental processes in plants and thus control plant growth.

The prenyllipids chlorophyll, carotenoids, and the benzoquinones plastoquinone-9 and α-tocopherol as well as the naphthoquinone phylloquinone K_1 occur exclusively in the chloroplasts[1]. They are components of the thylakoid membrane and are therefore of direct or indirect importance for the function of the photosynthetic apparatus. While plastoquinone-9 is known as an electron carrier of the photosynthetic electron transport chain near pigment system II, the site of action of phylloquinone K_1 and α-tocopherol is not yet fully understood. Recent results indicate that phylloquinone K_1 is a component of the photosynthetic electron transport chain with a position between Q and plastoquinone-9[2].

In our attempt to find out the factors that are responsible for the variability of the photosynthetic apparatus in nature, e.g. in sun and shade plants[3,4], we investigated the influence of phytohormones on the formation, composition and function of chloroplasts.

MATERIALS AND METHODS

Seedlings of Raphanus sativus L. (Saxa Treib) were grown on a 10 % van der Crone nutrient solution[5]. Phytohormones, when present, were added at the beginning of germination to the nutrient solution (IAA: 1 ppm = 5.7 μM, kinetin: 2 ppm = 9.3 μM).

The seedlings were illuminated with fluorescent light (white light: fluora 7.5 $J/m^2 \cdot$ sec; red light: emission maximum 660 nm, 1.0 $J/m^2 \cdot$ sec; blue light: emission maximum 450 nm, 1.5 $J/m^2 \cdot$ sec). Red and blue light were adjusted to the same light quanta density (5.5 $\mu Moles/m^2 \cdot$ sec).

The prenyllipids were extracted with acetone and transfered to petrolether. The components were measured spectrophotometrically after TC-separation [6,7]. The phototransformable protochlorophyll(ide) was measured from $77^{\circ}K$ in vivo absorption spectra of etiolated cotyledons. Hill activity (DCPIP and methylviologen reduction)[8] and P 700 measurements (chemical oxidation -reduction)[8] were carried out with chloroplasts isolated according to Nobel[9].

RESULTS

8 day old radish seedlings possess fully green cotyledons. When phytohormones like IAA and kinetin were present during growth the prenyllipid composition is markedly changed (Table 1). Higher amounts of chlorophyll a are formed, leading to a higher chlorophyll a/b ratio. The carotenoids are accumulated to a lower extent when the two phytohormones are present during growth, correspondingly the ratio chlorophylls to carotenoids (a+b/x+c) becomes higher.

It is generally agreed to that most of the chlorophyll b and of the xanthophylls (the main quantity of carotenoids) are associated with the photosyntheti light harvesting complex[10]. The lower chlorophyll b and carotenoid content in phytohormone treated seedlings therefore indicates a photosynthetic apparatus with fewer light harvesting complexes than in control plants.

TABLE 1

MODIFICATION OF THE LIGHT INDUCED CHLOROPLAST PRENYLLIPID SYNTHESIS UNDER THE INFLUENCE OF THE PHYTOHORMONES IAA AND KINETIN

µg in 100 8 day old radish plants grown for 1 day in the dark and 7 days in white light. The values are based on the data of [8,11,12]

	control	+ IAA	+ kinetin
chlorophyll a	2700	3500	3100
chlorophyll b	900	1100	900
chlorophyll a + b	3600	4600	4000
carotenoids (x + c)	572	508	480
P 700 / chlorophyll a + b	$44 \cdot 10^{-4}$	$52 \cdot 10^{-4}$	$62 \cdot 10^{-4}$
phylloquinone K_1	33	44	52
plastoquinone-9	299	590	414
α-tocopherol (+ α-tocoquinone)	353	282	148
a / b	3.0	3.2	3.4
a + b / x + c	6.3	9.1	8.3
a / P 700	169	147	124
a / K_1	82	79	62
a / PQ-9	9.1	5.9	7.5

IAA or kinetin treatment increase the level of phylloquinone K_1, plastoquinone-9, and of P 700 (lower a/K_1, a/PQ-9, and a/P 700 ratios). The benzoquinone α-tocopherol, on the other hand, is found in a lower concentration. Since phylloquinone K_1, like plastoquinone-9 and P 700, is a functional component of the photosynthetic electron transport chain[2], it appears that the number of electron transport chains is increased due to the action of IAA and kinetin which should give rise to a higher Hill activity.

A 77°K in vivo absorption spectrum of an etiolated radish cotyledo shows peaks for protochlorophyll(ide) at 650 and 637 nm. After phototransformation by one flash of light a protochlorophyll(ide) peak remains at 627 nm (non-phototransformable protochlorophyll(ide)), while the peaks at 650 and 637 nm disappear (phototransformable protochlorophyll(ide)) (Fig. 1). The ratio of the absorption at 650 nm to that at 627 nm indicates the relative amount of phototransformable protochlorophyll(ide). Kinetin when added to the growing plant leads to a higher amount of phototransformable protochlorophyll(ide) (Fig. 2). Correspondingly an increased chlorophyll formation can be expected upon illumination as has been demonstrated.

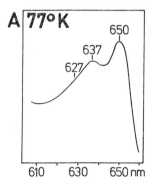

Fig. 1. $77°K$ in vivo absorption spectrum of an etiolated radish cotyledo.

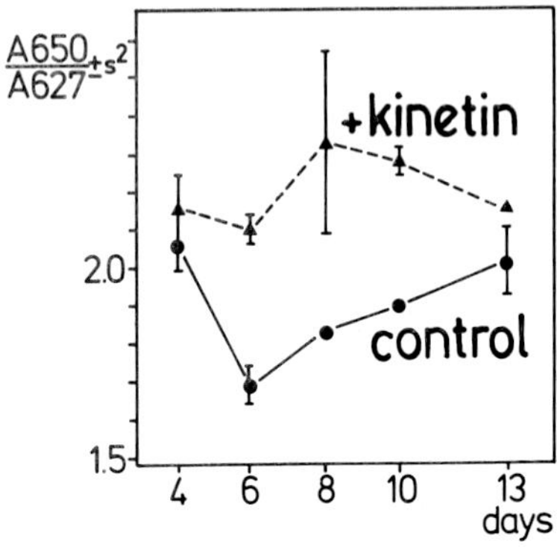

Fig. 2. The relative amount of phototransformable protochlorophyll(ide) in 4, 6, 8, 10, and 13 day old seedlings grown without (control) or with added kinetin (after [13]).

TABLE 2

HILL ACTIVITY OF CHLOROPLASTS FROM PHYTOHORMONE TREATED PLANTS

μMoles O_2/mg chlorophyll·h, Raphanus sativus L., growth condition: 1 day in darkness and 7 days in white light. Values based on [8].

	control	+ IAA	+ kinetin
DCPIP reduction			
2 d light	174	180	202
4 d light	88	94	122
7 d light	51	56	86
methylviologen-reduction			
2 d light	43	45	55
4 d light	22	25	31
7 d light	21	24	29

The changes in the prenylquinone levels of chloroplasts from phytohormone treated plants are correlated to an increased photosynthetic activity. The reduction of DCPIP and methylviologen (Hill activity) is markedly higher in all developmental stages examined than in chloroplasts from control plants (Table 2). Kinetin is more effective in this respect than IAA. While benzylaminopurine leads to changes like that of kinetin, gibberellic acid induces an earlier senescence[11] and abscisic acid prevents the light-induced formation of chlorophylls and prenylquinones[14]. The decrease of the initially high Hill activity with the illumination time is a normal developmental process and due to the increasing equipment of the photosynthetic membrane with chlorophyll. The prenyl-lipid pattern and Hill activity shown here for plants treated with the two phytohormones IAA and kinetin are similar to those of plants grown under high light intensity[3,4,15,16] and blue light[17].

TABLE 3
HILL ACTIVITY (DCPIP-REDUCTION OF CHLOROPLASTS FROM RADISH SEEDLINGS GROWN IN BLUE OR RED LIGHT WITHOUT (CONTROL) OR WITH ADDED KINETIN
μMoles O_2/mg chlorophyll·h, growth condition: 1 day in darkness and then illuminated with blue or red light.

	control	+ kinetin	increase
blue light[a]			
2 d light	235	289	23 %
4 d light	199	242	21 %
7 d light	136	171	26 %
red light[b]			
2 d light	160	193	20 %
4 d light	134	151	13 %
7 d light	91	156	71 %

[a] 450 nm maximum, 1.5 J/m^2·sec, 5.5 μMoles quanta/m^2·sec
[b] 660 nm maximum, 1.0 J/m^2·sec, 5.5 μMoles quanta/m^2·sec

In chloroplasts isolated from kinetin treated plants grown under blue or red light an increase of the Hill activity can also be seen. At day 8 after germination this effect becomes most pronounced for 'red light chloroplasts' (Table 3). This may be explained by a lack of endogenous cytokinins under red light which is overcome by adding kinetin. In blue light the increased Hill activity can also be explained by the formation of a higher level of endogenous cytokinins, which has been reported by several authors[18-22].

REFERENCES
1. Lichtenthaler, H.K. (1977) in Lipids and Lipid Polymers in Higher Plants, Tevini, M. and Lichtenthaler, H.K. eds., Springer, Berlin, pp. 231-258.
2. Lichtenthaler, H.K. and Pfister, K. (1978) in Photosynthetic Oxygen Evolution, Metzner, H. ed., Academic Press, London, pp. 171-194.
3. Lichtenthaler, H.K. (1971) Z. Naturforsch., 26b, 832-842.
4. Boardman, N.K. et al. (1974) in Proc. 3rd Intern. Congress on Photosynthesis, Elsevier Sci. Publ. Comp., Amsterdam, pp. 1809-1827.
5. Verbeek, L. and Lichtenthaler, H.K. (1973) Z. Pflanzenphysiol., 70, 245-258.
6. Lichtenthaler, H.K. et al. (1977) Physiol. Plant., 40, 105-110.
7. Hager, A. and Meyer-Bertenrath, T. (1966) Planta, 69, 198-217.
8. Buschmann, C. and Lichtenthaler, H.K. (1977) Z. Naturforsch., 32c, 798-802.
9. Nobel, P. (1967) Plant Physiol., 42, 1389-1394.
10. Thornber, J.P. (1975) Ann. Rev. Plant Physiol., 26, 127-158.
11. Straub, V. and Lichtenthaler, H.K. (1973) Z. Pflanzenphysiol. 70, 34-45.
12. Straub, V. and Lichtenthaler, H.K. (1973) Z. Pflanzenphysiol. 70, 308-321.
13. Buschmann, C. and Sironval, C. (1978) Planta, 139, 127-132.
14. Lichtenthaler, H.K. and Becker, K. (1970) Phytochemistry 9, 2109-2113.
15. Grahl, H. and Wild, A. (1975) in Environmental and Biological Control of Photosynthesis, Marcelle, R. ed., Dr.W. Junk, The Hague, pp. 107-113.
16. Wild, A. et al. (1975) in Environmental and Biological Control of Photosynthesis, Marcelle, R. ed., Dr.W. Junk, The Hague, pp. 115-121.
17. Kleudgen, H.K. and Lichtenthaler, H.K. (1974) Z. Naturforsch., 29c, 142-146.
18. Göring, H. and Mardanov, A.A. (1976) Biol. Rdsch., 14, 177-189.
19. Dörfler, M. and Göring, H. (1978) Biol. Rdsch., 16, (in press).
20. Göring, H. and Dörfler, M. (1979) J. exp. Bot., (in press).
21. Zeinalova, S.S. et al. (1967) Dokl. Akad. Nauk SSSR, 176, 955-958.
22. Fletcher, R.A. and Zalik, S. (1964) Plant Physiol., 39, 328-331.

ACKNOWLEDGEMENTS
This work was sponsored by a grant from the Deutsche Forschungsgemeinschaft.

HETEROGENEITY OF LIPIDS IN WHEAT FREEZE-DRIED CHLOROPLAST LAMELLAE

CLAUDE COSTES and RENE BAZIER
Laboratoire de Chimie biologique et de Photophysiologie , Institut National Agronomique (INRA) , Thiverval-Grignon , 78850 , (France)

ABSTRACT

A sequence of successive solvents extracted several compartments of lipids from Wheat lyophilized chloroplasts , at 20°C , namely : 1 , pentane ; 2 , benzene ; 3 , increasing ratio of pyridine in benzene ; 4 , increasing ratio of acetic acid in benzene ; 5 , acetone ; 6 , increasing ratio of water in acetone . Monogalactosyldiglycerides (MGDG) was dissolved into 2 pools , digalactosyldiglycerides (DGDG) , phosphatidylcholine (PC) and sulfolipid (SL) respectively into 3 pools each . Phosphatidylglycerol (PG) was the single lipid to not be fractionnated . The results allow to discuss the types of interactions between the different components of chloroplast lamellae .

INTRODUCTION

Different ways permit to study the molecular associations inside membranes . First , detergents ensure the breakdown of membranes into subparticles ; thus , subparticles with photosystem I or II activities have been separated in chloroplast lamellae : the ratios of each group of lipids to chlorophylls and carotenoids were measured in such subparticles [1-3] and the distribution of some proteins has been determined [4-9] . Another method consists in mild extraction of lipids from freeze-dried membranes . First applied on erythrocyte ghosts [10], and on mitochondria membranes [11] , we used this method on freeze-dried chloroplast lamellae [12,13] when it was demonstrated that lyophilisation did not disturb the heterogeneity of pigments [14] . This method gave information on the association of lipids through hydrogen bonding partly between themselves as galactolipides [13] , or with pigments or proteins [15] . The aim of the present work is to determine a new system of solvents for mild extraction .

MATERIAL AND METHODS

Chloroplasts were extracted according to Arnon et al [16] from young Wheat leaves (Triticum aestivum , cv Florence-Aurore) . Chloroplast pellets were freeze-dried overnight, giving chloroplast lamellae powder devoid of stroma ; 200 mg of this powder was put in a chromatographic column and treated successively by the following solvents (fig.1) :

(1) - pentane which disrupts the bonding between alkyl residues ;

(2) - a gradient of benzene (till 100%) in pentane ; benzene dissolves easily several pure lipids and can break some hydrogen bonding ;

(3) - a gradient of pyridine (till 20%, v/v) in benzene : this is a basic solvent which can extract cationic lipids and can disrupt hydrogen bonding ;

(4) - a gradient of acetic acid (till 4%, v/v) in benzene : this solvent can react with anionic lipids ;

(5) - a gradient of acetone (till 100%, v/v) in benzene, which can break residual hydrogen bonding ;

(6) - finally a mixture of acetone and water (till 20% water, v/v) solubilizes the residual lipids and pigments strongly linked to the membranes.

The sequence of solvents and their concentration gradient are shown in figure 1. Lipid analysis was performed through thin layer chromatography and gas chromatography of fatty acids as previously published [13].

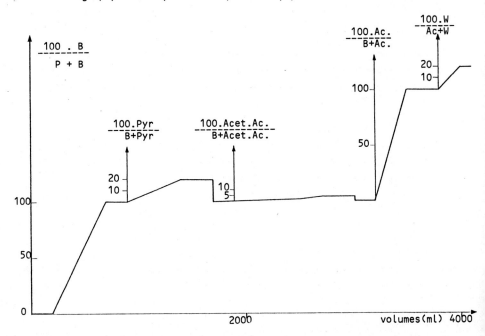

Fig. 1. Profile of the successive solvents applied to 200 mg of freeze-dried lamellae ; % are given in volumes. (P : Pentane, B : Benzene, Pyr. : Pyridine, Acet. Ac. : Acetic Acid, Ac. : Acetone, W : Water)

RESULTS

The first results were related to the ability of each solvent to dissolve each lipid from freeze-dried powder . It appeared that benzene extracted all the MGDG without any solubility of DGDG . For the gradients we determined the values of the content of pyridine or acetic acid in benzene able to dissolve the other lipids (figure 2) . We observed that with pyridine , the content of 20% (in vol.) in benzene permitted to dissolve a fraction of DGDG without extraction of appreciable amount of other lipids . With 4% (v/v) acetic acid in benzene all the PG was extracted , and well defined fractions of DGDG and of SL were dissolved from the freeze-dried lamellae previously treated with pure benzene .

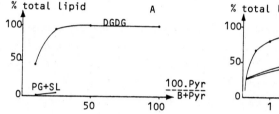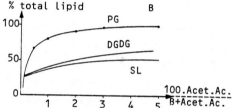

Fig. 2. Fractions of lipids extracted by several mixtures of pyridine (A) or of acetic acid (B) with benzene , from 50 mg of freeze-dried lamellae ; the extracted fractions are given as parts of total lipid .

The second group of results illustrated the heterogeneity of these lipids . MGDG was separated into two compartments respectively liberated with solvent 1 and 2 . DGDG was splitted into three compartments dissolved in solvents 3 , 4 and 5 . PC and SL were also fractionnated into three pools : but PC was more easily soluble (solvents 1,2 and 4) than SL (solvents 4,5 and 6) . PG was alone to constitute a single pool liberated with acidic solvent (4) . Chlorophylls a and b were present in all solvents , but the bulk was distributed between solvents 2 and 3 , one part with the bulk of MGDG , the second part with the bulk of DGDG (Table 1 - Figure 3) .

TABLE 1

MOLECULAR DISTRIBUTION OF SOLUTES AT 20°C (μmoles.g^{-1} freeze-dried chloroplasts)

Solvents n°	1	2	3	4	5	6
MGDG	7.0	106.0	-	-	-	-
DGDG	-	-	52.6	13.9	5.6	-
PC	0.5	5.2	-	2.7	-	-
PG	-	-	-	23.5	-	-
SL	-	-	-	3.4	8.1	9.0
Chlorophylls a and b	3.7	59.1	36.4	3.0	0.35	0.30

The last result was obtained when this mild extraction was performed on freeze-dried lamellae after heat denaturation of chloroplast proteins in boiling water for 10 minutes . In this experiment all the components became more easily extractible but it was striking to see that (Figure 4) :

- MGDG exhibited always two compartments , the second being always associated with almost half of the total chlorophylls ;
- an important part of DGDG remained in the third compartment but was almost completely devoided of chlorophylls which passed in solution in the first solvent ;
- PG and SL conserved almost the same pattern of extraction but appeared in easily soluble lipids .

CONCLUSIONS

All the results led us to the following interpretation .

1 - For MGDG , the main lipid of chloroplast , about 2/3 is not perturbed by heat denaturation of lamellae ; it seems to be associated mainly by hydrogen bonding , as the bulk of DGDG .

2 - An important part of chlorophylls (40 to 60%) appears to be associated with the bulk of MGDG ; this association would not be broken by heat denaturation of lamellae proteins .

3 - The second part of the bulk of chlorophylls (compartment 3) could be associated with lamellae proteins : it was liberated specifically after heat denaturation .

4 - PG could be bound to the lamellae structure mainly through ionic interactions : it was not possible to split this lipid in several compartments by extraction at 20°C (Figure 3) .

5 - PC and SL developped several types of interactions in the lamellae ; but SL is always associated with minor and strongly bound chlorophylls .

All these interpretation will be submitted to mathematical analysis in a next work .

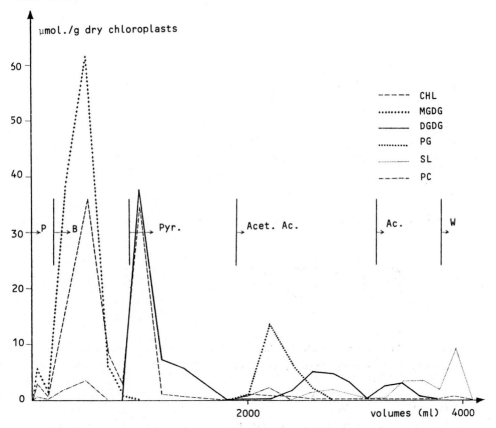

Fig. 3. Dissolution of chlorophylls and lipids from 200 mg of freeze-dried Lamellae along the profile of the successive solvents applied at 20°C (see Fig. 1.) - (P : Pentane , B : Benzene , Pyr. : Pyridine , Acet. Ac. : Acetic Acid , Ac. Acetone , W : Water)

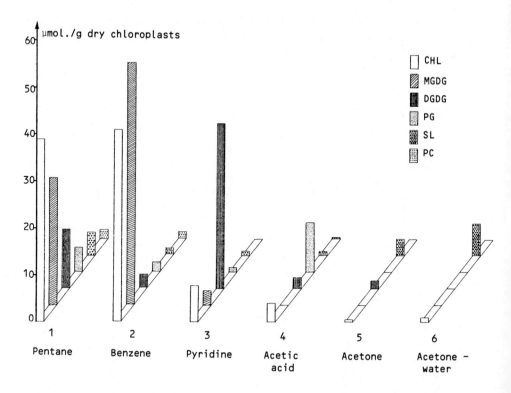

Fig. 4. Dissolution of chlorophylls and lipids from 75 mg of denaturated freeze-dried lamellae (heat treatment 10 minutes in boiling water) ; mild extraction was performed at 20°C .

REFERENCES

1. Vernon, L.P. , Shaw, E.R. and Ke, B. (1966) J. Biol. Chem., 241 ,4101.
2. Vernon, L.P. , Yamamoto,H.Y. and Ogawa, T. (1969) Proc. Natl. Acad.Sci. U.S., 63 , 911 .
3. Vernon, L.P. , Shaw, E.R. , Ogawa, T. and Raveed, D. (1971) Photochem. Photobiol., 14 , 343 .
4. Thornber, J.P. and Olson, J.M. (1971) Photochem. Photobiol., 14 , 329 .
5. Klein, S.M. and Vernon, L.P. (1974) Photochem. Photobiol., 19 , 43 .
6. Shiozawa, J.A. , Alberte, R.S. and Thornber, J.P. (1974) Arch. Biochem. Biophys., 165, 388 .
7. Kung, S.D. and Thornber, J.P. (1971) Biochim. Biophys. Acta, 253 , 285.
8. Bengis, C. and Nelson, N. (1975) J. Biol. Chem., 250 , 2783 .
9. Thornber, J.P. and Highkin, H.R. (1974) Europ. J. Biochem., 41 , 109 .

10. Haati, E.O. , Nanto, V. and Viikari, J. (1967) Acta Chem. Scandin., 21 , 2773 .
11. Sauner, M.T. and Levy, M. (1971) Biochim. Biophys. Acta , 241 , 97 .
12. Costes, C. and Bazier, R. (1972) in Proceed. 2nd Intern. Congress on Photosynthesis Research , publ. Junk, W. ; The Hague , p. 1635 .
13. Costes, C. , Bazier, R. and Lechevallier, D. (1972) Physiol. Veg. , 10 , 291 .
14. Deroche, M.E. (1969) Physiol. Veg. , 7 , 335 .
15. Costes, C. , Bazier,R. , Burghoffer, C. , Carrayol, E. and Deroche, M.E. , (1975) in Proceed. 3d Intern. Congress on Photosynthesis , ed. Avron, M. (Amsterdam) p. 2049 .
16. Arnon, D.I. , Allen, M.B. and Whatley, R.F. (1956) Biochim. Biophys. Acta , 20 , 449 .

LIPOXYGENASE ACTIVITY DISTRIBUTION IN YOUNG WHEAT CHLOROPLAST LAMELLAE

ROGER DOUILLARD and EDITH BERGERON
Laboratoire de Chimie biologique et de Photophysiologie (I.N.R.A.),
Institut National Agronomique Paris - Grignon, 78850 Thiverval-Grignon
(France)

ABSTRACT

Wheat shoot chloroplast lamellae have been purified by sucrose density gradient zonal centrifugation and do not seem to be contaminated by other membranes. Nevertheless the lipoxygenase activity peak is slightly shifted towards high density as compared with the chlorophyll peak. An interpretation of this shift is provided, based on the decrease in apparent density of the lamellae when their "age" increases.

INTRODUCTION

The lipoxygenases (E.C. 1.13.11.12) catalyse polyunsaturated (linoleic or linolenic) fatty acid hydroperoxydation in plant tissues or in various subcellular systems. This enzymic activity has not been related with any physiological process [1]. Some data suggest a chloroplastic localization of this activity [2-7] and we have also previously caracterized the lipoxygenase activity of Wheat chloroplast lamellae purified on a sucrose density gradient [8]. Nevertheless Wardale and Galliard [9] have shown that in the case of cauliflower floret plastids purified by sucrose density gradient centrifugation the lipoxygenase activity peak is slightly shifted towards high density as compared with the plastid marker enzyme triose phosphate isomerase peak. They concluded that lipoxygenase activity was bound to a fraction "very similar but not identical with intact plastids".

In the present work we show that in Wheat leave chloroplast lamellae an analogous shift can be observed between the lipoxygenase and chlorophyll (lamellae marker) peaks but that shift could be explained by a lamellae distribution in the gradient according with their apparent density directly correlated with their "age".

MATERIALS AND METHODS

<u>Chloroplast lamellae purification</u> . The chloroplast lamellae have been extracted from Wheat shoots (C.V. Florence Aurore) that have been grown under a 10 000 lx 15h/9h (light/dark) period since the fourth growth day . The sucrose density gradient centrifugation is carried out according with a previously described method [8] in a A type M.S.E. zonal rotor . When the centrifugation has been completed fractions of about 30 ml are collected : their sucrose content is determined using an ATAGO type 500 hand sugar refractometer and their weight measured . When needed the glycolate oxydase activity is immediately measured and aliquots of each fraction are frozen for delayed enzymatic activities and chlorophyll measurements .

<u>Analytical methods</u> . In order to get dry weight and nitrogen values a non frozen aliquot of the fraction is diluted and centrifuged . Then the lamellae pellet is resuspended and centrifuged two times for sucrose elimination and freeze dried for weight measurement , nitrogen determination by the Kjeldahl method and chlorophylls reading according with McKinney [10] .

The lipoxygenase activity has been measured with a Clark electrode according to our previous method [8] ; the succinate : cytochrome c oxidoreductase and NADPH : cytochrome c oxidoreductase activities according to Douce et al [11] ; and the glycolate oxydase activity by dichlorophenol indophenol bleaching according to Zelitch and Ochoa [12] .

<u>Result expression</u> . Some centrifugation profiles have been normalized in the following way : each fraction is described by its density in abscissa and by the percentage (P) of the total enzyme or product per density unit in the fraction so that :

$$P = \frac{100 \, A}{T} \cdot \frac{\Delta v}{\Delta \rho}$$

T : total enzyme or product in the gradient ; A : activity or concentration in the fraction ; Δv : fraction volume and $\Delta \rho$ the fraction density extent . The so build diagrams are independent of the unities used and of the slope and volume of the gradient . Their area is always 100 (dimensionless) .

RESULTS AND DISCUSSION

It appears in figure 1 that in a chloroplast lamellae density gradient centrifugation profile , the lipoxygenase activity peak is slightly shifted towards high density with regards to the chlorophyll peak . One can wonder

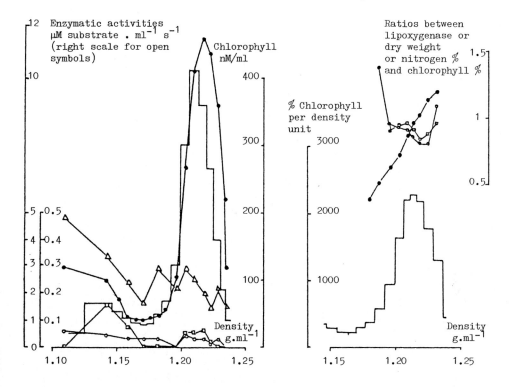

Fig. 1. (Left) Distribution of the chlorophyll and of the enzymatic activities associated with the chloroplastic lamellae centrifuged on a density gradient. ──chlorophyll. Enzymatic activities: •──•lipoxygenase ; △──△NADPH cytochrome c oxydoreductase ;□──□succinate cytochrome c oxydoreductase ;○──○glycolate oxydase.

Fig. 2. (Right) Distribution of the chlorophyll and ratios between the percentages per density unit of lipoxygenase•──•, dry weight○──○, nitrogen□──□ and the percentage per density unit of chlorophyll.

with Wardale and Galliard whether the lipoxygenase activity is bound, at least partly, to a non plastidial membranous structure. This point will now be checked.

<u>Purity criteria of the plastidial preparation</u>. As shown in figure 1 the succinate : cytochrome c oxydoreductase, NADPH : cytochrome c oxydoreductase and glycolate oxydase activities are very low compared to the lipoxygenase activity. Moreover they have no peak coincident with the chlorophyll or lipoxygenase peaks. Moreover figure 2 shows that the ratio between the

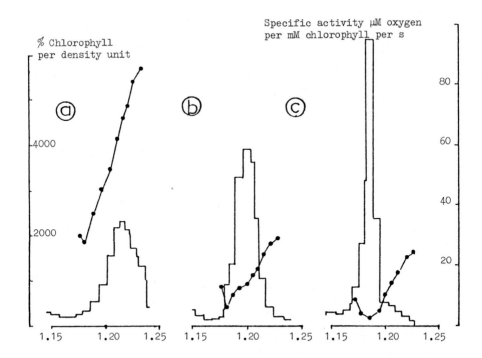

Fig. 3. Dependence of the lamellae chlorophyll and of the lipoxygenase specific activity distributions in density gradient centrifugation upon the shoot age . ──── Chlorophyll ;●──● specific lipoxygenase activity in μM of oxygen consumed per mM chlorophyll per s . Wheat shoot age ⓐ : 7 days ; ⓑ : 8 ; ⓒ : 12 .

lipoxygenase percentage per density unit and the same chlorophyll percentage increases with fraction density . This fact is indicative of a shift between the chlorophyll and the lipoxygenase peaks . On the other hand in the dry weight or nitrogen case the same ratios are practically constant in the chlorophyll peak vicinity . We are thus able to conclude that by the chlorophyll peak , the mitochondrial endoplasmic reticulum or peroxisomal contaminations are not significant and that there is no other contaminant discernible by its weight or nitrogen .

Influence of the shoot age upon lamellae and lipoxygénase activity distribution in centrifugation profiles . Let us compare in figure 3 , the mean values of the distribution peaks . It appears that the mean apparent density of the lamellae and the mean specific activity of the lipoxygenase decrease when the shoot age increases . Equivalently it can be said that the lamellae mean density is a measure of their "age" and that the specific activity decreases when the lamellae mean apparent density decreases . Let us compare now the fractions of a single centrifugation peak (figure 2 or 3) . Here it appears also that the lipoxygenase specific activity decreases when the apparent density of the fractions decreases .

The asymetrical distribution of the lipoxygenase activity in a chloroplast lamellae peak (figure 1) could thus be explained if one assumes that the lamellae are distributed according to their apparent density and consequently according to their "age" .

The "younger" they are, the higher their apparent density and the higher their lipoxygenase activity .

CONCLUSIONS

This interpretation implies that :

1) the lipoxygenase activity we previously described[8] is indeed bound to the chloroplastic lamellae ;

2) if the cauliflower floret results [9] are interpretable in the same way , the lipoxygenase activity would also be plastid bound ;

3) the younger the plastids , the higher their activity . One could wonder to which morphological or physiological phase is correlated this "young" feature : is it a proplastid phase or an organite division cycle phase [13] ? The question is open ;

4) the physiological meaning of the lipoxygenase activity is very likely correlated with a juvenile development phase of the plastidial membranous system .

REFERENCES

1. Galliard, T. (1975) in Galliard, T. and Mercer, E.I. ; Recent Advances in the Chemistry and Biochemistry of Plant Lipids , Academic Press , London , New York , San Francisco , pp. 319-357 .
2. Sisakyan, N.M. and Kobyakova, A.M. (1957) Biokhimija SSSR , 22 , 516-522 .
3. Holden, M. (1970) Phytochem., 9 , 507-512 .
4. Grossman, S. , BEN-Aziz, A. , Ascarelli, I. and Budowski, P. (1972) Phytochem., 11 , 509-514

5. Haydar, M. and Hadziyev, D. (1973) J. Sc. Food Agric., 24 , 1039-1053 .
6. Borisova, I.G. and Budnitskaia, E.B. (1975) Doklady Akademii Nauk SSSR , 225 , 439-441 .
7. Takagi, S. , Matsugami, M. and Moritoki, T. (1977) Scient. Reports Fac. Agric. Okayama Univ., 49 , 35-44 .
8. Douillard, R. and Bergeron, E. (1978) C.R. Acad. Sc. Paris , 286 , 753-755 .
9. Wardale, D.A. and Galliard, T. (1975) Phytochem., 14 , 2323-2329 .
10. McKinney, G. (1941) J. Biol. Chem., 140 , 315-322 .
11. Douce, R. , Mannella, C.A. and Bonner, W.D.Jr. (1973) Bioch. Bioph. Acta , 292 , 105-116 .
12. Zelitch, I. and Ochoa, S. (1953) J. Biol. Chem., 201 , 707-718 .
13. Kirk, J.T.O. and Tilney-Bassett, R.A.E. (1967) The Plastids . Freeman, W.H. and Company , London and San Francisco .

TRACER KINETIC ANALYSIS OF CHLOROPLAST PIGMENTS IN CHLORELLA PYRENOIDOSA

KARL H. GRUMBACH, GEORGE BRITTON AND TREVOR W. GOODWIN
Department of Biochemistry, University of Liverpool, P.O. Box 147,
Liverpool L69 3BX (England)

ABSTRACT

The incorporation of $[4,5-^3H_2]$-leucine (for 18 hours) and sodium $[2-^{14}C]$-acetate (for 40 minutes) into the photosynthetic pigments of the green alga Chlorella pyrenoidosa has been achieved and the fate of the ^3H and ^{14}C label over the subsequent 3 hours determined. Deuterium has been incorporated into α- and β-carotenes by Chlorella cells suspended in 60% deuterium oxide for 24 hours. After a further 24 hours in H_2O, deuterium was also present in lutein. The results of these experiments are in agreement with a precursor-product relationship between carotenes and xanthophylls and also suggest the existence of two metabolic pools of some compounds, e.g. β-carotene.

INTRODUCTION

Previous experiments[1] with the green alga Chlorella pyrenoidosa have shown that ^{14}C-label was rapidly incorporated into chloroplast pigments, chlorophyll and carotenoid, and into other prenyllipids, such as α-tocopherol and plastoquinone, during a 2-hour exposure to $^{14}CO_2$ but that much of the radioactivity was lost from some of these compounds within 4 hours when the algae were subsequently returned to an atmosphere containing unlabelled CO_2. Loss of label from α- and β-carotenes, concomitant with increased radioactivity in the xanthophylls was in line with a precursor-product relationship between carotenes and xanthophylls. We now report the results of further labelling experiments designed to give more information about the synthesis and breakdown of chloroplast pigments and other prenyllipids in Chlorella.

MATERIALS AND METHODS

Experiment 1. Incubation of Chlorella pyrenoidosa with $[4,5-^3H_2]$-leucine and sodium $[2-^{14}C]$-acetate. Chlorella pyrenoidosa (10 litres) was grown phototrophically for 7-10 days, at 23°C, then supplied with $[4,5-^3H_2]$-leucine (500μCi, specific activity 1.0 Ci/mmole) for a period of 18 hours. For the final 40 minutes of this 18 hour period sodium $[2-^{14}C]$-acetate (200μCi, specific activity 55mCi/mmole) was also supplied. The cells were then harvested, washed free from exogenous labelled substrate and reincubated in the light in unlabelled medium. After various time intervals aliquots were taken and the pigments and quinones isolated, purified[1,2] and assayed for ^{14}C and ^{3}H radioactivity by liquid scintillation. The results are shown in Figure 1.

Experiment 2. Labelling of Chlorella pyrenoidosa cells with deuterium from deuterium oxide. Chlorella pyrenoidosa (10 litres) was grown phototrophically for 7-10 days then harvested and resuspended for 24 hours in the light in medium (300 ml) prepared from 60% deuterium oxide. The cells were then harvested again; half were extracted, the other half resuspended in fresh medium prepared from 100% H_2O for a further 24 hours, again in the light, before extraction. The carotenoids were extracted from the two samples, purified and examined by mass spectrometry. An A.E.I. MS 12 instrument was used, with the direct insertion probe, at an ion source temperature of 180-200° and ionizing voltage 70eV.

RESULTS

The results of the first experiment in which Chlorella cells were incubated with [4,5-^{3}H$_2$]leucine for 18 hours and with [2-^{14}C]acetate for the final 40 minutes of this period are shown in Figure 1. Several things are apparent from this experiment. First the aminoacid leucine was incorporated into all the chloroplast prenyllipids, with reasonably efficiency. The pathway of leucine utilization is not known; it could be broken down to small molecules such as acetate or may be metabolized via 3-methylcrotonyl-CoA and 3-hydroxy-3-methyl-glutaryl-CoA. Nor is it known whether the major transformation of leucine into utilizable products occurs inside or outside the chloroplast. Secondly the incorporation of acetate was much more efficient than that of leucine. Similar levels of ^{14}C and ^{3}H radioactivity were incorporated into the various compounds even though the cells were exposed to leucine for 27 times as long as to acetate. Thirdly in most cases (α-carotene, lutein, chlorophyll b and plastoquinone-9) the ^{3}H:^{14}C ratio remained approximately constant over the 3 hour period after the supply of labelled substrates ceased. The molecules labelled from acetate and those labelled from leucine were therefore behaving similarly. In two notable cases, however, β-carotene and chlorophyll-a, this was not so. ^{14}C-Activity was lost to a greater extent than was ^{3}H activity. β-Carotene is the compound which was found to lose ^{14}C-label from ^{14}CO$_2$ most rapidly and is therefore turned over most rapidly.

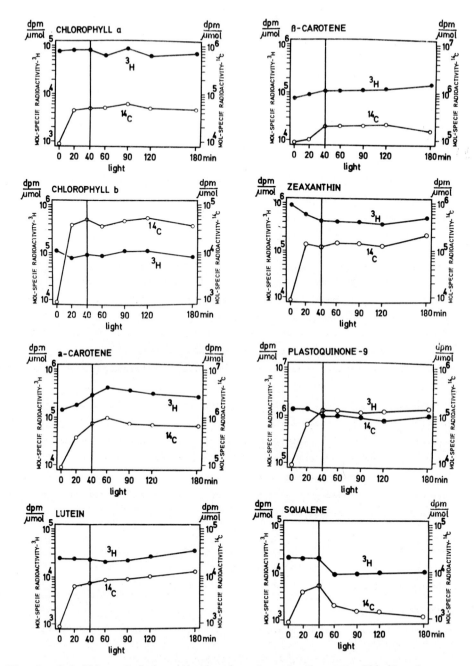

Fig. 1. Labelling of photosynthetic pigments and other prenyllipids of Chlorella pyrenoidosa in a tracer kinetic experiment with $[4,5-^3H_2]$-leucine and $[2-^{14}C]$-acetate as substrates.

A possible explanation for the result of the double-labelling experiment is that there are two pools of these compounds. There may be some selectivity in the way that acetate and leucine label the two pools. Alternatively label from any substrate may be rapidly incorporated into a metabolic pool that can turn over rapidly, but over a longer period of time the molecules become fixed, perhaps into the thylakoid membranes so the label is not easily lost. The longer time of incubation with leucine is sufficient to ensure that most of the ^{3}H label is in the 'fixed' molecules. The exposure to acetate is much shorter so more of the labelled products remain in the labile pool and are therefore susceptible to breakdown or available for use as precursors of other compounds. Thus chlorophyll a is a precursor of chlorophyll b and β-carotene of xanthophylls, especially zeaxanthin. The $^{14}C:^{3}H$ ratio of zeaxanthin actually increased during the 3 hour resuspension.

The second experiment showed that after incubation of light-grown Chlorella in D_2O solution for 24 hours, α- and β-carotenes both contained substantial amounts of deuterium (main species approximately D_{40}) indicating the extent of synthesis from CO_2 during this period. No deuterium was detected in any of the xanthophylls. After a further 24 hours in H_2O the proportions of deuterium-labelled α- and β-carotenes were considerably decreased but deuterium was now present in lutein. No deuterium was detected in zeaxanthin, antheraxanthin or violaxanthin.

Although this is only a preliminary result it nevertheless supports the idea of turnover of the carotenes and of a precursor-product relationship for α-carotene and lutein. Deuterium labelling from D_2O may thus be able to provide useful information on the biosynthesis, turnover and function of carotenoids and other prenyllipids in continuously light-grown algae, just as it has with mutant strains of <u>Scenedesmus obliquus</u> which change their carotenoid composition on transfer from darkness to light.

ACKNOWLEDGEMENTS

Financial support from the Royal Society, London and Deutsche Forschungsgemeinschaft is gratefully acknowledged.

REFERENCES

1. Grumbach, K. H., Lichtenthaler, H. K. and Erismann, K. H. (1978) Planta, 140, 37-43.
2. Britton, G. and Goodwin, T. W. (1971) Methods Enzymol., 18C, 654-701.
3. Britton, G., Lockley, W. J. S., Powls, R., Goodwin, T. W. and Heyes, L. M. (1977) Nature, 268, 81-82.

ULTRASTRUCTURE AND LIPID COMPOSITION OF CHLOROPLASTS OF SHADE AND SUN PLANTS

T. GUILLOT-SALOMON, C. TUQUET, M. de LUBAC, M.F. HALLAIS and M. SIGNOL
Laboratoire de Biologie Végétale IV, Université Pierre-et-Marie Curie,
12, rue Cuvier, 75005 Paris (France)

ABSTRACT

The results of a biometric and a biochemical analysis of chloroplasts isolated from leaves of shade and sun plants are reported. Three types of chloroplasts can be characterized depending on the organization of the plastid lamellae.
Type I chloroplasts from mesophyll cells of sun plants (spinach, barley, bean, maize) have well-developed grana and intergrana thylakoids. *Type II* chloroplasts, typical of shade plants (arum) present giant grana stacks and few interconnecting thylakoids. *Type III* chloroplasts from bundle-sheath cells of leaves from C_4-plants (maize) are characterized by an extensive development of stroma thylakoids with only few rudimentary grana.

The amounts of polar lipids are nearly similar in all plastid types. Only stricking differences appear in the fatty acid composition of phosphatidylglycerol where *trans*-3-hexadecenoic acid ($C_{16:1\ trans}$) is localized.

A strong correlation exists between the amounts of galactolipids per protein unit and the development of thylakoids and between the amounts of phosphatidylglycerol molecules containing $C_{16:1\ trans}$ and the percentages of appressed membranes. These results suggest that $C_{16:1\ trans}$ plays a specific role in grana stacking.

INTRODUCTION

It has been proved that the formation of thylakoids during plastid differentiation is correlated with important modifications in lipid composition[1,2,3]. Recent studies have shown a) that the differentiation of thylakoids is correlated with the synthesis of galactolipids and b) that grana stacking requires the presence of phosphatidylglycerol molecules (PG) containing *trans*-3-hexadecenoic acid ($C_{16:1\ trans}$)[4,5]. In this study, we have tried to extend these observations by comparing plastids isolated from different species of shade and sun plants.

Experiments were carried out on mesophyll chloroplasts with well-developed grana from plants grown under normal day-light (spinach, barley, bean, maize), on agranal bundle-sheath chloroplasts from maize, a C_4-plant, and on mesophyll plastids from a shade plant (arum) characterized by the presence of giant grana[6].

MATERIALS AND METHODS

Materials

Leaves of arum (*Arum maculatum* L.) grown under low-light intensity (below 2 W.m^{-2}) were collected in shady zones of woods near Paris. Seedlings of maize (*Zea mays* L., var. INRA 258), barley (*Hordeum vulgare* L., var. Cérès) and bean (*Phaseolus vulgaris* L., var. Commodore) were grown in a green house on moist vermiculite at 22°C, under normal light (12 W.m^{-2}) with a photoperiod of 14 hours. Leaves were harvested respectively after 14 (barley), 12 (maize) and 16 (bean) days, 3 hours after the beginning of illumination. Spinach leaves (*Spinacia oleracea* L.) were obtained from local market.

Chloroplasts were isolated in a sucrose solution and purified by centrifugation on a density gradient as described previously[7].

Biochemical analysis

Chlorophylls were measured according to Mackinney[8]. Polar lipids were extracted according to Folch et al.[9] and identified by two dimensional thin layer chromatography on silica gel[7]. Quantitative determinations of galactolipids, phospholipids and fatty acids were carried out as described previously[7].

Electron microscopy

For electron microscopy studies, samples were successively fixed by glutaraldehyde (6%) and OsO_4 (2%) in 0,15 M cacodylate buffer (pH 7,5) and then embedded in araldite. Sections were contrasted with uranyl acetate and lead citrate.

To control the purity of plastid preparations, 15 mM glutaraldehyde and 1% OsO_4 were successively added to suspension medium. The samples were embedded as the tissue samples.

Biometric analysis

Biometric studies of chloroplast ultrastructure were carried out as described previously[7]. All measurements refer to 1 μm^2 of plastid section. Percentages of stacked membranes (SM) were calculated from the length of granal partitions (LP) and the total length of thylakoids membranes (membrane density : MD) according to the relation : $SM = \frac{LP \times 2}{MD} \times 100$

RESULTS

Biometric analysis

The biometric analysis of plastid ultrastructures shows that membrane densities are very similar in all chloroplasts (20 to 30 $\mu m/\mu m^2$ section) (Table).

Whereas significant differences are seen in the percentages of stacked membranes. Three types of chloroplasts can thus be defined : *Type I* chloroplasts, from mesophyll cells of sun plants, display well-developed grana and intergrana thylakoid membranes (Fig. 1,2) with a percentage of stacked membranes in the range of 50 to 60% : *Type II* chloroplasts, from arum, show giant grana stacks and few interconnecting thylakoids (Fig. 3,4) with a percentage of appressed membranes higher than 80%. *Type III* chloroplasts of maize bundle sheaths have a very low level of appressed membranes (2%) and exhibit an extensive development of stroma thylakoids whereas only rudimentary stackings are observed (Fig. 5,6).

TABLE

BIOMETRIC AND BIOCHEMICAL ANALYSIS OF CHLOROPLASTS FROM DIFFERENT SPECIES OF SHADE AND SUN PLANTS

Plastid type	Plant species	MD	SM	P	CH	GL	PG	$C_{16:1}$	$\frac{GL / P}{MD}$
I	spinach	20 ± 8	54 ± 4	11,8	558	1285	189	40	5,45
	barley	25 ± 6	56 ± 10	5,7	420	820	180	31	5,75
	bean	22 ± 5	57 ± 6	9,8	1129	970	303	43	4,50
	maize mesophyll	32 ± 4	56 ± 4	12,0	1482	2155	374	30	5,65
II	arum	20 ± 8	83 ± 5	16,2	1077	1807	294	50	5,55
III	maize bundle sheath	20 ± 7	2 ± 1	12,5	1077	1600	335	8	6,40

MD : membrane density ($\mu m/\mu m^2$) ; SM : % of stacked membranes (mean values with their standard error) ; P : protein (pg/plastid) ; CH : chlorophyll $a + b$ (10^6 mol/plastid) ; GL : galactolipids (10^6 mol/plastid) ; PG : phosphatidylglycerol (10^6 mol/plastid) ; $C_{16:1}$: % of $C_{16:1\ trans}$ in the fatty acids of PG.

Biochemical analysis

Protein, chlorophyll, galactolipid and phosphatidylglycerol contents (Table) are specific of each plant material. No significant differences are observed in these components between the different plastid types. However the analysis of fatty acids of polar lipids has revealed stricking differences in the composition of PG. Amounts of $C_{16:1\ trans}$, localized in this phospholipid, are in the ranges of 30 to 40%, 50% and less than 8% of total fatty acids in type I, II and III plastids respectively.

If one takes into account the presence of a slight contamination (5%) by mesophyll chloroplasts in agranal chloroplast preparations[5], these results show that a relation exists between the levels of this particular fatty acid and the amounts of stacked membranes.

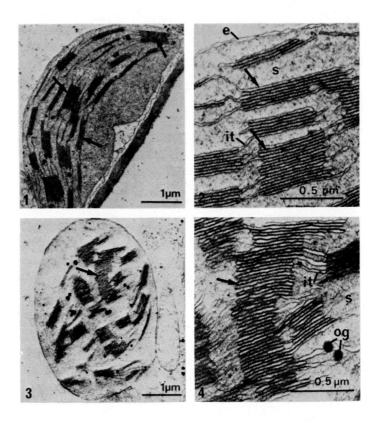

Figs 1 and 2. Type I chloroplasts (spinach) with well-developed grana (arrows) and intergrana thylakoids. Figs 3 and 4. Type II chloroplasts (arum) with giant grana stacks (arrows) and few interconnecting thylakoids.

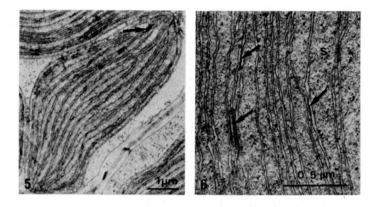

Figs 5 and 6. Type III chloroplasts (maize bundle-sheath) characterized by an extensive development of stroma thylakoids and few rudimentary grana (arrows). (e : chloroplast envelope ; it : intergrana thylakoid ; og : osmiophilic globule ; s : stroma)

Comparison of ultrastructure and lipid content

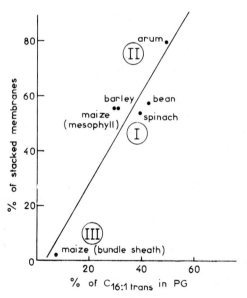

Fig. 7. Relation between % of stacked membranes and % of $C_{16:1}$ *trans* in PG molecules in different types of chloroplasts.

The ratios between the amounts of galactolipids per protein unit and the development of thylakoids (GL/P : MD) are similar in all plastid types (Table). Therefore, no relation can be observed between the amounts of grana, or of lamellae per granum, and the levels of $C_{16:1}$ *trans* in PG molecules[7]. However a strong correlation exists between the percentages of stacked membranes of $C_{16:1}$ *trans* in phosphatidylglycerol (Fig. 7). Type II chloroplasts, which contain the highest amount of granal stacking have the highest content of this component. Type III chloroplasts, very poorly stacked, show a low level of $C_{16:1\ trans}$. Values between those of types II and III are found for type I chloroplasts.

SUMMARY

For all plastid types, the comparison of ultrastructural and biochemical data suggests a strong correlation between a) the amounts of galactolipids per protein unit and the development of thylakoids and b) the amounts of phosphatidylglycerol molecules containing $C_{16:1\ trans}$ and the percentages of appressed membranes (grana stacks) within the plastid stroma.

The above results suggest that PG molecules containing $C_{16:1\ trans}$ could be specially involved in the cementing of granal thylakoids, whereas galactolipid molecules seem localized in equal amounts in all plastid membranes.

REFERENCES

1. James, A.T. and Nichols, B.W. (1966) Nature, 210, 372-375.
2. Leese, B.M. and Leech, R.M. (1976) Plant Physiol., 57, 789-794.
3. Tevini, M. (1977) in Lipids and lipid polymers in higher plants, Tevini, M. and Lichtenthaler, H.K. eds., Springer Verlag, Berlin, pp. 121-145.
4. Guillot-Salomon, T., Tuquet, C., Hallais, M.F. and Signol, M. (1977) Biol. Cell., 28, 169-178.
5. Tuquet, C., Guillot-Salomon, T., de Lubac, M. and Signol, M. (1977) Plant Sci. Letters, 8, 59-64.
6. Anderson, J.M., Goodchild, D.J. and Boardman, N.K. (1973) Biochim. Biophys. Acta, 325, 573-585.
7. Guillot-Salomon, T., Tuquet, C., de Lubac, M., Hallais, M.F. and Signol, M. (1978) Cytobiologie, 17, 442-452.
8. Mackinney, G. (1941) J. Biol. Chem., 140, 315-322.
9. Folch, J., Lees, M. and Sloane-Stanley, G.H. (1957) J. Biol. Chem., 226, 497-509.

CORRELATION BETWEEN PHOTOSYNTHESIS AND PLANT LIPID COMPOSITION.

KLAUS-PETER HEISE, GÖTZ HARNISCHFEGER
Lehrstuhl für Biochemie der Pflanze, Göttingen Univers.,
Untere Karspüle 2, 3400 Göttingen (Germany)

ABSTRACT/SUMMARY

Comparison of lipid composition and functional measurements made immediately after exposure of chloroplasts (1mg Chl./ml) to white light ($2 \cdot 10^5$ ergs $\cdot$ cm$^{-2} \cdot$ s^{-1}) show:
1. A linear correlation between the monogalactolipid:phosphatid-ratio (MGDG/PL) and the P/2e value.
2. An increase in neutral lipids (NL) and a simultaneous decrease of the P/2e-quotient, the amount being dependent on the time of illumination.

The latter result is corroborated by ^{14}C-incorporation from H ^{14}CO$_3^-$ into lipids of isolated chloroplasts. The increase of NL in plastid suspensions induced by short illumination periods is interpreted as an inhibited glycolipid- and phospholipid synthesis from diglycerides.

DCMU on the other hand seems to inhibit the described light induced lipid changes and to promote NL accumulation in chloroplast membranes [1-3]. The data, therefore, link changes in thylakoid lipid composition with the electron-transport and ATP-formation capability of the chloroplast.

INTRODUCTION

Early work on diurnal variations in the structure [4] and lipid composition [5,6] of chloroplast membranes suggested, that a rapid "turnover" of lipids takes place. In following up these studies we measured daily variations in leaf lipid content in relation to the function of isolated chloroplasts. The result, an apparent linear correlation between the MGDG/PL ratio of

---

ABBREVIATIONS: DGDG, digalactosyl diglyceride; MGDG, monogalactosyl diglyceride; NL, neutral lipids containing acyl glycerides and fatty acids; PL, total phospholipids.

spinach leaves and the P/2e value obtainable with the respective plastid preparation [7] lead to further investigation of the functional interaction between the photosynthetic apparatus and its' lipid environment in chloroplast membranes.

MATERIALS AND METHODS

Lipid composition of Jensen-Bassham type chloroplasts was altered by illuminating dense plastid suspensions (1mg Chl./ml) without electron acceptor for various amounts of time with white light ($2 \cdot 10^5$ ergs·cm^{-2}·s^{-1}) under mechanical shaking at 20°C. Functional measurements and lipid extraction were as close as possible in time to this light exposure (details see [7]). For comparison, dilute chloroplast suspensions were also used, all components for activity being present already during light exposure. The sample contained (µmol/ml): Tricine buffer (pH 8,0) 34,5; FeCy 1,38; Na_2HPO_4 3,45; $MgCl_2$ 3,45; ADP 3,45 and chloroplasts (20µg Chl./ml) in a total volume of 30 ml. In inhibition experiments DCMU was added to a final concentration of $5 \cdot 10^{-7}$M.

^{14}C-incorporation into the lipid fraction and CO_2-fixation of isolated Jensen-Bassham chloroplasts was measured according to Heise [8]. The lipids were extracted according to Bligh and Dyer [9]. Separation and analysis of the lipid extract was as described by Heise and Jacobi [10].

RESULTS AND DISCUSSION

By illumination of fairly concentrated suspensions (1mg Chl./ml) of Jensen-Bassham type plastids in absence of an electron acceptor a variation in lipid composition can be achieved similar to that described for the in vivo system (Table 1.). Hill activity and ATP-formation determined in a subsequent illumination period (3min.) under hypotonic conditions, necessary to circumvent limitations due to membrane barriers, show furthermore a linear correlation between the P/2e value and the MGDG/PL ratio (Fig. 1A). The data of Fig. 1B, calculated from the linolenic acid content of these lipid fractions, show essentially the same behaviour. Thus, the linolenic acid content of the lipids parallels their concentration.

TABLE 1

Correlation between light induced lipid changes in Jensen-Bassham type chloroplasts and their subsequently determined photosynthetic activity

time of illumination (min.)	lipid composition (µmol lipid/µmol Chl.)				photosynthetic activities		
	MGDG	DGDG	PL	NL	(1)	(2)	(3)
0	1,59 (2,42)	0,68 (0,93)	1,08 (1,15)	0,25	136	60	0,88
2	1,74 (2,59)	0,63 (0,51)	1,05 (1,20)	0,17	131	77	1,18
4	1,60 (2,34)	0,74 (0,65)	1,01 (1,08)	0,23	132	72	1,09
6	1,51 (2,38)	0,77 (0,75)	1,01 (1,09)	0,30	136	68	1,00
8	1,43 (2,25)	0,62 (0,54)	1,06 (1,14)	0,48	157	66	0,84
10	1,22 (1,97)	0,58 (0,49)	0,96 (1,03)	0,83	200	62	0,62

(1) e^--transport (+ADP+Pi) [µmol FeCy red./mg Chl.x h]
(2) photophosphorylation [µmol ATP/mg Chl.x h]
(3) P/2e-quotient
The α-$C_{18:3}$-ratio of the same lipids is given in the brackets.

This interpretation of the data has one drawback. For the determination of electron transport and photophosphorylation the plastids need to be shocked osmotically and are, thus, altered. Yet even more significant changes in the MGDG/PL-ratio were found, if dilute chloroplast suspensions (20µg Chl./ml) were illuminated under hypotonic and phosphorylating conditions for short periods (3 min.) of time in the presence of an electron acceptor (compare the first two rows in Table 2).

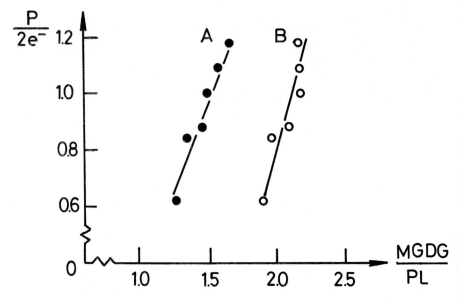

Fig. 1 A. Correlation between the ratio of MGDG/PL, determined directly, and P/2e. The data correspond to those of Table 1.
B. Same, but the MGDG/PL-ratio was calculated indirectly from the linolenic acid concentration in the same lipid fractions.

DCMU leads to a MGDG/PL-ratio that resembles that of the dark control, indicating possibly a "fixation" of light dependent lipid changes and thus of the membrane fluidity by DCMU. Furthermore DCMU seems to promote NL accumulation in the chloroplast membranes [2,3].

Table 1 argues that the decline in the P/2e-ratio appears to be mainly due to an increasing electron transport rate. This seems to give less importance to an inhibition of photosynthetic phosphorylation by products of lipid breakdown, which was observed by light ageing of brocken chloroplasts [11-13]. Some support for this, however, comes from experiments of ^{14}C-incorporation into the lipid fraction of isolated chloroplasts. The data show an increasing ^{14}C-accumulation in the NL-fraction only while the ^{14}C-uptake into the remaining membrane lipids decreases with duration of light [8]. The predominant part of the ^{14}C-label (80-90%) was found in the fatty acid residues of the NL-fraction.

TABLE 2
Influence of DCMU on lipid composition and photosynthetic activities of chloroplasts, in the presence of FeCy as acceptor.

time of illumination (min)	DCMU-concentration (mol/l)	MGDG/PL	Relative NL-proportion* (in %)	photosynthetic activities (1)	(2)	(3)
0 (Dark)	-	3,2 (4,9)	10,2	-	-	-
3	-	2,2 (3,7)	8,1	214	103	0,96
3	5x10^{-7}	3,0 (5,0)	11,5	91	44	0,97

(1), (2) and (3) see table 1. In brackets the α-C$_{18:3}$-ratio is given. * The total lipid content was taken as 100 %

A decrease in the P/2e value as given in table 1 is also correlated to a significant increase in NL content (figure 2).

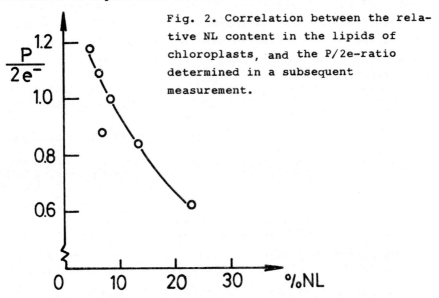

Fig. 2. Correlation between the relative NL content in the lipids of chloroplasts, and the P/2e-ratio determined in a subsequent measurement.

From fig. 3 one can infere, that ^{14}C-incorporation into neutral lipids seems to be connected to photosynthesis.
The ratio of the initial slopes of the time courses of ^{14}C-incor-

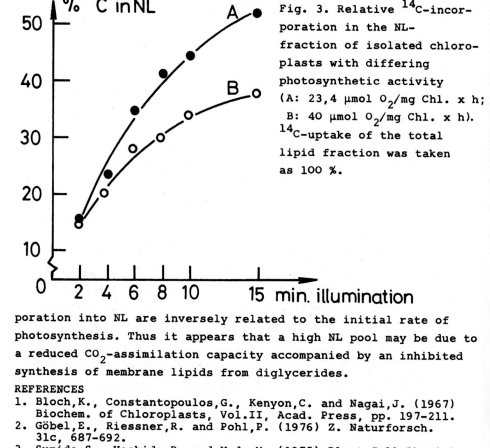

Fig. 3. Relative ^{14}C-incorporation in the NL-fraction of isolated chloroplasts with differing photosynthetic activity (A: 23,4 µmol O_2/mg Chl. x h; B: 40 µmol O_2/mg Chl. x h). ^{14}C-uptake of the total lipid fraction was taken as 100 %.

poration into NL are inversely related to the initial rate of photosynthesis. Thus it appears that a high NL pool may be due to a reduced CO_2-assimilation capacity accompanied by an inhibited synthesis of membrane lipids from diglycerides.

REFERENCES
1. Bloch,K., Constantopoulos,G., Kenyon,C. and Nagai,J. (1967) Biochem. of Chloroplasts, Vol.II, Acad. Press, pp. 197-211.
2. Göbel,E., Riessner,R. and Pohl,P. (1976) Z. Naturforsch. 31c, 687-692.
3. Sumida,S., Yoshida,R. and Ueda,M. (1975) Plant Cell Physiol. 16, 257-264.
4. Bartels,F. (1971) Protoplasma 72, 27-41.
5. Jarvis,M.C. and Duncan,H.J. (1975) Phytochemistry 14, 77-78.
6. Heise,K.-P. and Stottmeister,A. (1976) Ber. Dtsch. Bot. Ges. 89, 677-694.
7. Heise,K.-P. and Harnischfeger,G. (1978) Z. Naturforsch. 33c.■-■
8. Heise,K.-P. (1978) Z. Naturforsch. 33c.■-■
9. Bligh,E.G. and Dyer, W.J. (1959) Can. J. Biochem. 37, 911-917.
10. Heise,K.-P. and Jacobi,G. (1973) Z. Naturforsch. 28, 120-127.
11. Mc Carthy,R.E. and Jagendorf,A.T. (1965) Plant Physiol. 40, 725-735.
12. Heath,R.L. and Packer,L. (1968) Arch. Biochem. Biophys. 125, 189-198 and 850-857.
13. Hoshina,S. and Nishida,K. (1975) Plant Cell Physiol. 16, 465-474 and 475-484.

IS THE CHLOROPLAST ENVELOPE A SITE OF GALACTOLIPID SYNTHESIS ? YES !

JACQUES JOYARD, MARIANNE CHUZEL and ROLAND DOUCE
DRF/Biologie Végétale, CENG and USMG, 85 X, F 38041 GRENOBLE-Cedex (France)

SUMMARY

Owing to their envelope membranes, the chloroplasts are able to synthesize their own galactolipids : (1) the envelope is the site for the incorporation of galactose from UDP-galactose into galactolipids, (2) the envelope is able to incorporate sn-glycerol 3-phosphate into MGDG, (3) the envelope is able to incorporate fatty acids actively synthesized in the stroma from acetate into MGDG

INTRODUCTION

Plant galactolipids, monogalactosyldiglyceride (MGDG) and digalactosyldiglyceride (DGDG), represent about 80 % of the polar lipids in the chloroplasts of higher plants, and contain a high level of unsaturated fatty acids. The isolation in large quantities of envelope from spinach chloroplast, free of contamination from other cell membranes paved the way for a careful study of the localization of enzymatic activities involved in galactolipid synthesis within the chloroplast[1,2].

In this article, we shall summarize the different steps of the assembly of the three parts of the galactolipid molecules (galactose, glycerol, fatty acid) on the chloroplast envelope.

GALACTOSE INCORPORATION INTO GALACTOLIPIDS.

As already discussed[2], the envelope is the site of galactose incorporation into galactolipids[3]. Two distinct enzymes responsible for the synthesis of MGDG and DGDG are associated with the chloroplast envelope membranes[4]. The first enzyme or UDP-gal : diacylglycerol galactosyltransferase catalyzes the formation of MGDG:

$$\text{diacylglycerol} + \text{UDP-galactose} \rightarrow \text{MGDG} + \text{UDP}$$

In this case, UDP-gal is the galactosyl donor for galactosylation of a large endogenous pool of diacylglycerol[5]. The second enzyme is either a UDP-gal : MGDG galactosyltransferase[6-8] :

$$\text{MGDG} + \text{UDP-galactose} \rightarrow \text{DGDG} + \text{UDP}$$

or a galactolipid : galactolipid galactosyltransferase[9] :

$$\text{MGDG} + \text{MGDG} \rightarrow \text{DGDG} + \text{diacylglycerol}$$

In fact, both reactions seems to occur on the spinach chloroplast envelope.

The first galactosylation enzyme can be easily distinguished from the second enzyme : the maximum activity of MGDG synthesis is obtained around pH 8.0 ; in contrast, the second enzyme has its maximum activity around pH 6.5[4]. Triton X-100 (0.9 % by volume) is a strong inhibitor of the second enzyme but is almost without effect on the first galactosylation enzyme[10].

GLYCEROL INCORPORATION INTO GALACTOLIPIDS

Isolated chloroplasts are able to catalyse the incorporation of sn-glycerol 3-phosphate into phosphatidic acid and diacylglycerol[11,12]. We have shown that intact purified chloroplasts contain two acyltransferases[12]. One is a soluble enzyme (acyl CoA : sn-glycerol 3-phosphate acyltransferase) and catalyse the acylation of sn-glycerol 3-phosphate with oleoyl-CoA and palmitoyl-CoA[12,14]. The final product formed is lysophosphatidic acid :

sn-glycerol 3-phosphate + acyl-CoA → lysophosphatidic acid + CoA SH

This enzyme is probably loosely bound to the inner surface of the inner envelope membrane and becomes detached during the course of the envelope preparation. The enzyme catalyse the incorporation of saturated fatty acids in C-1 position of sn-glycerol 3-phosphate into lysophosphatidic acid[13], the maximum activity was obtained around pH 7.0[13].

The second acylase in chloroplast is an acyl-CoA : acyl-sn-glycerol 3-phosphate acyltransferase which is firmly and specifically bound to the envelope membranes (probably the inner) and forms phosphatidic acid[12]. Neither thylakoïd membranes nor stroma contain this enzyme[12] :

lysophosphatidic acid + acyl CoA → phosphatidic acid + CoA SH

This acyltransferase is probably specific for C-2 position. Furthermore, it is not known whether the two acyltransferases from chloroplasts can also use acyl-acyl carrier protein as acyl donor. However, in contrast to observations on *Euglena*[15], Shine et al[16] have shown that in spinach chloroplast the acyl-acyl carrier protein thioesters do not function as acyl donors.

The envelope membranes contain a specific alkaline phosphatidate phosphatase which hydrolyzes fairly rapidly phosphatidic acid formed from lysophosphatidic aci leading to the accumulation of diacylglycerol[12,17] :

phosphatidic acid → diacylglycerol + Pi

This enzyme has a maximum activity around pH 9.0, it is inhibited by cations (Mg^{2+}, Ca^{2+}) and is stimulated by EDTA[17]. It is specific for endogenous formed phosphatidic acid[17]. In the presence of acyl-CoA and sn-[^{14}C]-glycerol 3-phospha the envelope membranes accumulate labelled diacylglycerol[12]. Thus the addition o UDP-gal to the incubation medium induces a rapid decrease in the radioactivity incorporated into diacylglycerol which is accompanied simultaneously by a rapid synthesis of MGDG[12,17].

FATTY ACIDS INCORPORATION INTO GALACTOLIPIDS

The results described above show that the chloroplast envelope may catalyse the transfer of fatty acids to MGDG.

If UDP-gal and sn-glycerol 3-phosphate are synthesized in the cytoplasm of the cell[18], numerous investigations have revealed that isolated chloroplasts from various leaves, chromoplasts and proplastids possess the complete machinery for the biosynthesis of the hydrocarbon chain of fatty acids (for review, see 19,20). There is now considerable evidence that all the enzymes involved in fatty acid synthesis (fatty acid synthetase, stearoyl-ACP desaturase) are localized in the stroma phase[20]. Acyl-ACP synthesized by this multienzyme complex is converted to acyl-CoA by a switching system involving acyl-ACP thioesterase and one or several acyl-CoA synthetase[19]. The latter enzyme which catalyses the reaction :

$$\text{fatty acid} + \text{ATP} + \text{CoA SH} \rightarrow \text{acyl-CoA} + \text{AMP} + P \sim Pi$$

is specifically associated with the chloroplast envelope[12,21]. It is doubtful whether chloroplasts are capable of further desaturation of oleoyl-CoA and/or

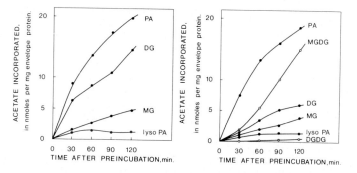

Fig. 1 : Incorporation of [^{14}C]-acetate into polar lipids of the chloroplast envelope. Soluble enzymes from intact and purified chloroplasts are incubated in the following medium : tricine-NaOH, 100 mM, pH 8.9 ; CoA SH, 50 µM ; ATP, 4 mM ; NADPH, 0.5 mM ; sodium bicarbonate, 1 mM ; dithiothreitol 1 mM ; Mn Cl$_2$, 1 mM ; sodium acetate, 2 mM, 7.5 µCi/1.2 ml (final volume) ; soluble proteins, 10 mg.
Left : After 1 hour of incubation , envelope membranes (1 mg protein) and sn-glycerol 3-phosphate 1 mM are added to the medium, the pH is lowered to 7.5 in order to be in the same conditions as for right curves. Note in this case the steady synthesis of phosphatidic acid and diacylglycerol.
Right : After 1 hour of incubation, envelope membranes (1 mg protein), sn-glycerol 3-phosphate 1 mM and UDP-gal 1 mM are added to the medium, the pH is lowered to pH 7.5 in order to allow galactolipids synthesis. Note that diacylglycerol is utilized for MGDG synthesis, small amounts of DGDG are also synthesized.
Abbreviations : PA, phosphatidic acid ; DG, diacylglycerol ; MG, monoacylglycerol; lyso PA, lysophosphatidic acid ; MGDG, monogalactosyldiglyceride ; DGDG, digalactosyldiglyceride. For lipids analysis, see ref. 5 and 12.

oleoyl-ACP although some blue-green algae, the progenitors of plastids, contain polyunsaturated fatty acids (e.g. C 18 : 3) (see ref. 22).

We have demonstrated that the chloroplast stroma, devoid of membrane fraction, is able to incorporate the radioactivity of [^{14}C]-acetate into saturated (C 16 : 0 and C 18 : 0) and mono-unsaturated (C 16 : 1 and C 18 : 1) fatty acids, but not in polyunsaturated fatty acids. The presence of ATP, coenzyme A, NADPH$_2$ and dithiothreitol in the incubation medium is required. The optimum pH is around 9.0.

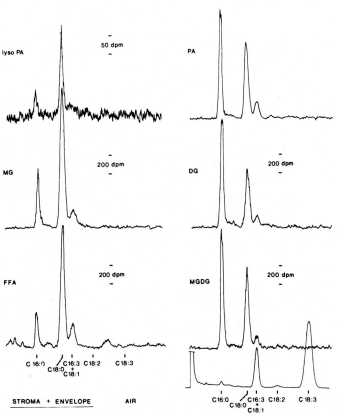

Fig. 2 : Labelled fatty acids distribution in envelope lipids after 3 hours of incorporation of [^{14}C]-acetate. Fatty acid methyl esters are separated by gas-chromatography, radioactivity is simultaneously detected in each lipid.
The mass spectrum is given for MGDG (lower right curve). Note that the major fatty acid labelled in lyso PA and MG (which derive from lyso PA by dephosphorylation) is C 18:0; in PA, DG and MGDG, the major labelled fatty acid is C 16:0. These spectrum show that PA, DG and MGDG present the same fatty acid pattern. Experimental conditions and abbreviations : see Fig. 1, for lipid analysis, see ref. 5.

When *sn*-glycerol 3-phosphate and envelope membranes are added to the incubation medium, the labelled fatty acids are incorporated very rapidly into lysophosphatidic acid, phosphatidic acid and diacylglycerol (Figure 1). C 18 fatty acids are preferentially transferred at the C-1 position of *sn*-glycerol 3-phosphate yielding 1-acyl-*sn*-glycerol 3-phosphate. C 16 fatty acids are then preferentially transferred at the C-2 position of 1-acyl-*sn*-glycerol 3-phosphate yielding 1,2-diacyl-*sn*-glycerol 3-phosphate (Figure 2).

The addition of UDP-gal to the incubation medium induced a fast decrease of [^{14}C]-acetate radioactivity incorporated into diacylglycerol which was accompanied simultaneously by a rapid synthesis of MGDG. Traces of DGDG are also labelled under these conditions (Figure 1). In this case, the labelled fatty acids incorporated into MGDG are essentially saturated (Figure 2).

Since galactolipids are rich in polyunsaturated fatty acids (except some DGDG species which contain a unique C 16:0 / C 18:3 combination, see ref. 22), MGDG and/or diacylglycerol molecules containing saturated fatty acids, specifically synthesized in the envelope, could constitute a direct substrate for one or several specific desaturases. *In vivo* experiments have led SAFFORD and NICHOLS[23], SIEBERTZ and HEINZ[24] and BOLTON and HARWOOD[25] to propose that *in vivo* desaturation occurs after the formation of MGDG. Our *in vitro* results are in agreement with this suggestion. Furthermore, we believe that desaturation is confined to envelope membranes. It is likely that the quinones found in the envelope membranes (LICHTENTHALER, JOYARD and DOUCE, unpublished data) may play an important role in the mechanism of the desaturation.

The question concerning the specific positioning of unsaturated fatty acids in galactolipids remain unanswered (see ref. 22). It is certain that both the desaturase(s) and acyltransferases involved in the acylation of *sn*-glycerol 3-phosphate are important in fatty acyl chain location. The specific acyl galactosyl diacylglycerol-forming activity found in the envelope[10] which catalyzes the transfer of acyl groups from the glycerol moiety of polar lipids to the galactose region of MGDG (see ref. 22) may also play an important role in inter-mixing all the polar lipid fatty acids in the chloroplast envelope.

CONCLUSION

The results presented here show that:

- the envelope is a site of assembly of the three parts of the galactolipid molecules (galactose, glycerol, fatty acids); it is clear that the envelope, and probably the inner membrane, contain the complete array of galactolipids biosynthetic machinery;

- the endoplasmic reticulum is not directly involved in the synthesis of galactolipids, the major structural lipids of chloroplasts.

In these circumstances, one can assume that:

- MGDG containing saturated and monounsaturated fatty acids, and specifically synthesized in the chloroplast envelope, should constitute a direct substrate for one or several specific desaturases.

REFERENCES

1. Douce, R., Holtz, H.B. and Benson, A.A. (1973) *J.Biol.Chem.* **248**, 7215-7222.
2. Douce, R. and Joyard, J. *this volume*.
3. Douce, R. (1974) *Science* **183**, 852-853.
4. Joyard, J. and Douce, R. (1976) *Physiol.Vég.* **14**, 31-48.
5. Joyard, J. and Douce, R. (1976) *Biochim.Biophys.Acta* **424**, 125-131.
6. Ferrari, R.A. and Benson, A.A. (1961) *Arch.Biochem.Biophys.* **93**, 185-192.
7. Ongun, A. and Mudd, J.B. (1968) *J.Biol.Chem.* **243**, 1558-1566.
8. Williams, J.P., Watson, G.R., Khan, M.U. and Leung, S.P.K. (1975) *Plant Physiol.* **55**, 1038-1042.
9. Van Besouw, A. and Wintermans, J.F.G.M. (1978) *Biochim.Biophys.Acta* **529**, 44-53 and *this volume*.
10. Heinz, E., Bertrams, M., Joyard, J. and Douce, R. (1978) *Z.Pflanzenphysiol.* **87**, 325-331.
11. Douce, R. and Guillot-Salomon, T. (1970) *FEBS Lett.* **11**, 121-124.
12. Joyard, J. and Douce, R. (1977) *Biochim.Biophys.Acta* **486**, 273-285.
13. Bertrams, M. and Heinz, E. (1976) *Planta* **132**, 161-168 and *this volume*.
14. Chuzel, M. (1978) 3rd cycle Thesis, University of Grenoble, France.
15. Renkonen, O. and Bloch, K. (1969) *J.Biol.Chem.* **244**, 4899-4903.
16. Shine, W.E., Mancha, M. and Stumpf, P.K. (1976) *Arch.Biochem.Biophys.* **172**, 110-116.
17. Joyard, J. and Douce, R. *in preparation*.
18. Heinz, E. (1977) in *Lipids and Lipid Polymers in Higher Plants*, Tevini, M. & Lichtenthaler, H.K. eds. Springer-Verlag, Berlin, pp. 102-120.
19. Stumpf, P.K. (1976) in *Plant Biochemistry*, 3rd edition, Bonner, J. & Varner, J.E. eds. Academic Press, New York, pp. 427-461.
20. Douce, R. and Joyard, J. (1979) *Adv.Bot.Res.* **7**, in press.
21. Roughan, P.G. and Slack, C.R. (1977) *Biochem.J.* **162**, 457-459.
22. Heinz, E., Siebertz, H.P., Linscheid, M., Joyard, J. and Douce, R., *this volume*.
23. Safford, R. and Nichols, B.W. (1970) *Biochim.Biophys.Acta* **210**, 57-64.
24. Siebertz, H.P. and Heinz, E. (1977) *Z.Naturforsch.* **32 C**, 193-205.
25. Bolton, P. and Harwood, J.L. (1978) *Planta* **138**, 223-228.

BUILDING UNITS OF PROLAMELLAR BODIES FROM ETIOPLASTS OF
AVENA SATIVA L.:
SAPONIN CONTENT AND REAGGREGATION EXPERIMENTS.

J. KESSELMEIER AND H.G.RUPPEL*
Dept. of Botany, University of Cologne, Gyrhofstr. 15,
D-5000 Cologne 41,
* Dept. of Biology, University of Bielefeld, D-4800 Bielefeld 1

SUMMARY

Isolated prolamellar bodies (PLBs) of etioplasts of Avena sativa contain large amounts of two steroidal saponins. By reaggregation experiments and by demonstrating that the saponin content depends on the crystalline character of PLBs it is shown that saponins are the main building units of PLBs.

INTRODUCTION

Completely dissolved prolamellar bodies of Avena sativa are able to reaggregate into tubules resembling native PLB - tubules.[1] Even eliminating proteins has no influence on the reaggregation rate and pattern[2]. So the tubular structure of the PLB cannot depend on proteins.

To answer the question which building units are responsible for tubular structure, acetone extracts of isolated and highly purified PLBs were examined concerning their lipid content and their capability to build up artificial PLB - like tubules.

MATERIALS AND METHODS

Isolation of purified PLBs and thylakoids. Isolation of purified etioplasts and chloroplasts from 6 - 7 days old seedlings of Avena sativa has been described by Lütz[3]. Prolamellar bodies or thylakoids were obtained on gradients consisting of 0 - 50 % urografin or 20 - 50 % sucrose[4].

Extraction of lipids, pigments and saponins. Insolated PLBs and thylakoids were extracted with acetone, twice with 80 % and once with 100 %.

Identification of saponins. Thin layer chromatography (TLC) was used on precoated silica gel plates (Macherey & Nagel) in the following solvent systems:

I.: $CHCL_3$:MeOH:acetic acid:H_2O = 170:30:20:6
II.: $CHCL_3$:MeOH:H_2O = 70:30:4
III.: ethyl acetate:MeOH:H_2O = 100:16,5:13,5

Preparative isolation of saponins and aglyca succeeded by column chromatography[4].
Sugars were determined by TLC and gas chromatography[4].
Saponins and their aglyca were identified by mass spectrometry[4].

Reaggregation experiments. Reaggregation experiments were carried out with the complete mixture of compounds from the acetone extract and with isolated saponins suspended in 0,5 % SDS, 0,05 m Tris / Borat, 0,1 % 2 - mercaptoethanol pH 8,5 [4].

Electron microscopy. Fractions of isolated or reaggregated tubules were controlled by electron microscopy after negative staining with 1 % Na-phosphotungstic acid, pH 7,0.

Thin sectioning of leaves was carried out according to conventional methods.

Determination of saponin concentration. Saponin concentration was measured by weighing isolated saponins or by scanning negatives of photographs of orcinol developed TLC - plates[9].

RESULTS AND DISCUSSION

Ruppel et al.[2] showed that proteins have no influence on tubular structure in vitro. Therefore we expected to find the building units in acetone extracts of isolated PLBs[4].

As main constituents of the acetone extract two saponins were identified (Fig. 1) with the Rf-values (solvent system II) of 0,3 and 0,2. They amounted to 48 - 50 % (d.w.) of the acetone extract! Both saponins contain rhamnose and glucuse.

By mass spectrometric analysis they were identified as Avenacosid A and B found by Tschesche et al. in extracts of green leaves in very small amounts[5-8]. Testing these saponins for their capability to reaggregate into tubular structures, we got PLB - like tubular complexes, resembling native PLBs (Fig. 2).

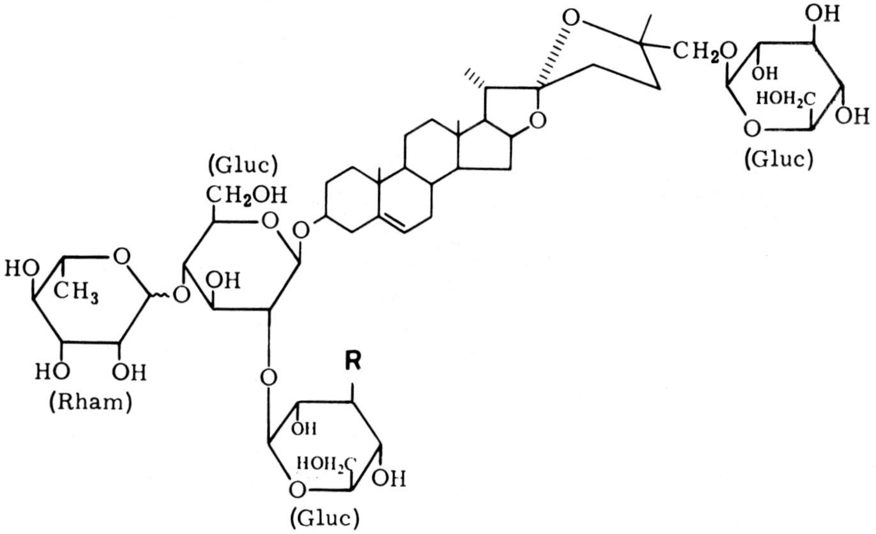

Fig. 1: Molecular structure of saponin A and B.
Saponin A: R = H ; saponin B: R = glucose.
(Reproduced with the permission of the editor of Zeitschrift für Pflanzenphysiologie.)

By a simple experiment we could demonstrate that the tubular structure of native PLBs in leaves of Avena sativa depends on the saponin content[9]. Dark cultivated 6 days old seedlings were exposed to light for 1, 4, 8, and 24 hours. After light treatment the plants were replaced to dark for the same time. Determination of the saponins after isolation of PLBs and thylakoids clearly showed that the PLB - structure in vivo depends on the saponin content. During greening the concentration is lowered and during re-etiolating it is raising again. This lowering and raising is directly coupled to the disintegration and rebuilding of PLBs in the leaves as shown by electron microscopic studies[9].

All these results - high concentration of saponins in PLBs, reaggregation capability and the relation between PLB - structure and saponin content - lead to the assumption that saponins are the main building units of prolamellar bodies.

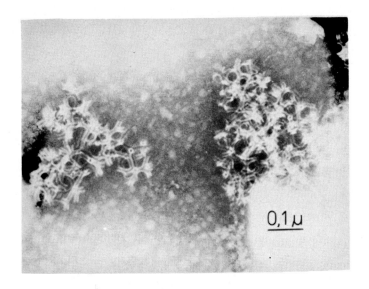

Fig.2: Reaggregated tubules, built up by purified saponins A and B, seen in the electron microscope after negative staining. (Reproduced with the permission of the editor of Zeitschrift für Pflanzenphysiologie.)

Reasons for the late discovery are certainly the high polarity (extraction with 80 % acetone!) and the fact that their concentration depends on the crystalline character and the occurrence of PLBs[9]. So we could show that barley PLBs contain lower concentrations of these two saponins. This corresponds with the non crystalline PLB - structure in barley etioplasts: only 20 % of barley PLBs are as crystalline as Avena PLBs[10].

A simular correspondence can be shown in the case of pea etioplasts.

ACKNOWLEDGEMENTS

This work was supported by the Deutsche Forschungsgemeinschaft through Sonderforschungsbereich 74. We wish to thank too Prof. Dr. H. Budzikiewicz, Köln, for mass spectrometric analysis and Prof. Dr. R. Tschesche, Bonn, for comparison material.

REFERENCES

1. Lütz, C. Kesselmeier, J. and Ruppel, H.G.:Z. Pflanzenphysiol. 85, 327 (1977).
2. Ruppel, H.G., Kesselmeier, J. and Lütz, C.: Z. Pflanzenphysiol. in press (Nov./Dec. 1978).
3. Lütz, C.: Z. Pflanzenphysiol. 75, 346 (1975).
4. Kesselmeier, J. and Budziekiewicz, H.: Z. Pflanzenphysiol. in press (1979).
5. Tschesche, R. and Schmidt, W.: Z. Naturforschg. 21 b, 896 (1966).
6. Tschesche, R., Tauscher, M., Fehlhaber, H.W. and Wulf, G.: Chem. Ber. 102, 2072 (1969).
7. Tschesche, R. and Lauven, P.: Chem. Ber. 104, 3549 (1971).
8. Tschesche, R. and Wiemann, W.: Chem. Ber. 110, 2416 (1977).
9. Kesselmeier, J. and Ruppel, H.G.: Z. Pflanzenphysiol. in press (1979).
10. Berry, D.R. and Smith, H.: J. Cell Sci. 8, 185 (1971).

ENDOGENOUS LIPOXYGENASE CONTROL AND LIPID-ASSOCIATED FREE RADICAL SCAVENGING AS MODES OF CYTOKININ ACTION IN PLANT SENESCENCE RETARDATION

YA'ACOV Y. LESHEM[*], SHLOMO GROSSMAN[*], ARYEH FRIMER[**] and JOSEPH ZIV[**]
Departments of Life Sciences[*] and Chemistry[**], Bar-Ilan University, Ramat-Gan (Israel)

ABSTRACT

Evidence is presented on hitherto unknown modes of action of cytokinin (CK) regulation of plant senescence pertaining to lipoxygenase, antioxidation and lipid associated free radical scavenging. The hormone significantly lowers lipoxygenase activity in senescing tissues and is mimicked by the endogenous lipid antioxidant α-tocopherol in its chlorophyll retaining and enzyme lowering effect. Experimentation with CK model compounds indicates that the active site of the molecule may be the α carbon atom of the amine from which hydrogens are abstracted resulting in formation of amide (e.g. $\phi\text{-}CH_2\text{-}NH_2 \rightarrow \phi\text{-}\overset{||}{C}\text{-}NH_2$), the abstracted hydrogens probably serving as free radical scavengers.

INTRODUCTION

The retardation of plant senescence by the phytohormone CK is well documented (Miller[1], Skoog & Armstrong[2], Kende[3] and Hall[4]). Recently Grossman & Leshem[5] have reported that salient feature of the hormone's anti-senescent effect in pea plants is a marked decrease in lipoxygenase levels which, during the course of senescence, either normal or induced by leaf detachment, invariably increase. It was also suggested that the mode of action may involve free radical scavenging (Grossman & Leshem[6]).

The present research has endeavoured to clarify the primary mode(s) of CK action in senescence physiology, i.e. the nature of the lipoxygenase-CK interaction, the mechanism of CK-induced chlorophyll retention[7,8] and the possible relationship to endogenous plant lipid antioxidation associated with α-tocopherol (Vitamin E) which may be membrane bound[9-12]. Furthermore, in an attempt to ascertain the possible involvement of free radical quenching phenomena and to understand and pinpoint the protective metabolism of the group of CK hormones (which are basically *amino*-purines) from an organic-chemical perspective, the *in vivo* effect of a free-radical generating system (X-XO) interacting with CK was looked into, and *in vitro* CK model compounds I-IV (Table 3) were reacted with the superoxide ($O_2^{\cdot-}$) free radical in aprotic solvents and reaction products characterized.

EXPERIMENTS AND RESULTS

CK and lipoxygenase. In experiment 1 plants were grown and CK applied as outlined elsewhere[5]. Lipoxygenase in this and all other experiments was assayed by the polarographic method described by Grossman et al.[13] Results are given in Table 1.

TABLE 1

CK (KINETIN 30 ppm) EFFECT ON LIPOXYGENASE IN SENESCING PEA FOLIAGE

Treatment	Specific Activity O_2 µl/mg protein/min
Control-intact leaves	0.44
CK treated detached leaves	1.50
Non CK treated detached leaves	2.90

From this table it is apparent that detachment markedly increased lipoxygenase specific activity and furthermore CK treatment of detached tissue significantly reduced the enzyme's activity.

Effect of CK and α-tocopherol. Plants were grown and treated as outlined above with the difference that foliage remained intact throughout. In this and the next experiment the α-tocopherol was applied as the light stable acetate emulsified in 0.01% Triton X-100. The latter was also added to all other treatments. Figure 1 clearly indicates that at concentrations applied the α-tocopherol has a similar effect to CK regarding lipoxygenase decrement.

CK, α-Tocopherol and chlorophyll retention: In trials 1 and 2 performed on peas plants were grown and treatments applied as outlined above, while in trial 3 on oat foliage the procedure of Gunning and Barkley[14] was followed. Table 2 presents results expressed as relative amounts of chlorphyll based on O.D. units at 663nm assessed according to Anderson & Brenner[15]. From this table it is apparent, that both CK and α-tocopherol enhance chlorophyll retention in both pea and oat tissue.

Interaction of free radicals, CK and lipoxygenase. Plants were grown as above and spray treated twice weekly. The free radical generating system comprised of xanthine oxidase (*ca* 2 units per spray treatment to 5x20 plants) acting on phosphate buffered (0.1M pH 7.4 and EDTA 10^{-4}M) xanthine (10^{-4}M). This system produces free radicals in general and particularly superoxide[16-18].

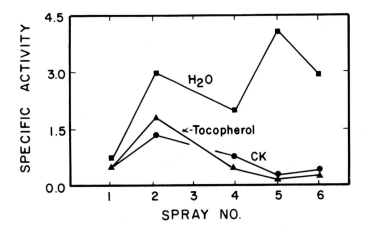

Fig. 1. Effect of CK (kinetin 30 ppm) and α-tocopherol acetate (10^{-5}M) on specific activity of lipoxygenase of intact pea plants.

TABLE 2

EFFECT OF CK AND α-TOCOPHEROL ACETATE ON CHLOROPHYLL RETENTION IN SENESCING PEA AND OAT FOLIAGE

Treatment	Trial 1 (Intact pea plants)	Trial 2 (Detached pea foliage)	Trial 3 (Detached oat blades)
H_2O-Control	100	100	100
Kinetin 30 ppm	128	120	174
α-Tocopherol acetate	170[a]	127[b]	178[b]
Level of statistical significance between control and treatment*	$p<0.05$	$p<0.05$	$p<0.01$

* Calculated by analysis of variance   a = 10^{-4}M;   b = 10^{-6}M

To ensure contact of the very labile free radicals with plant tissue, the enzyme was added to the substrate immediately prior to spraying. Control and CK treatments were buffered and contained EDTA as above. Results of preliminary experimentation still being carried out indicate a tendency of lipoxygenase to increase upon X-XO treatment, this increment apparently being lowered by CK. However, this yet remains to be substantiated.

In vitro reaction of cytokinin model compounds with the free radical $O_2^{\cdot -}$.
In general 2.5 mmole of aryl amine, 5.0 mmole of the crown ether 18-crown-6 and 10 mmole of KO_2 were dissolved in 20 ml of dry benzene and stirred for 24 hours at room temperature. The reaction mixtures were then neutralized (10% HCl), washed with saturated $NaHCO_3$ solution, dried ($MgSO_4$) and separated into their components by preparative TLC. Compounds and product yields are given in Table 3. It is clear from the results that the major product in all instances is the corresponding amide.

TABLE 3

REACTION OF ARYL AMINES (CK MODEL CYTOKININ COMPOUNDS) I-IV WITH $O_2^{\cdot -}$ (KO_2/18-CROWN-6)

Serial No.	Substrate	Product[a]	Isolated yields(%)[b] of Identified Primary Products[c]
I	Benzylamine $\phi - CH_2 - NH_2$	$\phi - \overset{O}{\underset{\|\|}{C}} - NH_2$	65
II	2-Furfurylamine furyl $- CH_2 - NH_2$	furyl $- \overset{O}{\underset{\|\|}{C}} - NH_2$	60
III	N-phenylbenzylamine $\phi - CH_2 - NH - \phi$	$\phi - \overset{O}{\underset{\|\|}{C}} - NH - \phi$	35
IV	N-benzyltoluidine $\phi - CH_2 - NH - \phi - CH_3$	$\phi - \overset{O}{\underset{\|\|}{C}} - NH - \phi - CH_2$	36

a. Identity determined by comparison of physical properties and spectral data with authentic samples. b. Based upon amount of reacted substrate.
c. Excluding further scission and oxidation products.

It is clear from the above results that the major products in all instances is the corresponding amide.

DISCUSSION

The data presented in Table 1 clearly indicate that during the process of senescing, lipoxygenase increases. These findings are in accord with those reported by Galliard[19]. The table also shows that the phytohormone CK significantly lowers the enzyme's activity. Since even after affinity chromatography purification of the enzyme the same trend exists[5], it is postulated that the lipoxygenase-CK interaction is not mediated by antioxidants since the latter are not bound by the affinity chromatography column.

Further insight into lipid-associated CK control of senescence pertaining to endogenous lipoxygenase levels, chlorophyll retention and the purported chloroplast-membrane association[9,10] with the natural antioxidant α-tocopherol is provided by Fig. 1 and Table 2. These indicate that the hormone considerably lowers lipoxygenase increment in aging tissues and in this respect is mimicked by α-tocopherol. The latter also has a like physiological effect upon chlorophyll retention in senescing foliage of both peas and oats (Table 2).

The apparent similarity of physiological effect between CK and α-tocopherol poses a paradox since Grossman & Leshem[5] have shown that the CK lipoxygenase interaction appears to be independent of the presence of endogenous antioxidants. The two diverging lines of experimental evidence may be accommodated by the interpretation of compartmentation of CK action in senescence. One compartment could possibly be the chloroplast membrane which is associated with α-tocopherol (Lichtenthaler & Calvin[9], Lichtenthaler[10], and Barr & Crane[20]) and hence possesses antioxidative action. A second compartment may be the plasmalemma, where CK could, as a free radical scavenger, serve a direct source of quenching of free radicals formed by the action of lipoxygenase on PUFA chains (Galliard[19], Veldink et al.[21], and Gurr & James[22]). A like mode of action at least in phenomenology of membrane damage caused by photooxidation of linolene acid has been reported by Van Hasselt[23] & de Kok & Kuiper[24]. Another possibility is that at the plasmalemma compartment CK acts through an indirect effect on the lipoxygenase, modifying its active site and thus decreasing its specific activity. This may be inferred from the experiment[5] in which the lipoxygenase was purified by affinity chromatography procedure and was found to be less active after CK-treatment.

If, as the above results imply, the CK acts as a free radical scavenger, as seen from Table 3, the probable mechanism of scavenging involves an initial hydrogen abstraction from the CK molecule, and the conversion of an amine to an amide, as shown below:

$$\begin{array}{c} Ar \\ | \\ H_2C \\ | \\ HN \\ | \\ R \end{array} \xrightarrow{O_2^-} \begin{array}{c} Ar \\ | \\ HC\cdot \\ | \\ HN \\ | \\ R \end{array} \longrightarrow \begin{array}{c} Ar \\ | \\ HC-OOH \\ | \\ HN \\ | \\ R \end{array} \xrightarrow{-H_2O} \begin{array}{c} Ar \\ | \\ C=O \\ | \\ HN \\ | \\ R \end{array}$$

According to this mechanism the function of the aryl or vinyl group in cytokinins is to activate the α-methylene hydrogens towards abstraction by delocalizing the resulting free radical. Further experimentation on the X-XO system and the elucidation of the nature of its interaction with CK will provide further *in vitro* information on the above hypothesis.

ACKNOWLEDGEMENT

We thank K. Tal, Jeanette Barnes, Margalit Bergman and Tsipora Goldweitz for their technical aid.

REFERENCES

1. Miller, C. (1961) Ann. Rev. Pl. Phys. 12, 395-408.
2. Skoog, F. and Armstrong, D.C. (1970) Ann. Rev. Pl. Phys. 21, 359-384.
3. Kende, H. (1971) Int. Rev. Cytol. 31, 301-339.
4. Hall, R.H. (1973) Ann. Rev. Pl. Phys. 24, 415-444.
5. Grossman, S. and Leshem, Y. (1978) Phys. Pl. 43, 359-362.
6. Grossman, S. and Leshem, Y. (1977) Pl. Phys. (Sup.) 59, 415.
7. Richmond, A. and Lang, A. (1957) Science 125, 650-651.
8. Osborne, D.J. and McCalla, C.C. (1961) Pl. Phys. 36, 219-221.
9. Lichtenthaler, H.K. and Calvin, M. (1964) Biochim. Biophys. Acta 79, 30-40.
10. Lichtenthaler, H.K. (1967) Ber. ptsch. Bot. Ges. 82, 483-497.
11. Lucy, J.A. (1972) Ann. N.Y. Acad. Sci. 203, 4-11.
12. Mead, J.F. (1976) in Free Radicals in Biology, Pryor, W., ed., Academic Press, N.Y., Vol. 1, 59-61.
13. Grossman, S., Ben Aziz, A., Budowski, P., Ascarelli, I., Gertler, A., Birk, Y. & Bondi, A. (1969) Phytochemistry 8, 2287-2293.
14. Gunning, B.E.S. and Barkley, W.F. (1963) Nature 199, 262-265.
15. Anderson, C.R. and Brenner, M.L. (1977) Pl. Phys. (Sup.) 59, 74.
16. Fridovich, I. (1976) in Free Radicals in Biology, Pryor, W., ed., Academic Press, N.Y., Vol. 1, 239-277.
17. Nishikimi, M. (1975) Biochem. Biophys. Res. Commun. 63, 463-468.
18. Lehninger, A.L. (1975) Biochemistry. Worth Publishers, N.Y., p. 471.
19. Galliard, T. (1975) in Recent Advances in the Chemistry and Biochemistry of Plant Lipids, Galliard, T. and Mercer, E., eds., Academic Press, N.Y. p. 335.
20. Barr, R. and Crane, F.L. (1977) Pl. Phys. 59, 433-436.
21. Veldink, G.A., Vliegenthart, J.F.G. & Boldingh, J. (1977) in Prog. Chem. Fats & Other Lipids, Vol. 15, Pergamon Press, pp. 131-166.
22. Gurr, M. & James, A.T. (1973) Lipid Biochemistry, Chapman & Hall, N.Y., pp. 1-231
23. Van Hasselt, P.R. (1974) Act. Bot. Neerl. 23, 159-169.
24. de Kok, L.J. and Kuiper, P.J.C. (1977) Phys. Pl. 39, 123-128.

INTERACTIONS OF CHLOROPHYLL A WITH CHLOROPLAST LIPIDS IN MIXED MONOLAYER FILMS

CONNY LILJENBERG and EVA SELSTAM
Department of Plant Physiology, Botanical Institute, University of Göteborg,
Carl Skottsbergs Gata 22, S-413 19 Göteborg (Sweden)

ABSTRACT

Surface pressure-area measurements of purified chlorophyll a and monogalactosyl diglyceride in mixed monolayers were performed with an automatic recording surface film balance at a constant compression rate at $20^\circ C$. Parts of chlorophyll as phytol and geranylgeraniol were studied pure and in mixed films with monogalactosyl diglyceride. All components studied formed stable monolayer films. Chlorophyll and monogalactosyl diglyceride containing mostly α-linolenic acid showed miscibility. Phytol was immiscible with synthesized monogalactosyl diglyceride containing only stearic acid. Mixed monolayers of phytol and monogalactosyl diglyceride isolated from barley containing 83 mole percent α-linolenic acid showed a remarkably strong interaction. The results are discussed as a model for the localization of chlorophyll a in the thylakoid membrane.

INTRODUCTION

The membranes of the thylakoids in the chloroplast are the sites of the photosynthetic apparatus for the conversion of solar energy to chemical energy. The chlorophylls and other photosynthetic pigments are situated in these membranes. It is not clear whether chlorophyll in the thylakoid membrane is bound to lipids or to proteins or to both.

From lipid analysis data of isolated thylakoid membranes[1] it seems reasonable to assume that the galactodiglycerides constitute the main components of the lipid double layer in the thylakoid membrane of higher plants. A characteristic feature of these galactodiglycerides is that they contain high amounts (c. 90-95 %) of α-linolenic acid.

It seemed worthwhile to examine a chlorophyll - lipid model and to study possible association between the pigment and the membrane lipid.

In the present investigation chlorophyll a and monogalactosyl diglyceride were chosen for these experiments. In addition, studies have been performed with a simplified pigment - lipid complex where those structural parts of the pigment which could be expected to be critical for an association have been chosen. Monogalactosyl diglycerides with different fatty acid composition have been tested. The interactions between pigment - lipid have been measured with a surface balance.

Abbreviations: TLC, thin layer chromatography; MGDG, monogalactosyl diglyceride; 16:0, hexadecanoic acid; 16:1, cis-9-hexadecenoic acid; 18:0, octadecanoic acid; 18:1, cis-9-octadecenoic acid; 18:2, cis,cis-9,12-octadecadienoic acid; 18:3, all-cis-9,12,15-octadecatrienoic acid.

MATERIAL AND METHODS

Prenols and chlorophyll a. Commercial phytol was purified according to Liljenberg and Odham[2]. Geranylgeraniol (a gift from F. Hoffman La Roche & Co., Basel) consisted of a mixture of trans- and cis-form of C_2 (85 % trans).

Chlorophyll a was isolated from blue-green algae. The acetone extract from the algae were run twice through a polyethylene powder column (The British Drug Houses Ltd)[3] and chlorophyll a was then further purified on TLC. The system used was 0.25 mm cellulose plates developed in light petrol (b.p. 60-80°C) - acetone-2-propanol 80:20:0.5 (by volume). Chlorophyll was determined both with absorption spectrophotometry[4] and through measurements of the magnesium content with atomic absorption spectroscopy. The analyses gave almost identical results.

MGDG. As a source of MGDG with high degree of α-linolenic acid ($MGDG_{18:3}$) dark grown leaves of barley (Weibull's Ingrid) were used. The lipids were extracted according to Folch et al.[5]. MGDG was roughly separated first on a DEAE-cellulose column in acetate form[6] and then on a silicic acid column (Silic AR, CC-7, Mallinckrodt)[7]. MGDG was further purified on prewashed[8] silica gel H (Merck) plates eluted with chloroform-methanol-7N ammonium hydroxide 65:25:4 (by vol.)[9]. The fatty acid distribution of isolated MGDG was determined by gas chromatography and found to be: 16:0 5.6, 16:1 1.4, 18:0 4.0, 18:1 1.3, 18:2 4.7 and 18:3 83.0 mole percent.

MGDG containing stearic acid in both 1- and 2-position of the glycerol (MGDG 18:0) was synthetized according to Wehrli and Pomeranz[10].

Solvents. Chloroform, methanol, light petrol and acetone were distilled prior to use. The spreading solvent n-hexane (Merck) and methylene chloride (Fisher), both reagent grade, were taken directly from freshly opened bottles.

Surface balance technique. The surface pressure was measured with a glass plate suspended from a torsion wire (Wilhelmy technique). The surface pressure was measured continuously during compression of the surface film[11]. The small deflection of the plate was sensed by a forced transducer and recorded synchronously (the apparatus was built by Dr. Håkan Löfgren at the Department of Medical Chemistry, University of Göteborg). The glass plate, the barriers and the glass trough were washed with freshly prepared conc. dichromic sulphuric acid and rinsed several times with water before the measurements. The spreading solution was deposited on the subphase (phosphate buffer of pH 8.0; 10^{-3}M) from a mikropipette after the surface had been swept clean and the barriers positioned. The water used for buffer preparations and rinsing procedures was deionized, double-distilled and filtrated (Millipore Molsheim, France). Before polyunsaturated lipids and chlorophyll a were spread the apparatus was covered and flushed with nitrogen. After the lapse of three minutes for complete evaporation of the solvent and formation of the monolayer the film was compressed by movement of the barrier at a constant rate (24 $cm^2 \cdot min^{-1}$). All experiments with chlorophyll a were performed in dim green light. Pressure-area measurements were made at 20.0 $\pm$ 0.5°C and each experiment was repeated at least three times. Spreading solutions were prepared by diluting known concentrations of different pure lipids and by mixing aliquots of known concentrations of the lipids in n-hexane. Concerning $MGDG_{18:0}$ this lipid was incompletely dissolved in n-hexane and therefore this lipid, pure and in mixtures, was dissolved in methylene chloride.

RESULTS

$MGDG_{18:3}$ was run against chlorophyll a, phytol and geranylgeraniol, and $MGDG_{18:0}$ against phytol. Geranylgeraniol, a precursor of phytol, has been found to be coupled to chlorophyll pigments in higher plants[12,13,14]. All tested substances formed stable compressible monolayers on 10^{-3}M phosphate buffer. Dashed curves in the Figures were calculated for ideal mixtures by assuming that the total area of the mixed film is $n_x A_x + n_y A_y$ where the n's are the moles of the two components and the A's their molecular areas in pure films at the same surface pressure.

Figure 1 shows surface-pressure-area curves for pure $MGDG_{18:3}$, chlorophyll a and a mixture of the two components. The experimentally found values of the 6:1 mixture (mole/mole) of $MGDG_{18:3}$ and chlorophyll a showed good agreement with theoretical values indicating together with collapse pressure data (not shown in Figure), interaction between the two components.

Figure 3 shows surface pressure-area curves for pure phytol, $MGDG_{18:3}$ and the mixture MGDG/phytol 3:1. The mixture showed a strong negative deviation from theoretical values indicating that the complex occupied much less area than calculated. If, in a similar experiment, $MGDG_{18:0}$ was substituted for $MGDG_{18:3}$ the mixture 3:1 did instead show a positive deviation from theoretical values (Figure 2). The mixture of geranylgeraniol and $MGDG_{18:3}$ 1:3 showed a negative deviation from theoretical curves (Figure 4) indicating association between the two substances.

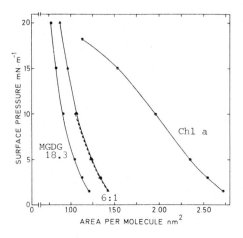
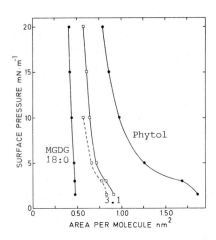

Fig. 1. Surface pressure-area curves for MGDG$_{18:3}$, chlorophyll a and the mixture 6:1 (mole/mole). Fig. 2. Surface pressure-area curves for MGDG$_{18:0}$, phytol and the mixture 3:1.
Solid lines - experimental curves. Dashed lines - calculated curves assuming ideal mixing. ●, pure substance; ▲, MGDG$_{18:3}$/chlorophyll a 6:1; □, MGDG$_{18:0}$/phytol 3:1.

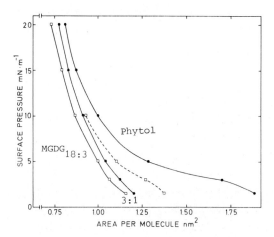
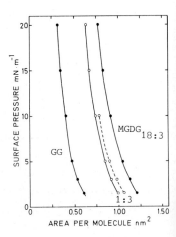

Fig. 3. Surface pressure-area curves for MGDG$_{18:3}$, phytol and the mixture 3:1. Fig. 4. Surface pressure-area curves for geranylgeraniol (GG), MGDG$_{18:3}$ and the mixture 1:3.
●, pure substance; □, MGDG$_{18:3}$/phytol 3:1; ○, GG/MGDG$_{18:3}$ 1:3.

DISCUSSION

$MGDG_{18:3}$ showed a cross-sectional surface area of 0.83 nm^2 at 10 $mN \cdot m^{-1}$ which agree well with values for galactodiglycerides from pelargonium leaves with a different fatty acid composition[15]. The experimentally found values of the 6:1 mixture (mole/mole) of $MGDG_{18:3}$ and chlorophyll a showed good agreement with theoretical values indicating together with collapse pressure data interaction between the two components (Figure 1).

The 3:1 mixtures between on one hand phytol and $MGDG_{18:3}$ and between geranylgeraniol and $MGDG_{18:3}$ on the other both showed negative deviation from theoretical values (Figures 3 and 4). This applies especially for the mixture with phytol where phytol seemed to have a strong condensing effect of the complex. In fact, the complex occupied less area than the free $MGDG_{18:3}$ alone. This effect is interpreted as an association between phytol and $MGDG_{18:3}$.

The difference of interaction between phytol and $MGDG_{18:3}$, on one hand, and between chlorophyll a and $MGDG_{18:3}$, on the other, might be explained by steric hindrance for the phytyl tail of the chlorophyll a molecule to interact with the fatty acid chains of the $MGDG_{18:3}$ in this water-air system.

The mixture 3:1 of phytol and $MGDG_{18:0}$ showed positive deviation from ideal behavior (Figure 2). Moreover, when running this mixture a typical pressure-area curve for phytol was followed by a characteristic $MGDG_{18:0}$ curve. This might indicate that phytol and $MGDG_{18:0}$ were not miscible.

It is difficult to draw conclusions about the thylakoid membrane from the studied model system. The phytyl tail of the chlorophyll molecule and especially free phytol interact strongly with the main lipid component of the thylakoid membrane, the monogalactosyl diglyceride in a monolayer model system. If part of the total chlorophyll of the thylakoid membrane is associated to the lipid bilayer the chlorophyll might stabilize the membrane by reducing the mobility of the fatty acid chains, protect the unsaturated chains and also reduce the permeability of the membrane.

REFERENCES

1. Bahl, J., Francke, B. and Moneger, R. (1976) Planta, 129, 193-201.
2. Liljenberg, C. and Odham, G. (1969) Physiol. Plant., 22, 686-693.
3. Anderson, A.F.H. and Calvin, M. (1962) Nature, 194, 285-286.
4. Mackinney, G. (1940) J. Biol. Chem., 132, 91-109.
5. Folch, J., Lees, M.C., Stanley, G.H.S. (1957) J. Biol. Chem., 226, 497-509.
6. Nichols, B.W. and James, A.T. (1964) Fette Seifen. Anstrichmittel, 66, 1003-1006.

7. Rouser, G., Kritchevsky, G., Simon, G. and Nelson, G.J. (1967) Lipids, 2, 37-40.
8. Schlotzhauer, P.F., Ellington, J.J. and Schepartz, A.I. (1977) Lipids, 12, 239-241.
9. Nichols, B.W. (1964) in New Biochemical Separations, James, A.T. and Morris, L.J. eds., van Nostrand, London, pp. 321-337.
10. Wehrli, H.P. and Pomeranz, Y. (1969) Chem. Phys. Lipids, 3, 357-370.
11. Shah, D.O. (1970) in Advances in Lipid Research, Paoletti, R. and Kritchevsky, D. eds., Academic Press, New York and London, 8, 347-431.
12. Liljenberg, C. (1974) Physiol. Plant., 32, 208-213.
13. Wellburn, A.R. (1976) Biochem. Physiol. Pflanzen, 169, 265-271.
14. Schoch, S., Lempert, U. and Rüdiger, W. (1977) Z. Pflanzenphysiol., 83, 427-436.
15. Oldani, D., Hauser, H., Nichols, B.W. and Phillips, M.C. (1975) Biochim. Biophys. Acta, 382, 1-9.

THE GALACTOLIPID AND PIGMENT COMPOSITION OF THE THYLAKOID MEMBRANES FROM NATURALLY DIFFERENTIATING CHLOROPLASTS OF AVENA SATIVA L.

R.O. MACKENDER
Department of Botany, The Queen's University of Belfast, Northern Ireland

ABSTRACT

Thylakoid membranes have been isolated from plastids at different stages in their natural development from proplastids to chloroplasts, and their galactolipid and pigment composition determined. Molar ratios between each lipid or pigment and chlorophyll a have been determined as well as the actual quantities of each, (on a per plastid basis), for each of the different stages of thylakoid proliferation. The conclusion as to the method of membrane assembly depends on the method of data analysis.

INTRODUCTION

There is a considerable literature concerning the composition biochemistry and biophysics of the thylakoid membranes of mature plastids. There are also numerous reports of investigations into changes in the composition of greening etiolated tissue[1,2] and the occassional report of changes in plastids isolated from greening etiolated tissue[3] and of whole greening algae[4] and of membrane fractions isolated from the algae[5]. However there is relatively little known about changes in the composition of thylakoid membranes during their proliferation during natural plastid development, and even less on the method/mechanism of membrane assembly.

The monocotyledon leaf is proving to be a very suitable system for the study of natural plastid development in higher plants[6,7]. Analyses have been carried out on membranes isolated from each of 9 sequential segments cut along the shoots of oat seedlings. This represents (on the basis of the chl/plastid[7]) at least seven stages in plastid development. The aim of this investigation was to determine the lipid and pigment composition of the membrane during its proliferation and to try and deduce the mechanism by which it did this.

MATERIALS AND METHODS

The shoots of 9-day old seedlings of Avena sativa L. var Mostyn which were grown in trays of sand at 20°C on 16 h photoperiod at light intensity of 1500 ft. candles were used in all these analyses. The shoots were harvested and cut into 9 one-centimetre segments - each segment contained plastids at a different stage of development, and these segments have been identified numerically (1-9) with segment 1 containing the youngest tissue and 9 the oldest.

Membranes were isolated by homogenizing 15-20 g of tissue with 100-150 mls of an ice cold medium containing 200 mM D(-) sorbitol, 5 mM $MgCl_2$, 0.2% $^w/v$ BSA buffered at pH 7.4 with 50 mM Na_2HPO_4/KH_2PO_4, in either an Atomix blender (segments 3-9; 4 x 3 sec bursts at full speed) or with a mortar and pestle (segments 1 and 2). In both cases the resultant brei was filtered through 4 layers each of cheesecloth and nylon plankton net (20µ mesh) and the filtrate centrifuged at 1500 g for 10 minutes at ca. $4°C$. The pellets were washed by resuspending in the same phosphate buffer containing 5 mM $MgCl_2$ and recentrifuging; this was repeated twice. The final pellets were resuspended in a small volume of the same medium and separate aliquots were taken for chlorophyll, carotenoid and acyl lipid analysis.

Chlorophyll was extracted into 80% acetone and determined spectrophotometrically according to Arnon[8]. Carotenoids were extracted into 80% acetone. The extract was then partioned against pet.ether (40-60 bp) and against diethyl ether (peroxide free) twice, after first diluting the extract with 100 mM NaCl solution (ca. 20% by volume). The combined pet ether and ether extracts were washed with 100 mM NaCl, shaken with dry NaCl, decanted and taken to dryness under a stream of O_2 free nitrogen. The carotenoids were separated by TLC (Kieselgel H; Merck) using pet ether (60-80 bp): ethyl acetate : diethylamine 58:30:12 v/v/v as the developing solvent. After chromatography the spots were eluted into ethanol and the pigments quantified spectrophotometrically using the following extinction coefficients ($E_{1\ cm}^{1\%}$) and absorption maxima[9]: β carotene 2620 (453 nm); Lutein 2550 (445 nm); Violaxanthin 2550 (443 nm); Neoxanthin 2243 (439 nm).

Lipids were extracted as described previously[10] except that a 20 fold excess of C:M 2:1 v/v was added to the membrane suspension. Lipids were separated by TLC using $C:M:H_2O$ 65:25:4 v/v/v and were located with iodine vapour and quantified as described previously[10].

RESULTS

Thylakoid membranes isolated from sequential segments cut along the shoots of 9-day-old oat seedlings have been analysed and their acyl lipid and pigment compositions determined. The leaves were cut into 9 segments, giving at least 7 different stages of plastid development (thylakoid proliferation) - 7 because the 3 segments in the centre of the leaf have more or less identical chl/plastid.

The ratios between the different pigments and lipids and chl (a) in the thylakoids from each segment are given in Table 1. In general these show (although there are exceptions) that between segments 4 and 9 (inclusive) the ratios are more or less constant, indicative of a stepwise insertion of the component molecules into the new membrane.

TABLE 1

THE MOLAR RATIOS OF THE GALACTOLIPIDS AND PIGMENTS TO CHLOROPHYLL A IN THYLAKOID MEMBRANES AT DIFFERENT STAGES OF PROLIFERATION

LIPID or PIGMENT RATIO	LEAF SEGMENT NUMBER								
	1	2	3	4	5	6	7	8	9
					molar ratios				
Chlorophyll b:a	0.45	0.34	0.35	0.35	0.34	0.34	0.34	0.35	0.34
MGDG : chl a	12.69	6.97	4.38	4.20	3.93	4.20	4.28	4.28	4.20
DGDG : chl a	7.59	4.11	2.14	1.87	1.79	1.70	1.79	1.87	1.79
β Carotene : chl a	0.17	0.13	0.08	0.10	0.09	0.08	0.11	0.10	0.11
Lutein : chl a	0.25	0.25	0.16	0.17	0.15	0.15	0.16	0.15	0.13
Violaxanthin : chl a	0.13	0.06	0.06	0.04	0.06	0.06	0.05	0.05	0.05
Neoxanthin : chl a	0.11	0.06	0.06	0.06	0.06	0.05	0.04	0.04	0.05
MGDG : DGDG	1.67	1.71	1.93	2.15	2.17	2.31	2.35	2.38	2.28

This type of addition is not so evident during the early stages of thylakoid proliferation, seemingly the chlorophyll is added after the other molecules. If this data is recalculated so that the number of each constituent molecular species is determined relative to a unit number of chlorophyll b molecules at each stage, it is possible to determine the molecular composition of this "membrane unit", and it would seem that each unit contains the following number of molecules: Chl a: 292; Chl b: 100; MGDG: 1230; DGDG: 540; β carotene: 28; lutein: 45; Violaxanthin: 15; Neoxanthin: 15.

Using this data and that for the chlorophyll conc per plastid, it has been possible to calculate the precise number of molecules of each molecular species in the thylakoid membranes at each stage of plastid development. These calculations show (Table 2) that between the leaf base and the leaf tip the lipids and pigments increase as follows: chl (a) x 16, chl (b) x 13, β carotene x 10, lutein x 9, violaxanthin x 6, neoxanthin x 8, MGDG x 5 and DGDG x 4. The fact that these increases are not constant would not support the idea of a single step membrane synthesis. However if the increases are taken between segment 4 and segment 9 (thus eliminating the early stages of thylakoid development) there is a greater constancy in the fold of increase - and is about 1.7.

Thus the data as it has been presented so far is seemingly consistent with a stepwise incorporation of the component molecules into the membrane (except in the early stages of development), during its proliferation.

TABLE 2

THE MOLECULAR COMPOSITION (MOLECULES PER PLASTID) OF THYLAKOID MEMBRANES AT DIFFERENT STAGES OF PROLIFERATION

LIPID / PIGMENT	LEAF SEGMENT								
	1	2	3	4	5	6	7	8	9
				Molecules per plastid x 10^{10}					
Chlorophyll a	5.1	10.5	29.6	47.2	50.5	52.7	60.0	63.4	80.2
Chlorophyll b	2.3	3.6	10.6	16.6	18.6	17.9	20.5	21.9	27.2
β Carotene	0.8	1.3	2.5	4.6	4.5	4.3	6.6	6.5	8.7
Lutein	1.3	2.7	4.9	8.0	7.6	7.6	9.7	9.5	10.6
Violaxanthin	0.7	0.7	1.9	1.9	2.9	2.9	3.1	3.0	4.1
Neoxanthin	0.5	0.7	1.7	2.8	2.9	2.4	2.5	2.8	3.8
MGDG	64.2	73.1	129.8	198.1	198.6	220.9	256.8	271.2	336.9
DGDG	38.4	43.1	63.4	88.2	90.5	89.4	107.4	118.5	143.6
Total chlorophyll	7.4	14.1	40.2	63.8	69.1	70.6	80.5	85.3	107.4
Total carotenoid	3.3	5.4	11.0	17.3	17.9	17.2	21.9	21.8	27.2
Total galactolipid	102.6	116.2	193.2	286.3	289.1	310.3	364.2	389.7	480.5

TABLE 3

THE CHANGES IN THE AMOUNTS OF CHL A, MGDG AND DGDG IN THE THYLAKOID MEMBRANES DURING THYLAKOID PROLIFERATION, AND THE RATIOS BETWEEN THEM

LEAF SEGMENT	LIPID/PIGMENT per plastid pmoles x 10^{-3}			Change in lipid/pigment concentration			Δ MGDG/ Δ chl a	Δ DGDG/ Δ chl a
	Chl a	MGDG	DGDG	Δ chl a	Δ MGDG	Δ DGDG		
1	84	1066	638	-	-	-	-	-
2	174	1213	715	90	147	73	1.63	0.81
3	492	2155	1053	318	942	338	2.96	1.06
4	783	3289	1464	291	1134	411	3.90	1.41
5	839	3297	1502	56	8	38	0.14	0.68
6	873	3667	1484	34	370	18	10.88	-
7	996	4263	1783	123	596	299	4.84	2.43
8	1052	4508	1967	56	240	184	4.28	3.28
9	1332	5594	2384	280	1091	417	3.90	1.49

Δ = change in quantity of lipid or pigment between successive leaf segments

However the picture becomes much less clear if, instead of expressing the ratios between the total amounts of each membrane component, the ratio between the increases in the number of molecules between consecutive stages in development, is calculated. This is shown in Table 3 for chl (a), MGDG and DGDG. The actual increases in the numbers of each of these molecules between successive developmental stages have been calculated and the Δ MGDG : Δ chl a and the Δ DGDG : Δ chl a determined. It is immediately apparent that neither of these series of ratios show the same pattern or constancy during the sequence of development as those shown in Table 1 (Similar calculations with each of the carotenoids show the same inconsistencies). These results are not consistent with a "single step" mechanism of membrane biogenesis.

DISCUSSION

Before discussing these results three general points should be made. First, although 9 day old plants were used (one day older than had been used to determine the chl/plastid), the shoots of the two batches were of the same length and the growth rate was similar. The difference was in the time taken for the coleoptile to emerge. Secondly, these analyses did not distinguish between the different membranes of the grana fret system i.e. the stroma lamellae, the end granal membranes and the partition membranes. Each may or may not have its own composition which may change during development. Thus recorded changes may be a reflection of the relative proportions of these membranes and or changes in their individual compositions. Thirdly, the method of data analysis is important: if it is intended (as it was here) to compare ratios between total amounts of the membrane components (e.g. MGDG : chl a) at different stages of development and to use them as an indicator of the mechanism of membrane assembly, then it should be remembered that unless the increases in one or other of the components is large the ratios will appear to remain fairly constant. To give due emphasis to small changes in the relative increases of the membrane components it is necessary to use difference data (Table 3).

Taking all these points into consideration, the truest picture of the mechanism of thylakoid membrane assembly is probably given in Table 3, which suggests a multistep mechanism of assembly. The constancy of the ratios (particularly in the latter stages of thylakoid proliferation) shown in Table 1 is a reflection of the small increases in components relative to the large amounts already present.

The only data with which this can be compared is that of analyses of differentiating plastids from maize leaves[11]. It should be noted that those analyses were of suspensions containing broken and intact plastids, and, in the suspensions isolated from older tissue, bundle sheath and mesophyll plastids (mainly mesophyll). Bearing these points in mind, it should be noted that the lipid concentrations per plastid were very similar, but the molar ratios between MGDG and DGDG and total chlorophyll were higher in oats than in maize. However in both cases there is a dramatic decrease in these ratios between the first two stages of plastid development examined.

Comparisons with the results from greening etiolated tissues are difficult, especially in the light of recent findings that etioplasts themselves show a developmental sequence and have a changing composition[12,13].

Finally, it should be noted that the molecular proportions of the galactolipids and pigments reported here are very similar to those reported for spinach[14].

REFERENCES

1. Tremolieres, A. and Lepage, M. (1971) Pl. Physiol, 47, 329-334.
2. Tevini, M. (1977) in Lipids and Lipid Polymers in Higher Plants. Tevini, M. and Lichtenthaler, H.K. eds., Springer-Verlag, Berlin Heidelberg, pp. 121-145.
3. Bahl, J., Franke, B. and Moneger, R. (1976) Planta, 129, 193-201.
4. Goldberg, I. and Ohad, I. (1970) J. Cell Biol., 44, 563-571.
5. Petrocellis, B de., Siekevitz, P. and Palade, G.E. (1970) J. Cell Biol., 44, 618-634.
6. Leech, R.M., Rumsby, M.G., Thomson, W.W., Crosby, W. and Wood, P. (1971) in Proc. 2nd Int. Cong. Photosynthesis Research, Forti, G., Avron, M. and Melandri, A. eds., pp. 2479-2488.
7. Taylor, J.A. and Mackender, R.O. (1977) Pl. Physiol., 59(6), 10.
8. Arnon, D.I. (1949) Pl. Physiol., 24, 1-15.
9. Davies, B.H. (1976) in Chemistry and Biochemistry of Plant Pigments 2nd Ed. Goodwin, T.W. ed., pp. 38-165.
10. Mackender, R.O. and Leech, R.M. (1974) Pl. Physiol., 53, 496-502
11. Leese, B.M.L. and Leech, R.M. (1976) Pl. Physiol., 57, 789-794.
12. Robertson, D. and Laetsch, W.M. (1974) Pl. Physiol., 54, 148-160.
13. Mackender, R.O. (1978) Pl. Physiol., in press.
14. Lichtenthaler, H.K. and Park, R.B. Nature, Lond., 198, 1070-1072.

CHANGES IN THE MITOCHONDRIAL LIPIDS OF MANGO FRUITS AT STORAGE TEMPERATURES INDUCING "CHILLING INJURY".

P. MAZLIAK and O. KANE

Laboratoire de Physiologie cellulaire, ERA 323, 12 rue Cuvier, 75005 Paris (France)

ABSTRACT

A marked decrease in the oxidation capacities of the mitochondria of chilling injured mangoes was correlated with a remarkable lowering of the molar ratio palmitoleic/palmitic acid, in the lipids of these mitochondria.

INTRODUCTION

Mangoes are tropical fruits which are grown in roughly the same regions of the world as those where bananas are grown. Bananas are easily transported towards the temperate countries because these fruits are well resistant to storage. On the contrary mangoes - the production of which is enormous : 6 million tons a year - are poorly resistant to storage and less than 0.1% of the production is presently exported. In an effort to find better storage conditions for the transport of mangoes, we have been studying the metabolism of mango fruits stored at different temperatures.

MATERIALS AND METHODS

Fruits. Mango fruits (*Mangifera indica*, cv. Amélie) have been grown in Mali (Africa), harvested at a mature stage and immediately sent by air to the laboratory in France. The green fruits have been stored, on one hand at 20 and 12°C (both these temperatures allowing normal ripening), on the other hand at 8 and 4°C, both these relatively low temperatures inducing chilling injury.

Mitochondria. Considerable difficulty was encountered during the initial attempts to isolate an actively respiring mitochondrial fraction from mango fruits because of the highly acidic nature of the fruit pulp and also because of the abundance of tannins in the cell vacuoles. These difficulties have been partially overcome by using polyvinylpyrrolidone in the grinding medium as recommended by Hulme et al.[1]. The succinate oxidation rates measured polarographically with our mitochondrial preparations were of the same order of magnitude as those reported by previous authors[2], although we had not added any cytochrome c in our respiratory medium. Analysis of mitochondrial fatty acids was by means of GLC as described previously[3].

RESULTS

Figure 1 shows the variations during storage of the oxidative capacities of the mitochondria isolated from fruits stored at four different temperatures. In the case of fruits ripening normally, at 20 or 12°C, the succinate oxidation capacities <u>increased</u> during the first ten days and then <u>decreased</u> slowly. A marked increase was observed at 20°C, corresponding to a marked climacteric rise in the respiration of entire fruits. At lower temperatures, the fruit presented chilling injury symptoms after about 10 days of storage and the succinate-oxidation capacities of isolated mitochondria did not show any increase during storage but instead a progressive decrease.

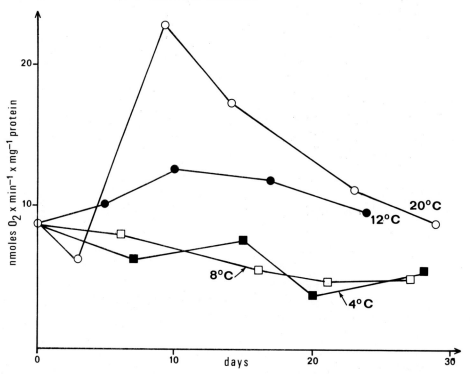

Fig. 1. Variations, during storage, of the succinate-oxidation capacities of mitochondria isolated from mango fruits stored at 4, 8, 12 and 20°C. Each point is the average of three independent assays.

To see if this decrease was related to some alteration of the mitochondrial lipids during the storage of the fruits at low temperature, analyses of mitochondrial fatty acids were realized. The major fatty acids found in the mitochondria of ripe fruits were <u>palmitoleic</u> (36.7 mol/100 mol), <u>palmitic</u> (23.7%), <u>oleic</u> (18.1%), <u>linoleic</u> (7.5%) and <u>linolenic</u> (6.7%) acids. Figure 2 shows the

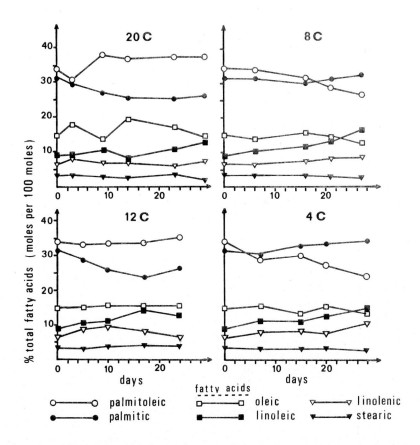

Fig. 2. Variations, during storage, of the fatty acid compositions of mitochondria isolated from mango fruits stored at 4, 8, 12 and 20°C.

-variations of these mitochondrial fatty acids during storage. Fatty acids having 16 carbon atoms were found to present the most remarkable evolution and figure 3 shows clearly that in the case of healthy fruits, ripening at 20 or 12°C, the molar ratio palmitoleic/palmitic acid increased with ripening. This increase was primarily due to a fall in the proportion of palmitic acid, thus suggesting an active desaturation of this acid during storage. On the contrary, when chilling injury developed on fruits, at 8 and 4°C, the molar ratio palmitoleic/palmitic acid continuously decreased, thus suggesting an inhibited activity of palmitate desaturase during storage. This result, indicating a lower degree of unsaturation in the mitochondrial lipids of fruits stored at low temperature is the contrary of what has been generally observed with other plant tissues.

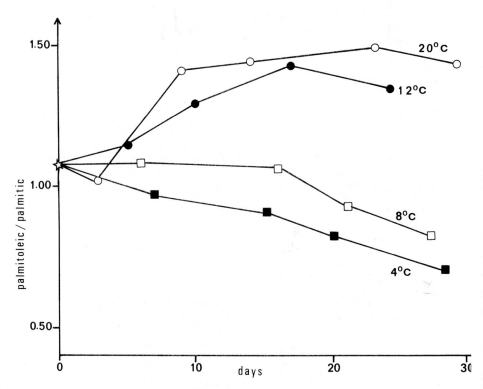

Fig. 3. Variations during storage, of the molar ratio palmitoleic/palmitic acid in the mitochondrial lipids from mango fruits stored a 4, 8, 12 and 20°C.

In summary, we have been able to correlate a marked decrease in the succinate-oxidation capacities of the mitochondria of chilling-injured mangoes with a remarkable lowering of the molar ratio palmitoleic/palmitic acid in the lipids of these mitochondria. It is tempting to attribute the lowering of the oxidative capacities of the mitochondria to some modification of the fluidity of their membranes ; this modification, in turn, could result from the changes in fatty acid compositions that we have observed during storage.

REFERENCES

1. Hulme, A.C., Jones, J.D. and Wooltorton, L.S.C. (1964) Phytochemistry, 3, 173-188.
2. Baqui, S.M., Mattoo, A.K. and Modi, V.V. (1974) Phytochemistry, 13, 2049-2055.
3. Mazliak, P., Oursel, A., Abdelkader, A.B. and Grosbois, M. (1972) Eur. J. Biochem., 28, 399-411.

A PROPOSED MODEL FOR SUDDEN FLUIDITY CHANGES

VERN McMAHON

Division of Biochemistry, University of Wyoming,
Laramie, Wyoming, 82071, USA.

ABSTRACT

Cyanidium caldarium will dramatically alter the fatty acids associated with specific lipid components as a response to changes in its culture temperature. Only one fatty acid, α-linolenic acid (18:3), decreases to less than 1% at the highest culture temperature. The remaining fatty acids of palmitic, oleic and linoleic acids are not greatly affected by culture temperature. Since 18:3 is inducible with a high to low shift in culture temperature, it would appear to be a major controlling factor in sudden fluidity changes of membrane lipids. With these observations a model is proposed. The model consists of the prospect that the Δ 15-desaturase(s) works on a membrane lipid (e.g., phosphatidylcholine—18:2 or a chloroplastic lipid of sulfoquinovosyldiglyceride—18:2) as its substrate and that desaturation occurs *in situ* to 18:3. The enzyme is viewed as being embedded in its substrate; a membrane which is high in phosphatidylcholine or another membrane which is high in sulfoquinovosyldiglyceride.

INTRODUCTION

Many organisms appear to be capable of altering the kinds of fatty acids associated with membrane lipids. It is generally known that fatty acid compositions may be greatly affected by the growth temperature. For the most part, cells grown at high temperature possess fatty acids with fewer numbers of double bonds totally than those cells grown at low temperature which have greater unsaturation in their fatty acids. The insertion of increasing numbers of double bonds into fatty acids must alter membrane fluidity if their melting points are markedly different [stearic, oleic, linoleic and linolenic acids have melting points of 69.6°, 13°, -5° and -11°C, respectively[1]]. Therefore, it is most evident that membrane fluidity is primarily controlled by specifically altering fatty acids associated with specific membrane lipids in such an appropriate way as to maintain the same relative degree of fluid character in membrane lipids at each of various growth temperatures. Or, the relative degree of fluidness or rigidity may be viewed as the cell's response to achieve a given homeostasis for optimal membrane function.

MATERIALS AND METHODS

The thermophilic alga, *Cyanidium caldarium*, which had been obtained from an acid, hot springs in Yellowstone National Park was grown on a defined medium, using a Magnaferm fermentor as previously described [2,3]. The various lipid techniques of extraction, separation, identification and quantitation are also previously described [3].

RESULTS

Cyanidium caldarium will alter its fatty composition very dramatically as a function of growth temperature, as can be seen in figure 1. The most unusual change is found with 18:3 which is the most abundant fatty acid (47%) at a 20°C growth temperature and nearly absent (<1%) at a 50°C growth temperature. All other fatty acids do not follow such a dramatic change; indeed, these all increase in total amounts upon increasing the culture temperature (e.g., from 20° to 50°C growth temperatures). Higher culture temperatures than 50°C result

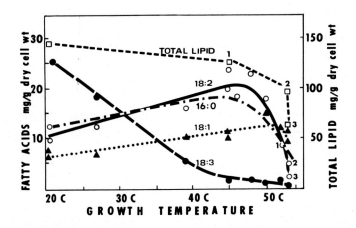

Figure 1. Effect of growth temperature upon fatty acids and total lipid. Cells were cultured at each of the growth temperatures for a period of two weeks before harvesting.

in an accelerated loss of fatty acids at or near the thermal death point (53°) of this clone-isolate of cells.

Since 18:3 is nearly absent at high temperature (>50°C) and is so markedly influenced by manipulation of growth temperature, we have pursued the conditions for its synthesis for most probably the Δ 15-desaturase plays a significant role in rapid changes of membrane fluidity. We have found that this desaturase is inducible with a high to low temperature shift in growth temperature. Also cycloheximide was found to inhibit completely 18:3 synthesis[3].

In other experiments[3], we report that 1-^{14}C-acetate only enters 18:3 to any extent with a high to low temperature shift in culture temperature. The appearance of the ^{14}C-label in 18:3 is first found in phosphatidylcholine and then nearly as quickly in the plant "sulfolipid", sulfoquinovosyldiglyceride. All other lipids such as monogalactosyldiglyceride, phosphatidylethanolamine and phosphatidylglycerol do not quickly become labelled and after a six-hour incubation; these lipids contain about one-fourth the label of 18:3.

The possibility exists that two Δ 15-desaturases are located in separate compartments such as the chloroplast and endoplasmic reticulum.

The following model is proposed for the role that 18:3 and its desaturase play in sudden changes in high growth temperature to lower temperature. The model is appropriate for the regulation of membrane fluidness in which cells are attempting to adjust membrane fluidity by altering specific fatty acids of specific membrane lipids. The model consists of the prospect that the Δ 15-desaturase works on phosphatidylcholine-18:2 as its substrate and that desaturation occurs *in situ* to 18:3. The enzyme is viewed as being imbedded in its substrate; a membrane which is high in phosphatidylcholine. We also propose that a separate Δ 15-desaturase exists in the chloroplast in which its substrate is sulfoquinovosyldiglyceride-18:2 and again encompasses the same concepts of the above model with the enzyme being imbedded in its substrate.

While other authors [4,5,6] have proposed that phosphatidylcholine-18:1 could be the substrate for the Δ 12-desaturase in which the product would be phosphatidylcholine-18:2, the conceptual extension of the data of this paper is consistent with their proposals.

SUMMARY

Our data show that the synthesis of 18:3 is especially temperature dependent. Its synthesis is inducible. ^{14}C-labelling experiments show that 18:3 accumulates rather quickly on two lipids. A model is presented for sudden fluidity changes.

ACKNOWLEDGEMENTS

This work was supported by the Wyoming Agricultural Experiment Station and is published as Journal Article No. SR 902.

REFERENCES

1. Markley, K.S. (1960) Nomenclature, Classification and Description of Individual Fatty Acids. In: K.S. Markley, ed., Fatty Acids and Their Chemistry, Properties, Production and Uses. Part I. Interscience Publ., Inc., New York.

2. Adams, B.L., McMahon, V.A. and Seckback, J. (1971) J. Biochem. and Biophys. Res. Commun. 42: 359-365.

3. Bedord, C.J., McMahon, V.A. and Adams, B. (1978) B. Arch. Biochem. Biophys. 185: 15-20.

4. Abdelkader, A.B., Cherif, A., Demandre, C., and Mazliak, P. (1973) Eur. J. Biochem. 32: 155-165.

5. Roughan, P.G. (1975) Lipids 10: 609-614.

6. Talamo, B., Chang, N., and Bloch, K. (1973) J. Biol. Chem. 248: 2738-2742.

HAS trans- 3-HEXADECENOIC ACID A ROLE IN GRANAL STACKING?

M.P. PERCIVAL, J. WHARFE, P. BOLTON, A.O. DAVIES, R. JEFFCOAT,* A.T. JAMES *
and J.L. HARWOOD
Department of Biochemistry, University College, P.O. Box 78, Cardiff CF1 1XL
(United Kingdom)
* Unilever Research, Colworth House, Sharnbrook, Bedford MK44 1LQ (United
Kingdom)

ABSTRACT

The level of trans-3-hexadecenoic acid was not depleted in a mutant chlorophyll b-less barley which had severely reduced granal stacking. In addition treatment of broad bean plants with various light regimes, which resulted in large fluctuations in the amount of the acid, did not produce corresponding changes in the amount of stacking. These results, together with data from other workers, lead us to conclude that trans-3-hexadecenoic acid probably does not play a role in granal stacking.

INTRODUCTION

trans-3-Hexadecenoic acid is, apparently, a ubiquitous constituent of the photosynthetic tissues of plants. It is well known that it is localised in the chloroplast, is virtually absent from etiolated tissues and increases rapidly during greening[1,2]. Similar results have been reported for Chlorella vulgaris when the acid appears after heterotrophically grown algae are transferred to phototrophic conditions[3]. In addition, trans-3-hexadecenoic acid is located almost exclusively at the 2-position of phosphatidylglycerol[4,5], the main chloroplastic phospholipid.

Because of these facts and the unique nature of its trans double bond, the function of the fatty acid has excited some speculation. Recently, Tuquet et al.[6] in extending previous work on the lipid composition of bundle sheath and mesophyll cells from maize[7] noted that trans-3-hexadecenoate was present in much higher amounts in the mesophyll cells. They also found, in contrast to other workers[1,8,9], that the acid was present in much higher amounts in granal rather than stromal lamellae. As a result of these observations, they suggested that trans-3-hexadecenoic acid plays a direct role in granal stacking.

One possible difficulty in equating lipid compositions between different cell types could be that alterations are introduced because of the widely differing metabolism of the cells. Accordingly we have approached the problem by considering only one cell type. Changes in the chloroplast structure in these cells are caused, firstly, by using the chlorophyll b-less barley mutant which exhibits poor stacking and, secondly, by subjecting broad bean plants to different light conditions.

MATERIALS AND METHODS

Plant material. Barley seeds were germinated in John Innes seed compost at $22°$ in an illuminated growth cabinet by using a 16h light/8h dark cycle. Illumination was 100 (photosynthetically active wavelengths) $\mu E/m^2$ per s.

Broad bean seeds were germinated under normal daylight conditions (May-June) for 18 days post-emergence. Plants for dark treatment were transferred into total darkness for 7 days after which some were returned to daylight conditions for a further 4 days. Samples were taken at each change in growth conditions from the second leaf pair and the fifth leaf pair which were, respectively, about 3 days old and 13 days old at the start of the treatment. Etiolated tissue was obtained from seedlings grown in total darkness for 7 days post-emergence.

Lipid extraction and separation. Lipids were extracted, purified by thin layer chromatography and quantitated by previously published methods[10].

Electron microscopy. This was carried out on barley as previously described[10]. Broad bean leaf sections (~1mm square) were immersed in freshly-prepared 0.1M-phosphate buffer (pH 6.8) containing glutaraldehyde (3%, w/v) for 2h at $0°$. After washing in buffered 0.5M-sucrose, samples were post-fixed with aqueous osmium tetroxide (2%, w/v). Sections were embedded in Spurr's low viscosity resin[11] after ethanolic dehydration. Thin sections (60-100nm) were cut on a Sorvall Porter-Blumm UM MT_2 microtome, stained with lead citrate and photographed under a Joel JEMM 7 electron microscope.

Carbon dioxide fixation. Carbon dioxide uptake was measured by the method of Etherington[12] using a Grubb Parsons SB_2 infra-red gas analyser.

RESULTS

The mutant barley was analysed for chlorophyll a and b contents and contained a ratio of over 500 for the two chlorophylls. Representative electron micrographs of leaf tissue from normal and mutant barley plants were then examined to ascertain the relative morphology. As previously observed for chlorophyll b-less barley mutants[13], the lamellar structure of the chloroplast appeared to be disorganised in comparison with normal plants. Chloroplast sections from both leaf types were compared by a number of criteria (Table 1). Highly significant differences were found between the chloroplasts of normal and chlorophyll b-less mutant barley. In particular the halving of the number of lamallae per granum in the mutant emphasised the lack of organisation. The results compare very well with those previously found by Goodchild et al.[13].

TABLE 1
COMPARISON OF THE CHLOROPLASTS AND trans-3-HEXADECENOATE CONTENT OF NORMAL AND CHLOROPHYLL b-LESS MUTANT BARLEY

	Normal	Mutant	Significance (p)
Grana/chloroplast section	42.6±4.3	29.8±0.6	<0.001
Single lamellae/chloroplast section	9.3±0.4	33.8±1.4	<0.001
Lamellae/granum	5.6±0.3	2.8±0.2	<0.001
Phosphatidylglycerol (% leaf acyl lipids)	8.1±0.5	9.6±0.9	n.s.
trans-3-Hexadecenoate (% total PG F.A.'s)	26.9±0.5	25.7±0.9	n.s.

Results are expressed as means S.D. (n=6). Significance was analysed by Student's t test.

In contrast to the large differences in chloroplast morphology which were found, the levels of both phosphatidylglycerol and trans-3-hexadecenoate in the mutant barley leaves were the same as in normal tissue.

It had been found that placing normally-grown broad bean plants in the dark followed by re-illumination under normal daylight conditions specifically caused large changes in the levels of trans-3-hexadecenoate in the leaves[5]. These changes have been followed in some detail, together with a simultaneous analysis of the chloroplasts by electron microscopy.

Control sections viewed by electron microscopy showed normal chloroplast structures with starch being produced and stored in granules. The older (expanded) leaves contained chloroplasts which had more granal stacking than the young (expanding) leaves and this was paralleled by the higher absolute levels of trans-3-hexadecenoate in the former[5]. When the plants were dark-treated no change at all was seen in the structure of the chloroplasts from older leaves. In those from younger leaves some distortion of shape and the appearance of small lamellar bodies was seen. Granal stacks, however, were still present in undiminished amounts. The lamellar body may have been formed by the reversion of previously formed structures or by the synthesis of new membranes in the dark. We are inclined to the latter opinion especially since it is known that, for example, young barley plastids show prolamellar bodies even in green tissues if rapidly expanding young leaves are examined at dawn. On re-exposure of the dark-treated bean plants to normal daylight periods, starch is re-synthesised and the morphology is comparable with untreated controls. Etiolated broad bean leaves contained prolamellar bodies with an extremely regular lamellar arrangement. Only one prolamellar body per etioplast was ever observed and a few vesicular inclusions and osmiophilic granules were usually seen within the organelle.

The acyl lipid content and fatty acid distribution of the bean leaves was determined (Table 2). The level of phosphatidylglycerol as a percentage of total lipids remained fairly constant during the various treatments. In contrast, the percentage of its fatty acids which were present as trans-3-hexadecenoic acid declined markedly during dark treatment in both young and old leaves. On re-exposure of the plants to daylight, the amount of the acid increased beyond control levels, especially in the younger leaves. The results confirmed the original observations made with broad beans and light exposure[5].

Thus in broad bean plants there was no obvious correlation between the level of trans-3-hexadecenoic acid and the amount of granal stacking. The wide fluctuations in the levels of the acid which can be induced in expanded leaves were without any obviously corresponding morphological change. In expanding leaves an 82% reduction in the level of the acid is not accompanied by any loss in granal stacking.

TABLE 2

CHANGES IN LEAF trans-3-HEXADECENOIC ACID IN BROAD BEANS GROWN WITH DIFFERENT LIGHT REGIMES

Leaf	Growth conditions	trans-3-Hexedecenoate (μmol/g fresh wt. (% control))
Young	1. Normal daylight	100
	2. As (1) + dark	18
	3. As (2) + normal daylight	213
Mature	1. Normal daylight	100
	2. As (1) + dark	38
	3. As (2) + normal daylight	109

There are further lines of evidence against a connection between trans-3-hexadecenoic acid and granal stacking. For example, work in our laboratory and in others[1,6,8,9] have shown that, although there is usually more trans-3-hexadecenoate in granal membrane fractions, there are also appreciable quantities in stromal lamellar preparations. If the acid were only concerned with granal stacking then one would not expect this. Furthermore, it has been shown[1] that exposure of etioplasts to intermittent light allowed synthesis of trans-3-hexadecenoic acid to normal levels but without the normal development of the thylakoids.

As a result of all this evidence we conclude that the function of trans-3-hexadecenoic acid is probably not connected with granal stacking. Certainly, that cannot be its exclusive function. We have shown a linear correlation between rates of carbon dioxide fixation and the level of trans-3-hexadecenoate in dark/light treated broad beans. In addition, the development of photosystem II activity, measured by DCMU-sensitive oxygen uptake of plastids in the light, correlated with the increase in the fatty acid.[2] We would suggest, therefore, that trans-3-hexadecenoate has a connection with the metabolic reactions of photosynthesis. Whether the acid is needed for efficient photosynthesis, or is merely formed as a by-product, needs further experimentation.

SUMMARY

The results of experiments with chlorophyll b-less barley, broad bean plants grown under different light conditions, chloroplast subfractionation and the greening of barley etioplasts are all against the idea that trans-3-hexadecenoic acid is involved in granal stacking. Instead, we suggest that it may be connected with the reactions of photosynthesis, though its actual role is, at present, unclear.

ACKNOWLEDGEMENTS

M.P.P. and A.O.D. were in receipt of SRC/CASE awards. The financial support of the ARC and the SRC is also gratefully acknowledged. We are also grateful for the skillful assistance of Mr. M. Stubbs of Unilever Research, Colworth for help in the electron microscopy.

REFERENCES

1. Bahl, J., Franke, B. and Moneger, R. (1976) Planta, 129, 193-201.
2. Sellden, G. and Selstam, E. (1976) Physiol. Plant, 37, 35-41.
3. Nichols, B.W., Stubbs, J.M. and James, A.T. (1967) in Biochemistry of

Chlorplasts, Goodwin, T. W. ed., vol. 2, Academic Press, New York and London, pp. 670-690
4. Haverkate, F. and Van Deenen, L. L. M. (1964) Biochim. Biophys. Acta, 84, 106-108
5. Harwood, J. L. and James, A. T. (1975) Eur. J. Biochem., 50, 325-334.
6. Tuquet, C., Guillot-Saloman, T. D., Delubac, M. and Signol, M. (1977) Plant Sci. Lett., 8, 59-64.
7. Bishop, D. G., Andersen, K. S. and Smillie, R. M. (1971) Biochim. Biophys. Acta, 231, 412-414.
8. Douce, R., Holz, R. B. and Benson, A. A. (1973) J. Biol. Chem., 248, 7215-7222.
9. Allen, C. F., Good, P., Trosper, T. and Park, R. B. (1972) Biochim. Biophys. Res. Commun., 48, 907-913.
10. Bolton, P., Wharfe, J. and Harwood, J. L. (1978) Biochem. J., 174, 67-72
11. Spurr, A. R. (1969) J. Ultrastruc. Res., 26, 31-43.
12. Etherington, J. R. (1967) Annals Bot., 31, 653-660.
13. Goodchild, D. J., Highkin, H. R. and Boardman, N. K. (1966) Exp. Cell Res., 43, 684-688.

LIPIDS AND LIPOLYTIC ACTIVITIES IN SPINACH PLASTIDS DURING SEASONAL DEVELOPMENT

ROLF FREY and MANFRED TEVINI
Botanical Institutes of the Universities of Kaiserslautern and Karlsruhe, D-6750 Kaiserslautern, D-7500 Karlsruhe, Kaiserstrasse 12, F.R.G.

ABSTRACT

Plastids of four spinach leaf stages were analyzed according to their size, chlorophyll-, protein-, galacto- and phospholipid content. After in vitro aging of the isolated thylakoids the observed lipid changes were found to be due to the activity of several lipolytic enzymes.

INTRODUCTION

During the development from proplastids to gerontoplasts the plastids always modify their morphological structure and biochemical composition. This change is caused by anabolic, metabolic and catabolic processes relating to the photosynthetic apparatus.

We have good information about the biochemical components of plastids, especially about the phospho- and glycolipids during light induced chloroplast development[1]. But there is hardly anything known about the alteration of the thylakoid lipid contents during a period of vegetation and the enzymatic activities involved. These problems will be discussed in this paper.

MATERIALS AND METHODS

Spinach was grown in the garden. The investigations were made after the following four phases of development:

Stage I   : 5 weeks after sewing (small light-green leaves)
Stage II   : 8 weeks after sewing (fully expanded green leaves)
Stage III : 10 weeks after sewing (green-yellow leaves)
Stage IV   : 12 weeks after sewing (yellow-green leaves)

Chloroplasts were isolated after a modified method of Park[2] and Pon 1961. Thylakoid material could be separated from the stroma material by centrifuging broken chloroplasts at 13000 x g for 20 minutes and washing the pellet (thylakoid material). A Coulter Counter was used to count the chloroplasts and to determine their size. The lipids were extracted by the method of Bligh[3] and Dyer 1959. The method of Roughan[4] and Batt 1969 was used to determine quantitatively the lipids after two-dimensional TLC on silica gel H (Merck 5721) in chloroform : methanol : acetic acid : aqua dest. 65 : 25 : 15 : 3 and in aceton : acetic acid 85 : 25. Chlorophylls were measured by the method of Arnon 1962[5], soluble proteins by the method of Lowry 1951[6]. Insoluble proteins were determined by weighing after extraction of pigments, lipids and water-soluble components. The incubation experiments were made in 0.2 M acetate buffer pH 4.6 and in a potassium buffer pH 7.5. The buffers contained 40 mM calcium chloride. The incubation temperature was 25°C, the incubation time 2 hours. The incubation was stopped by 1.5 M TCA.

RESULTS

a) <u>Distribution of the plastid size</u>. The size of plastids is quite different in the four investigated phases of development. Chloroplasts of young spinach leaves are 4 - 9 μm in diameter without any detectable maximum of size distribution. Fully expanded green leaves and green-yellow leaves contain plastids which are in the main about 4 μm in diameter whereas gerontoplasts of yellow-green leaves are mostly about 6 μm in diameter (Fig. 1).

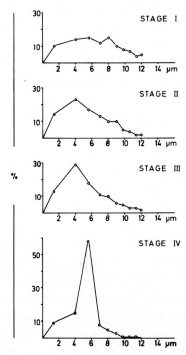

Fig. 1. Distribution of the plastid size of plastids deriving from leaves in four different phases of development.

b) <u>Chlorophylls and Proteins</u>. The chlorophyll content increases from the early developmental stage up to a maximum in plastids of fully expanded green leaves (stage II) and decreases when degeneration sets in. The soluble and insoluble proteins show the same behaviour (Table 1).

TABLE 1
CONTENTS OF PROTEINS (SOLUBLE AND INSOLUBLE) AND CHLOROPHYLLS IN PLASTIDS DERIVING FROM LEAVES IN FOUR DIFFERENT PHASES OF DEVELOPMENT
Data in µg / 1 x 10^6 Plastids

stage of development	I	II	III	IV
soluble proteins	6.4	46.1	14.2	4.7
insoluble proteins	21.0	47.0	15.0	3.2
chlorophylls	1.7	2.5	0.4	0.04

c) <u>Galacto- and phospholipids of thylakoids</u>. According to the chlorophylls and proteins the amount of the galactolipids MGDG and DGDG in thylakoid fractions of plastids increases from the first phase of development to the second and decreases with the initial degeneration of the plastids. In fully expanded green leaves the proportion of MGDG to DGDG is 1.6 and much higher than in other stages (Table 2). The main components of the phospholipids are as expected PG and PC. However, only in the thylakoid fraction of young plastids the contents of PC are higher than those of PG. PA can be measured in thylakoids of all four investigated stages. The value of PI is remarkably high in gerontoplasts.

TABLE 2
LIPID CONTENT OF THYLAKOIDS FROM PLASTIDS IN 4 DIFFERENT STAGES OF DEVELOPMENT
Data in µg/ 10^6 Plastids

	Stage I	Stage II	Stage III	Stage IV
MGDG	4.1	15.2	4.1	3.8
DGDG	3.8	9.6	4.6	4.6
SL	0.6	3.5	0.4	0.5
PG	0.1	2.8	0.7	0.6
PC	0.15	1.4	0.5	0.5
PE	0.05	0.2	0.1	0.3
PI	tr	0.5	0.05	0.9
PA	0.2	0.5	0.25	0.23
LPC	0.01	-	-	0.03
MGDG:DGDG	1.1	1.6	0.9	0.8

d) <u>Galacto- and phospholipids of aged thylakoids.</u> For studying the lipolytic enzyme activities in vitro we used the already described thylakoid fractions (Table 2) and incubated them for 2 hours at pH 7.5. As seen in Table 3 the percentage amounts of all glycolipids are more or less lower in aged thylakoids than in the controls. The extent of the decrease of the galactolipids is very high in stage I- and stage IV-thylakoids. In young plastids we found a high amount of PG whereas in stage IV-plastids the amount of PC is increased. This is due to the activity of Phospholipase D transphosphatidylation[7]. In chloroplasts of stage II and III PG, PC and PA are hydrolized to a high extent.

TABLE 3
PERCENTAGE AMOUNTS OF LIPIDS IN THYLAKOIDS AFTER 2 HOURS AGING
Incubation at pH 7.5, control directly after the isolation = 100 %

	Stage I	Stage II	Stage III	Stage IV
MGDG	27.3	37.9	89.8	67.9
DGDG	57.4	81.2	84.3	37.8
SL	36.7	98.6	65.0	68.0
PG	440.0	75.0	14.3	35.0
PC	113.3	45.7	16.0	168.0
PE	96.0	120.0	0.0	106.7
PA	30.0	32.0	60.0	126.1
PI	0	96.0	180.0	91.1
LPC	0	-	-	0

The results indicate that the thylakoids obviously contain enzymes for the lipid metabolism; in the main hydrolytic enzymes can be observed. They can be regarded as a phospholipase A which can hydrolyze one fatty acid from the lipid molecule, a phospholipase D and a lipolytic acylhydrolase which can liberate both fatty acids from galacto- and phospholipids. These enzymes can also be observed in a crude phospholipase D homogenate made of stroma and envelope material[7]. Besides these lipolytic activities there are enzyme activities in the thylakoids which can cause an increase of certain lipids during incubation such as PG, PC, PA and PI. A localization and characterization of these activities is described elsewhere[7].

SUMMARY

During their seasonal development from proplastids to gerontoplasts the plastids are submitted to a morphological and physiological change. Their size alters from about 4 µm in diameter in the early investigated stages of development to about 6 µm in diameter in the phase of plastid degeneration (Fig. 1). The chlorophylls similar to soluble and insoluble proteins reach their highest level in plastids isolated from fully expanded green leaves (Table 1).

The galactolipids MGDG and DGDG and the phospholipids PG, PC and PA obtain their highest amounts in thylakoids of ripe green plastids (stage II, Table 2). In contrast to this the phospholipids PE, PI and LPC reach their highest values in thylakoid fractions of yellow-green plastids. All investigated thylakoid fractions show remarkable amounts of PA. Lipolytic enzyme activity changing the contents of phospho- and galactolipids can be observed in vitro after two hours aging in all the thylakoid fractions studied (Table 3). Acylhydrolases, Phospholipase D and galactolipases are working in vitro with different activities in the different thylakoids isolated from physiologically different plastids.

ACKNOWLEDGEMENTS

This work was supported by the Deutsche Forschungsgemeinschaft. We are very grateful to Mrs. U. Widdecke for her valuable assistance.

REFERENCES

1. Tevini, M. (1977) Lipids and Lipid Polymers in Higher Plants, Tevini, M. and Lichtenthaler, H.K., eds., Springer, Heidelberg, pp. 121-140.
2. Park, R.B. and Pon, N.G. (1961) J. Mol. Biol. 3, 1-10.
3. Bligh, E.G. and Dyer, W.J. (1959) Can. J. Biochem. Physiol. 37, 8, 911-917.
4. Roughan, P.G. and Batt, R.D. (1968) Analytical Biochem. 22, 74-88.
5. Arnon, D.J. (1949) Plant Physiol. 24, 1-15.
6. Lowry, O.H. et al. (1951) J. Biol. Chem. 193, 265-275,
7. Frey, R. (1977) Thesis in: Karlsruher Beiträge zur Entwicklungsphysiologie der Pflanzen, Tevini, M., ed., University of Karlsruhe.

THE CONVERSION OF 9-D- AND 13-L-HYDROPEROXYLINOLEIC ACID BY SOYBEAN LIPOXY-GENASE-1 UNDER AEROBIC CONDITIONS

JAN VERHAGEN, GERRIT A. VELDINK, JOHANNES F.G. VLIEGENTHART AND JAN BOLDINGH
Laboratory for Organic Chemistry, University of Utrecht, Croesestraat 79, 3522 AD Utrecht (The Netherlands)

ABSTRACT

A novel pathway for the conversion of hydroperoxylinoleic acid by soybean lipoxygenase-1 is described. During the reaction oxygen is consumed (0.7 mol/mol 13-L-LOOH[*]; 0.33 mol/mol 9-D-LOOH) which may explain that in the presence of oxygen much more polar material is formed than in its absence. Besides, epoxy-hydroxy- and oxodienoic fatty acids are formed. However, under aerobic conditions no 13-oxo-tridecadienoic acids are produced from 13-L-LOOH.

INTRODUCTION

Lipoxygenase-1 from soybeans catalyses the formation of fatty acid hydroperoxides from poly-unsaturated fatty acids (e.g. linoleic acid) and molecular oxygen. When oxygen is depleted another type of reaction starts viz. the coupled conversion of the fatty acid and the fatty acid hydroperoxide. In recent years the mechanism of these reactions has been studies with various techniques e.g. EPR and optical spectroscopy[1,2]. It has been established that, depending on the reaction conditions, different enzyme species can occur[3]. Following an investigation of the kinetics of the anaerobic reaction[4] a mechanism has been proposed that accounts for both the spectroscopic and kinetic evidence.

Recently, two novel pathways for lipoxygenase-catalysed conversions of hydroperoxylinoleic acid have been discovered[5,6]. When lipoxygenase-1 is incubated with either 13-L- or 9-D-hydroperoxylinoleic acid approximately 40% of the hydroperoxide is converted into oxodienes absorbing at 285 nm. Under such anaerobic conditions also more polar compounds are formed, the main constituent of which was shown to be *threo*-11-hydroxy-*trans*-12,13-
-epoxy-9-*cis*-octadecenoic acid.

[*]13-L-LOOH: 13-L-hydroperoxylinoleic acid
 9-D-LOOH: 9-D-hydroperoxylinoleic acid

If, however, oxygen is admitted to a system containing lipoxygenase and hydroperoxylinoleic acid the reaction rate is considerably lower and a partly different set of products are formed. This paper describes the latter reaction in more detail.

MATERIALS AND METHODS

Lipoxygenase-1 was isolated from soybeans according to Finazzi-Agrò et al.[7], spec. activity: 230 µmol linoleic acid oxygenated per min per mg. 13-L-LOOH and 9-D-LOOH were prepared as described previously[5] and purified by high performance liquid chromatography[4]. Spectrophotometric measurements were carried out in a 1 cm pathlength cuvette at 25 C. The reaction was started by the addition of 200 µg of enzyme to 2.5 ml of a hydroperoxide solution in airsaturated 0.1 M sodiumborate buffer (pH 9.0). The change in absorbance at 285 nm or 234 nm was recorded with a Cary 118C spectrophotometer. Spectra were taken before and after reaction. Oxygen uptake was measured with a Clark oxygen electrode connected to a GME Oxygraph model KM (GME, Middleton, Wis., U.S.A.). The measuring cuvette was completely filled (2.4 ml) with air-saturated 0.1 M sodiumborate buffer (pH 9.0) and sealed with a rubber cap.

Preparative experiments were carried out by incubating 10 mg 13-L-LOOH or 9-D-LOOH with 32 mg lipoxygenase-1 in 400 ml air-saturated 0.1 M sodiumborate buffer (pH 9.0) at 25 C (LOOH: 80 µM, oxygen: 240 µM, enzyme: 0.8 µM). After completion of the reaction (9-D-LOOH: approx. 2.5 h, 13-L-LOOH approx. 17 h) the solution was acidified to pH 2 and extracted with diethylether. The extract was washed with water, dried and concentrated. The residue was esterified with diazomethane and subjected to thinlayer chromatography on 0.25 mm pre-coated silicagel plates (Silicagel 60 F-254, 20x20 cm, E. Merck, A.G., Darmstadt, Germany). Separation of the products was performed with the solvent system hexane-diethylether (60:40, $^v/v$). The compounds were located by spraying with phosphomolybdic acid (5% in ethanol, $^w/v$). 2,4-Dinitrophenylhydrazine (0.4% in 2 M HCl, $^w/v$) was used to identify carbonyl compounds and an acidified ferrous thiocyanate (0.2 g NH_4SCN in 15 ml acetone plus 10 ml of an acidified solution of $FeSO_4$ in water, 4% ($^w/v$))to identify hydroperoxy-compounds.

RESULTS AND DISCUSSION

Fig. 1 shows progress curves for the conversion of 13-L-LOOH by lipoxygenase-1 starting at an oxygen concentration of 5 µM.

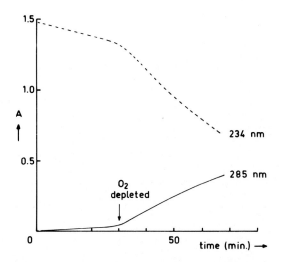

Fig. 1. Progress curves for the conversion of 13-L-LOOH by lipoxygenase-1 recorded at 234 nm and 285 nm. 13-L-LOOH: 60 µM, initial oxygen concentration: 5 µM, enzyme: 0.8 µM, 0.1 M sodiumborate buffer (pH 9.0), t = 25 C. Measurements were carried out as described previously (5).

From these curves recorded at 234 and 285 nm it appears that oxygen inhibits the rate of formation of oxodienoic acids (absorbing at 285 nm) as well as the rate of conversion of hydroperoxide itself. Moreover, the yield of oxodienoic acids per mol converted 13-L-LOOH was found to be reduced in the presence of oxygen. Apparently, oxygen reacts with intermediate radicals which leads to the formation of products other than oxodienoic acids. Therefore, the existence of a lag period in the anaerobic conversion of 13-L-LOOH by lipoxygenase-1 can be attributed to the inhibition by traces of oxygen, which are slowly consumed. In the presence of 240 µM of oxygen the conversion of hydroperoxide as well as the formation of oxodienoic acids were indeed found to proceed much slower than in the absence of oxygen (Table 1).

No reaction was observed with heat-inactivated lipoxygenase-1 (5 min at 85 C in a 2% sodiumdodecylsulphate solution). Fig. 2 shows ultra-violet spectra before and after the conversion of 13-L- and 9-D-LOOH. It appears that from 13-L-LOOH under aerobic conditions considerably less oxodienoic acids are formed than under anaerobic conditions (8% and 41% respectively), whereas with 9-D-LOOH the difference is less pronounced (see Table 1).

TABLE 1

COMPARISON OF THE AEROBIC AND ANAEROBIC CONVERSIONS OF 13-L-LOOH AND 9-D--LOOH BY LIPOXYGENASE-1

Anaerobic measurements were carried out as described previously (5) and aerobic experiments as described in Materials and Methods. 13-L-LOOH or 9-D-LOOH: 80 µM, oxygen in the aerobic experiments: 240 µM, enzyme: 0.8 µM (spec. act.: 230 µmol linoleic acid converted per min per mg) and 1.0 µM (spec. act.: 160 µmol linoleic acid converted per min per mg) in the aerobic and the anaerobic experiments respectively.
Buffer: 0.1 M sodiumborate (pH 9.0), t = 25 C.

	spec. act. nmol LOOH $min^{-1}mg^{-1}$	A_{285} formed %	A_{234} remaining %	O_2 uptake µM	spec. act. nmol O_2 $min^{-1}mg^{-1}$
13-L-LOOH	12	41	16	-	-
9-D-LOOH[a]	266	36	12	-	-
13-L-LOOH	3.0	8	16	56	2.3
9-D-LOOH[b]	20	26	15	26	6.6

[a] anaerobic
[b] aerobic

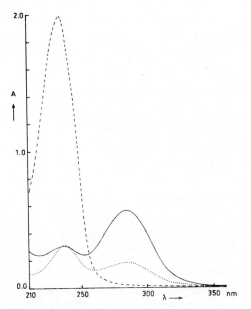

Fig. 2. Ultra-violet spectra before (-----) and after the aerobic conversion of 80 µM 13-L-LOOH (·····) and 80 µM 9-D-LOOH (———) by 0.8 µM lipoxygenase-1 in air-saturated 0.1 M sodiumborate buffer (pH 9.0) at 25 C.

The extent of the oxygen uptake during the reaction was found to be proportional to the concentration of the hydroperoxides, but it appeared to be different for the two isomeric hydroperoxides: 0.7 mol oxygen per mol 13-L-LOOH and 0.33 mol oxygen per mol 9-D-LOOH were consumded (Table 1). Fig. 3 shows thinlayer chromatograms of the products formed in the aerobic conversion of 13-L-LOOH and 9-D-LOOH by lipoxygenase-1.

To obtain information about the relative amounts of the products formed from 13-L-LOOH, incubations were carried out with (U-^{14}C)-labelled 13-L-LOOH as substrate. The results are given in Table 2.

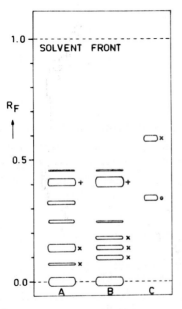

Fig. 3. Thin-layer chromatograms of the products formed in the aerobic conversion of 13-L-LOOH (A) and 9-D-LOOH (B) by lipoxygenase-1.
C: methyl-esters of 13-L-LOOH and 9-D-LOOH (lower band) and linoleic acid (upper band).
The compounds were located by spraying with phosphomolybdic acid. The compounds were visible under UV light unless indicated with: X
+: band coloured orange when sprayed with dinitrophenylhydrazine.
O: band coloured red when sprayed with ferrous thiocyanate.

TABLE 2
CONVERSION OF 13-L-[U-^{14}C]-LOOH BY LIPOXYGENASE-1 UNDER AEROBIC CONDITIONS
Experiments were carried out as described previously for anaerobic conditions[5]. Incubation time: 24 h. [13-L-(U-^{14}C)-LOOH]: 76 µM, [enzyme]: 4 µM, [oxygen]: 240 µM. buffer 0.1 M sodiumborate (pH 9.0). t = 25 C.

R_F value	% of total radioactivity	tentative assignment
origin	63.1	oxidative polymers
0.15	16.9	epoxy-hydroxy-octadecenoic acid
0.24	7.7	unknown
0.32	2.3	unknown
0.42	8.5	13-oxo-octadecadienoic acid
0.46	1.5	unknown

From the absence of 13-oxo-tridecadienoic acids it can be concluded that oxygen interferes with the chain-fission reaction which is observed in the anaerobic conversion. However, 13-oxo-octadecadienoic acid (R_F=0.42) appears to be formed in both the aerobic and the anaerobic reaction. The lower yield under aerobic conditions observed in the spectrophotometric experiments (Table 1) was confirmed by the results of the experiments with (U-^{14}C)-labelled 13-L-LOOH (Table 2). Considerably more material (63%) can be found at the origin under aerobic conditions suggesting that in the presence of oxygen oxidative polymerization might have occurred (0.7 mol oxygen consumed per mol 13-L-LOOH).

In case of 9-D-LOOH the product-pattern observed in the aerobic reaction does not differ markedly from that of the anaerobic reaction, but also here material has accumulated at the origin, probably for the same reason as with 13-L-LOOH as substrate.

ACKNOWLEDGEMENTS

The authors wish to thank Mrs. G.P.M. Rijke-Schilder for skilful technical assistance. This investigation was supported in part by the Netherlands Foundation for Chemical Research (SON) with financial aid from the Netherlands Organization for the Advancement of Pure Research (ZWO).

REFERENCES

1. De Groot, J.J.M.C., Garssen, G.J., Veldink, G.A., Vliegenthart, J.F.G., Boldingh, J. and Egmond, M.R. (1975) FEBS Lett. 56, 50-54.
2. De Groot, J.J.M.C., Veldink, G.A., Vliegenthart, J.F.G., Boldingh, J., Wever, R. and Van Gelder, B.F. (1975) Biochim. Biophys. Acta 377, 71-79.
3. Egmond, M.R., Fasella, P.M., Veldink, G.A., Vliegenthart, J.F.G. and Boldingh, J. (1977) Eur. J. Biochem. 76, 469-479.
4. Verhagen, J., Veldink, G.A., Egmond, M.R., Vliegenthart, J.F.G., Boldingh, J. and Van der Star, J. (1978) Biochim. Biophys. Acta 529, 369-379.
5. Verhagen, J., Bouman, A.A., Vliegenthart, J.F.G. and Boldingh, J. (1977) Biochim. Biophys. Acta 486, 114-120.
6. Verhagen, J. (1978) Thesis, University of Utrecht, The Netherlands.
7. Finazzi-Agrò, A., Avigliano, L., Veldink, G.A., Vliegenthart, J.F.G. and Boldingh, J. Biochim. Biophys. Acta 326, 462-470.

LIPID METABOLISM DURING THE REGREENING OF THE CHAETOPHORALEAN GREEN ALGA
FRITSCHIELLA TUBEROSA IN AXENING CULTURE

MICHAEL WETTERN

Institut für Allgemeine Botanik und Botanischer Garten der Universität Hamburg, Jungiusstr. 6, D-2000 Hamburg-36 (F.R. Germany)

INTRODUCTION

The filamentous and branched green alga *Fritschiella tuberosa* IYENGAR, growing by means of apical cells, is able to produce secondary carotenoids in N-limited cultures as other green algae do[1]. This process of aging is accompanied by variations in morphology and fine structure and results in the production of thick-walled akinetes filled with cytoplasmic orange-red lipid globules. After transferring these orange-red algae to a complete nutrient medium, the changes occuring during regreening under continuous white light illumination and axenic conditions were examined.

REGREENING

Only a few hours after the start of the regreening process the orange-red akinetes germinated with green protuberances, followed by mitotic divisions and cytokinesis leading to long and thin-walled filaments. Two days later the alga exhibits the characteristics of the logarithmically growing *Fritschiella tuberosa* thallus in liquid culture.

Variation in pigment content

Whereas the synthesis of chlorophyll a starts after a lag phase of 6 hours, an increase of the chlorophyll b content could only be observed after 12 hours, and the leval of total carotenoids remains unchanged for 20 hours after the start of the experiment (Fig. 1). Then the ratio of chlorophyll a plus chlorophyll b to carotenoids reaches the value 1, in this stage the alga is called regreened, although traces of secondary carotenoids (astaxanthin, echinenone, cantaxanthin, adonixanthin and fritschiellaxanthin[1,2]) can still be observed 24 hours after the beginning of the experiment. From the start of regreening the ratio of chlorophyll a to chlorophyll b increases from 0.9 to 3.2 within 20 hours, then it remains constant. During an experiment of 108 hours duration the chlorophyll a content increased about 30fold, that of chlorophyll b about 9fold and that

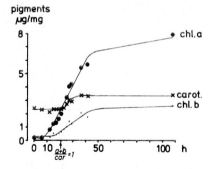

Fig. 1 Change in the pigment content during the regreening of *Fritschiella tuberosa*

of total carotenoids only to a slight extent.

Variation in total lipid and triacylglycerol content

After a lag phase of about 12 hours the total lipid content (percentage of dry weight) and the content of triacylglycerol begins to decrease (Fig. 2). Both show a strong decrease within a following period of 30 hours. It is likely that the decrease of the total lipid content is caused only by the decrease of the triacylglycerol content.

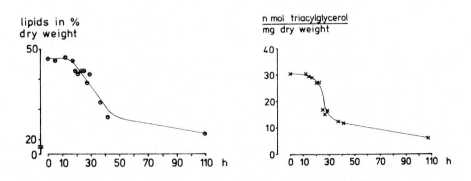

Fig. 2 Change in the total lipid and triacylglycerol content during the regreening of *Fritschiella tuberosa*

Variation in glycolipid and phospholipid content

The synthesis of monogalactosyl diacylglycerol, which is the most abundant glycolipid, starts after a lag phase of about 16 hours (Fig. 3). At that time a marked decrease of triacylglycerol takes place. Therefore, it may be assumed that the triacylglycerol, localized in the cytoplasm, is partially utilized in the synthesis of glycolipids, the biosynthesis of which is localized in the chloroplast membranes[3].

Phospholipids are sythesized in the beginning of the experiment though to a different degree. The most conspicuous variation in the phospholipid content is exhibited by phosphatidyl choline. This lipid reaches its maximal concentration after 12 hours, before the synthesis of monogalactosyl diacylglycerol starts. This might indicate the donor function of phosphatidyl choline for unsaturated C-18 fatty acids in the biosynthesis of galactolipids in *Fritschiella tuberosa* as it was also stated for *Cucurbita pepo*, *Zea mays*, and *Vicia faba*[6,7].

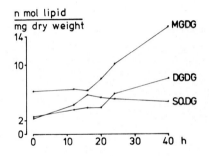 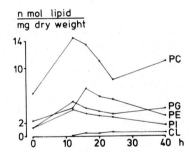

Fig. 3 Change in the glycolipid and phospholipid content during the regreening of *Fritschiella tuberosa*

Variation in relative fatty acid content

The main fatty acids of *Fritschiella tuberosa* are linoleate, α-linolenate, oleate, palmitate and the hexadecatrienoic and -tetraenoic acid (Fig. 4). No acids with a greater number of carbon atoms than the α-linolenic acid could be detected. With the time elapsed since the start of the regreening process, the relative concentration of α-linolenic acid increased, whereas that of palmitic acid increased after a lag phase of about 12 hours. An evident decrease of oleic and linoleic acid could be observed after about 20 hours.

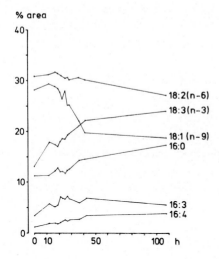

Fig. 4 Change in the relative fatty acid content during the regreening of *Fritschiella tuberosa*

Variation in molar ratios

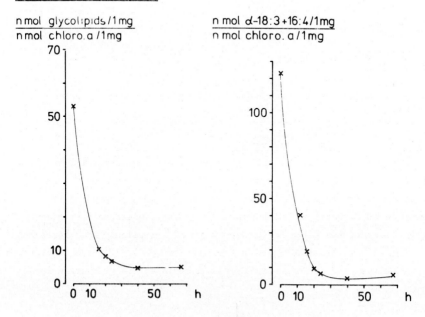

Fig. 5 Change in the molar ratios of glycolipid to chlorophyll a and α-linolenic acid and hexadecatetraenoic acid to chlorophyll a during the regreening of *Fritschiella tuberosa*

The diagrams in figure 5 show results indicating the possible funktion of glycolipids and their unsaturated fatty acids on the stabalization of chlorophyll a as it was postulated for $Euglena^8$ and $Pisum\ sativum^9$. For Fritschiella tuberosa, 30 hours after the beginning of regreening a relation of four to six molecules of glycolipids forming the matrix to one molecule chlorophyll a can be calculated. At the same time a constant molar ratio of α-linolenic and hexadecatetraenoic acid to chlorophyll a was found to be six to eight.

FINAL REMARKS

As a result of the lipid transformation during the regreening of Fritschiella tuberosa two phases can be defined.

The first phase commences with the start of regreening. This results in the synthesis of α-linolenic acid and chlorophyll a which is reflected by changes in the activity of the enzymes involved in their synthesis. Phosphatidyl choline participates earlier in the regeneration than do the glycolipids.

The second phase probably requires the induction of enzyme proteins which leads to a decrease in the triacylglycerol content wihtin 12 hours after the onset of regreening. A parallel increase in monogalactosyl diacylglycerol is also observed.

The stated results, especially the molar ratios, mark the end of the regreening process in Fritschiella tuberosa after approximately 30 hours. Until that time the insertion of all chloroplast lamellae lipids necessary for the formation of intact thylakoids is completed. After this time the chloroplast merely enlarges. Therefore, in Fritschiella tuberosa the regeneration during the regreening comes to an end about the same time as it was observed in $Chlorella\ fusca^{10}$. On the other hand the capability and the duration of the regreening certainly depends on the physiological condition of the algae. In older cultures regreening is retarded increasingly. There might be a "point of no return"[11], when regreening will be impossible.

REFERENCES

1. Weber, A. (1975) Arch. Microbiol., 102, 45-52
2. Buchecker, R., Eugster, C.H. and Weber, A. (in press) Helv. Chim. Acta
3. Douce, R. (1974) Science, 183, 852-853
4. Roughan, P.G. (1975) Lipids, 10, 609-614
5. Slack, C.R. and Roughan, P.G. (1975) Biochem. J., 152, 217-228

6. Trêmolières, A. and Dubaq, J.P. (1976) in abstracts of the symposium "Lipids and Lipipolymers in Higher Plants", Karlsruhe (F.R.Germany)

7. Williams, J.P., Watson, G.R. and Leung, S.P.G. (1976) Plant Physiol., 57, 179-184

8. Rosenberg, A. (1967) Science, 157, 1191-1196

9. Trêmolières, A., Jaques, R. and Mazliak, P. (1973) Physiol. Vég., 11, 239-251

10. Porra, R.J. and Grimme, L.H. (1974) Arch. Biochem. Biophys., 164, 312-321

11. Huber, D.J. and Newman, D.W. (1976) J. Exp. Bot., 27, 490-511

A PHOSPHOLIPID EXCHANGE PROTEIN FROM THE ENDOSPERM OF GERMINATING CASTOR BEAN SEEDS

TOSHINORI TANAKA and MITSUHIRO YAMADA
Department of Biology, University of Tokyo, Komaba, Meguro, Tokyo 153 (Japan)

ABSTRACT

When ^{14}C-labeled microsomes prepared from the endosperms of 4-day old castor bean seedlings were incubated with unlabeled mitochondria, the loss of ^{14}C-lipids from microsomes was accompanied by the increase of ^{14}C-lipids in mitochondria. The addition of castor bean cytosol from the same tissues markedly enhanced the lipid transfer. The lipid transfer factor was fractionated in the supernatant obtained by acidifying the cytosol to pH 5.1 and subsequently salted out by 75 % saturated ammonium sulfate. Further purification of the factor resulted in a good yield of the lipid transfer protein, the specific activity of which was 70 folds as high as that of potato tuber lipid exchange protein in Sephadex eluate step. The purified protein transferred phosphatidyl choline specifically. The reverse transfer from mitochondria to microsomes was as high as the transfer from microsomes to mitochondria and also specific for phosphatidyl choline. Thus, the purified protein is a phosphatidyl choline exchange protein; an apparent molecular weight was about 20,000 by gel filtration method. Castor bean phosphatidyl choline exchange protein was able to transfer phospholipids of castor bean microsomes to mitochondria from the other species such as potato tuber, califlower inflorescence, pumpkin hypocotyl and rat liver, but not to *Avena* etioplasts. The castor bean phospholipid exchange protein transferred phospholipids from potato tuber microsomes to potato tuber mitochondria.

INTRODUCTION

Recent studies[1-3] confirmed that the endoplasmic reticulum is the site of phospholipid synthesis in endosperms of germinated castor bean seeds. Castor bean mitochondrial membranes are rich in phosphatidyl choline and phosphatidyl ethanolamine, but this organelle does not contain the enzymes responsible for their syntheses. On the other hand, feeding experiments with ^{14}C-choline, ^{14}C-acetate and ^{3}H-glycerol in castor bean endosperms show that the label first appeared in the endoplasmic reticulum and subsequently in mitochondria and glyoxysomes[4,5]. These results suggest that a phospholipid exchange protein(s), as in animal tissues[6] and potato tubers[7] functions between castor bean organelles

This paper reports the presence of a phospholipid exchange protein in castor bean cytosol and the properties of the partially purified protein.

MATERIALS AND METHODS

Castor bean endosperm halves from 4-day old seedlings were fed with [1-^{14}C]-acetate(1 μCi/g fresh tissues) at 30°C for 4 hr and ground with three volumes of grinding medium consisting of 0.4 M sucrose, 1 mM EDTA, 1 mM $MgCl_2$, 10 mM KCl, 10 mM mercaptoethanol and 50 mM phosphate buffer(pH 7.5), and squeezed through two layer of nylon mesh. The homogenate was centrifuged at 10,000xg for 20 min and the supernatant obtained was further centrifuged at 105,000xg for 1 hr to give a microsomal pellet, which was suspended with the griniding medium(1×10^6 cpm/5 mg protein/ ml). Crude mitochondria were obtained by 10,000xg centrifugation for 20 min from unlabeled endosperm halves. The pellet suspension was put on a sucrose gradient(35 to 60 % w/w, 40 ml) and centrifuged at 22,000 rpm for 4 hr. Purified mitochondrial fraction was collected from 42 % sucrose layer, washed and suspended with the grinding medium(10 mg protein/ml).

Unlabeled mitochondria(5 mg protein) and ^{14}C-labeled microsomes(1 mg protein) were suspended with or without the lipid exchange protein in a total volume of 3 ml at 30°C for 15 min. The mixture was then diluted with 5 ml of cold grinding medium and centrifuged at 10,000xg for 20 min. The supernatant was used as microsomal fraction and the pellet as the mitochondrial fraction after washing. Contamination of the isolated mitochondria with microsomes(cross-contamination) was about 7 % of the lipid exchange rate obtained, according to the exchange assay with ^{3}H-leucine-labeled microsomes and unlabeled mitochondria[7]. Lipids were extracted from both fractions by the procedure of Bligh and Dyer[8]. Lipid exchange rate of ^{14}C-lipids(%) transferred from microsomes to mitochondria was corrected for both cross contamination and lipid loss in the organelle fractionation, according to Kader[7].

RESULTS AND DISCUSSION

Fig. 1 shows time course when ^{14}C-labeled microsomes were incubated with unlabeled mitochondria in the absence of castor bean cytosol. The transfer of ^{14}C-lipids from microsomes to mitochondria was increased with time and temperature of the incubation. Of interest is that the lipid transfer occurs even at 0°C, although the similar was obtained in the lipid transfer between potato tuber's organelles[7]. However, the lipid transfer from microsomes to mitochondria was markedly facilitated by the addition of castor bean cytosol prepared from endosperms of 4-day old seedlings as shown in Table 1.

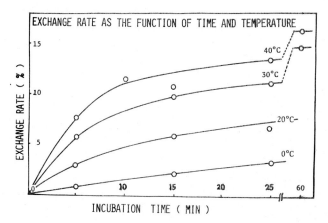

Fig. 1 Lipid transfer from microsomes to mitochondria in the absence of castor bean cytosol

Table 1 Effect of castor bean cytosol on the lipid transfer from microsomes to mitochondria[9]

Fraction added(mg protein)	Exchange rate(%)	Specific exchange rate[a]
None	8.5	
105,000xg supernatant(6.1 mg)	12.0	0.57
105,000xg supernatant(12.1 mg)	14.8	0.52
105,000xg supernatant(17.8 mg)	17.8	0.51

[a] Exchange rate / mg protein

Table 2 Purification of lipid transfer factor from castor bean cytosol [9]

Purification Step	Specific[a] Activity	Recovery %	Purification Factor	Specific Activity[a] of Potato PLEP[b]
105,000g Sup	2.4	100	1	
pH 5.1 Sup	6.2	56.2	2.1	
$(NH_4)_2SO_4$ ppt	32.1	20.2	13.4	
Sephadex G 100 eluate	72.6	7.3	30.3	1.3
DEAE-Cellulose eluate				5.2

[a] μg phospholipids transferred/mg protein/ 15 min
[b] Phospholipid exchange protein purified from potato tubers by Kader[7]

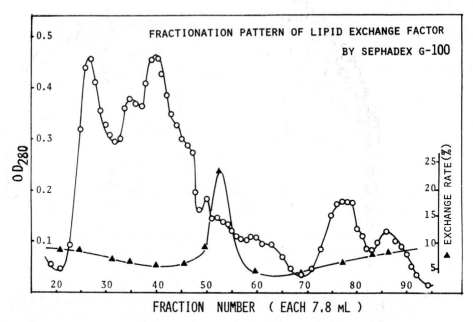

Fig. 2 Fractionation pattern of castor bean lipid transfer protein from Sephadex G-100 column[9]. The ammonium sulfate fraction was dialyzed, put on the column(4 X 50 cm) and eluted with 50 mM phosphate buffer(pH 7.5) containing 10 mM mercaptoethanol.

Thus, the lipid transfer factor contained in castor bean cytosol was purified according to the procedure of Wirtz and Zilversmit[10]. As shown in Table 2, the factor was fractionated in the supernatant after removing acid(pH 5.1) precipitate and subsequently salted out with 75 % saturation of ammonium sulfate. Ammonium sulfate fraction was further purified by gel filtration of Sephadex G-100. The lipid transfer protein was eluted from the column as a sharp peak, as shown in Fig. 2. The specific activity of purified lipid transfer protein increased from 2.4 in the 105,000xg supernatant to 72.6 in the Sephadex eluate; the specific activity was 70 times as high as that of potato's phospholipid exchange protein in Sephadex eluate step[7]. There was no inhibitory factor which was observed in lipid exchange between potato tuber organelles.

Recently Morgan and Moore reported a phospholipid transfer protein from castor bean endosperms. This protein transferred phosphatidyl choline from microsomes to mitochondria, but not from mitochondria to microsomes[11]. Fig. 3 shows that our protein from castor bean endosperms functions for the reversible exchange between microsomes and mitochondria, as is the cases of potato tuber[7] and manmalian organs such as rat liver and bovine heart[6]. The decrease of ^{14}C-lipids in microsomes was accompanied by the increase of that in mitochondria and the decrease of ^{14}C-lipids in mitochondria by the increase of that in microsomes, when the purified protein added was increased.

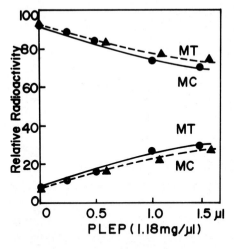

Fig. 3 Effect of castor bean phospholipid transfer protein on the rates of lipid transfer between microsomes(Mc) and mitochondria(Mt)[9].

Solid line: Decrease of ^{14}C-lipids in microsomes accompanied by increase of ^{14}C-lipids in mitochondria
Dotted line: Decrease of ^{14}C-lipids from mitochondria followed by increase of ^{14}C-lipids in microsomes
In the reaction mixture labeled microsomes or mitochondria(1 mg Protein) and unlabeled mitochondria or micorsomes were used.

Of interest is which lipid class is transferred by castor bean lipid exchange protein. The previous paper demonstrated that the lipid transfer was restricted to phospholipids among glycerolipids in microsomes[13]. Furthermore, Fig. 4 shows that castor bean lipipid exchange protein specifically exchanged phosphatidyl choline between microsomes and mitochondria, since the increase of

Fig.4 Effect of castor bean lipid exchange protein on the exchange of pnospholipids between microsomes and mitochondria[9].

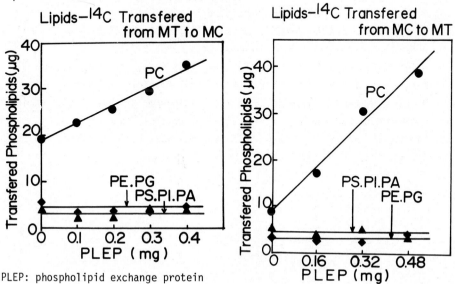

PLEP: phospholipid exchange protein

phospholipid exchange protein added resulted only in the increment of phosphatidyl choline from microsomes to mitochondria and vice versa.

Table 3 Organelle specificity of castor bean phospholipid exchange protein[9]

Lipid acceptor from castor bean microsomes	Specific exchange rate
Castor bean mitochondria	5.8 - 6.2
Potato tuber mitochondria	4.6 - 5.8
Cauliflower mitocondria	4.1 - 4.6
Pumpkin hypocotyl mitochondria	2.6 - 3.6
Rat liver mitochondria	0.4 - 1.8
Avena etioplasts	0

Table 4 indicates that castor bean phospholipid exchange protein possesses the capacity to transfer phospholipids from castor microsomes to various mitocondria, although Avena etioplasts did not serve as lipid acceptor. Furtheremore, castor bean phospholipid exchange protein transferred phospholipids from potato tuber microsomes to potato mitochondria; specific exchange rate(4.5-5.1).

REFERENCES
1. Lord, J.M. (1976) Plant Physiol., 57, 218-223.
2. Moore, T.S. (1974) Plant Physiol., 54, 164-168.
3. Moore, T.S., Lord, J.M., Kagawa, T. and Beevers, H., (1973) Plant Physiol., 52, 50-50-53.
4. Kagawa, T., Lord, M.J. and Beevers, H. (1972) Plant Physiol., 51, 61-66.
5. Donaldson, R.P. (1976) Plant Physiol., 57, 510-515.
6. Wirtz, K.W.A. amd Van Deenen, L.L.M. (1977) TIBS, 2, 49-51.
7. Kader, J.C. (1975) Biochim. Biophys. Acta, 380, 31-44.
8. Bligh, E.G. and Dyer, W.J. (1959) Can. J. Biochem. Physiol., 37, 911-917.
9. Tanaka, T. and Yamada, M. (1979) Plant Cell Physiol., in press
10. Wirtz, K.W.A and Zilversmit, D.B. (1969) Biochim. Biophys. Acta, 193, 105-116.
11. Moogan, M.C. and Moore, T.S. (1977) Plant Physiol. supplement p80.
12. Yamada, M., Tanaka, T., Kader, J.C. and Mazliak, P.(1978) Plant Cell Physiol. 19, 173-176.

LIPID COMPOSITION AND ANALYSIS

CHANGES IN PHOSPHOLIPID LEVELS DURING COLD STRATIFICATION AND GERMINATION OF PINUS PINEA SEEDS.

M.T. ALSASUA and E. PALACIOS-ALAIZ
Departamento de Bioquímica, Facultad de Farmacia, Ciudad Universitaria. Madrid (Spain)

ABSTRACT

Examination of four different methods for the extraction of pine seeds phospholipids showed that water-saturated n-butanol was the most efficient system.

The qualitative phospholipid composition was similar in dry and germinated seeds. The level of total lipid phosphorus decreased markedly with the cold stratification. The individual phospholipid, phosphatidyl choline showed significant increase during stratification and germination at 4º C. Phosphatidyl glycerol and diphosphatidyl glycerol were depleted with the low temperature treatment.

INTRODUCTION

Our previous investigations into the main metabolic changes in Pinus pinea seeds had led us to consider the lipid content and triglyceride breakdown during cold stratification[1]. Because of this metabolic activity and the often correlated low temperature response and membrane permeability [2,3], we have, therefore, been studying the phospholipid composition of pine seeds kept at low temperature and during germination. We have also compared various solvents systems with respect to efficiency of phospholipid extraction.

MATERIALS AND METHODS

Seeds were stratified at 4º C as described elsewhere[4].

Sterilized seeds were germinated in dark at 28º C on vermiculite soaked with water in covered cristallizing dishes.

Samples were harvested after 15 days of stratification, for the stratification tests.

The germination tests were made with samples harvested after

7 days of incubation at 28º C in previously stratified seeds and with samples incubated for 16 weeks at 4º C. The morfological characteristics of both lots of germinating seeds were similar.

Extration and analysis

Analysis was performed either on the whole seeds or on their dissected parts, endosperm and embryo.

Total lipids were extracted from the tissue by four different procedures, as outlined in table 1.

Lipids extracts were evaporated[5] and purified by partition chromatography on Sephadex LH-20[6].

Total lipid phosphorus was determined as inorganic phosphorus[7]. Separation of polar lipids into classes was achieved by two-dimensional TLC[8], using as solvent systems, chloroform-methanol-30% (w/v) ammonia (65:25:5 v/v) for the first dimension and chloroform-acetone-methanol-acetic acid-water (38:32:5:10:2 v/v) for the second dimension.

Confirmation of phospholipid identy was made by co-chromatography[9,10,11]. A modification of the method of Rouser et al[11], was used to determine the quantities of individual phospholipids.

RESULTS

TABLE 1.- TOTAL LIPID PHOSPHORUS EXTRACTION BY VARIOUS SOLVENTS

Method	Total Lipid Phosphorus mgrs. P/100 grs. Dry Weight		
	Whole Seed	Endosperm	Embryo
I	8,24	7,64	12,95
II	11,36	8,00	25,62
III	10,00	7,90	21,10
IV	37,45	31,55	40,47

Method I: 1) Boiling 2-propanol. 2) Propanol-2-ol:chloroform (1:1) v/v. 3) Chloroform.
Method II: 1) Boiling 2-propanol:water (80:20) v/v. 2) Propanol-2-ol:chloroform (1:1) v/v. 3) Chloroform.
Method III: 1,2,3) Chloroform:Ethanol: Water (200:95:5) v/v.
Method IV: Boiling water-saturated-n-butanol.

Boiling water-saturated-n-butanol, was the most effective solvent for lipid phosphorus extraction. The effectiveness of this procedure in extracting the total phospholipids from wheat has been reported by Laidman et al[8].

When pine seeds were stratified at 4º C, embryo and endosperm lost a considerable part of their total lipid phosphorus content. (Fig. 1).

There were no substantial qualitative differences in phospholipid composition of ungerminated, stratified and germinated pine seeds (Table 2).

TABLE 2.- CHANGES IN PHOSPHOLIPID COMPOSITION DURING COLD STRATIFICATION AND GERMINATION OF PINUS PINEA SEEDS.

Phospholipid	Phospholipid Content (% Total Lipid Phosphorus)					
	Endosperm			Embryo		
	Dry Seed	Stratified Seed	Germinated Seed	Dry Seed	Stratified Seed	Germinated Seed
P C	29,07	52,93	44,00	14,07	45,78	33,75
P I	17,17	16,66	12,50	22,17	22,61	18,25
P E	11,27	11,19	13,83	9,40	13,25	12,00
P A	T.	2,36	T.	5,50	2,32	14,75
P G	8,90	6,86	1,85	14,92	4,02	3,00
D P G	11,17	1,20	13,41	14,32	2,25	7,25
L P C	11,49	1,44	T.	8,99	3,01	T.
Others	10,93	7,36	12,90	10,63	6,76	10,00

Abbreviations:
PC: phosphatidylcholine; PI: phosphatidylinositol; PE: phosphatidylethanolamine; PA: phosphatidic acid; PG: phosphatidylglycerol; DPG: diphosphatidylglycerol; LPC: lysophosphatidylcholine; T.: traces.

The proportion of PC showed significant increase during cold stratification and there was a significant depletion of the more saturated lipids as PG and DPG during the same 15 days period of moistured low temperature treatment. These changes in individual phospholipid levels may account for the membrane permeability and low temperature adaptation of this chilling-resistant seed.

The level of total lipid phosphorus increased during germination (fig. 1) and synthesis of some individual phospholipids

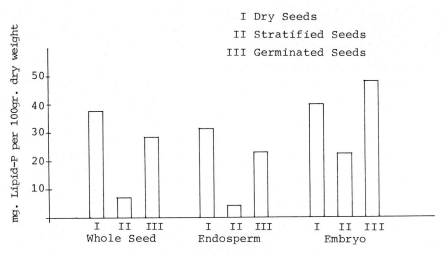

Fig. 1. Changes in total lipid phosphorus during cold stratification and germination.

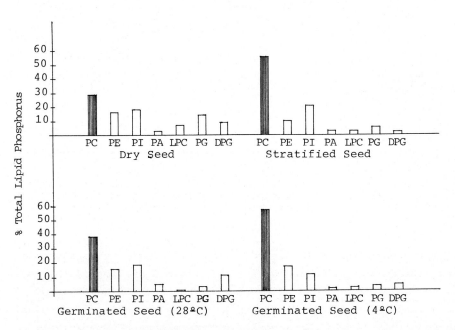

Fig. 2. Phospholipid contents in pine seeds.

were detected in germinated seeds as compared with stratified seed (Fig. 2).

During germination at low (4º C) and optimal temperature (28º C) there were significant changes in the proportion of some phospholipids. (Fig. 2). The PC/PE ratio of the 4º C seeds was about 30 % higher than that of the morphologicaly equivalent 28º C germinated seeds. A clearer difference was appreciated in the PC/DPG ratio, which suggests differential synthesis of individual phospholipids correlated with low temperature adaptation.

SUMMARY

The ability of three solvent systems to extract phospholipids from pine seeds were investigated. Water saturated n-butanol was consedered to be the most efficient system for the extraction of total phospholipid from embryo, endosperm and the whole seed.

The use of this extraction procedure inactivates degradative enzymes and there was no evidence of artifact formation due to no enzymic hydrolysis.

The level of total lipid phosphorus decreased substantially during cold stratification. The qualitative phospholipid composition was similar in ungerminated, stratified and germinated seeds, at 4º C and 28º C respectively.

Phosphatidyl choline showed significant increase during cold stratification and germination at 4º C.

Phosphatidyl glycerol and diphosphatidyl glycerol decreased with the low temperature treatment.

Low temperature germination of seeds was correlated with a higher PC/PE and PC/DPG ratio.

REFERENCES

1. M. Sanz-Muñóz, E. Palacios-Alaiz et al. (1971) Proceedings of the XIIIth International Congress of Refrigeration. Washington DC. vol. 3, 335-344.
2. Akabane, N. (1961), Rep. Hokkaido Prefectural Agri. exp. Sth 9, 47.
3. De la Roche, I.A., Andrews, C.J., et al. (1972). Can. J. Bot. 50, 2401-2409.

4. Palacios-Alaiz, E. and Sanz Muñóz, M. (1974). It. J. of Biochem. 23, 1.
5. Folch, J. and Lees, M. (1951). J. Biol. Chem. 191, 807.
6. Elingboe, J., Nyström, E. et al. (1969). Methods in Enzymology 14, 317-329.
7. Dryer, R.L., Tamnes, A.R., et al. (1957). J. Biol. Chem. 225, 177.
8. Colborne, A.J. and Laidman, L. (1975). Phytochemistry 14, 12.
9. Skidmore, W.D. and Entenman, C. (1962). J. Lipid Res. 3, 471.
10. Voskovsky, V.E. and Kostetsky, E.Y. (1968). J. Lipid Res. 9, 396.
11. Rousser, G., Kritchnevsky, G., et al. (1967). "Lipid Chromatographic Analysis" (Marinetti, G.V. ed. p.p. 99-162 Dekker, New York).

PYRROLIDIDES IN THE STRUCTURAL ANALYSIS OF LIPIDS BY MASS SPECTROMETRY

BENGT Å. ANDERSSON

Department of Structural Chemistry, Faculty of Medicine, University of Göteborg, Fack, S-400 33 Göteborg 33 (Sweden)*

ABSTRACT

Pyrrolidides of fatty acids offer a complement to the commonly used methyl ester derivative in the structural analysis of long-chain fatty acids by mass spectrometry. A number of examples are given where mass spectra of fatty acid pyrrolidides are discussed.

INTRODUCTION

The methyl ester is the most commonly used derivative in the structural analysis of long-chain fatty acids by mass spectrometry (MS) since the pioneering work of Ryhage and Stenhagen[1]. However, there are limitations in the MS analysis of methyl esters, for instance in distinguishing between positional isomers of unsaturated fatty acids. To overcome these limitations many derivatives have been suggested and this article will illustrate the potential of a derivative, in which the carboxyl group of the fatty acid has been converted to a pyrrolidide.

Vetter et al[2] proposed the use of pyrrolidides for determination of the position of double bonds in fatty acids and later Andersson continued these studies in a series of papers[3-12] which include different types of fatty acids. Based on these studies several papers have appeared which describe the use of pyrrolidides in the structural analysis of naturally occurring fatty acids[4,7,8,13-17].

MATERIALS AND METHODS

Preparation of pyrrolidides. There are several methods to convert fatty acids into pyrrolidides: e.g. the acid chloride method, carbodiimide coupling and aminolysis. Free fatty acids are used in the first two methods and fatty acid esters are required for aminolysis. Quantitative yield is obtained using the latter

---

*Present address: Department of Biochemical Ecology, University of Göteborg, Kärragatan 6, S-431 33 Mölndal (Sweden).

method and there are different ways to catalyze the reaction. Pyrrolidine is used both as solvent and reagent, and to promote the reaction Jordan and Port[18] use sodium methoxide, Vetter et al[2] acetic acid and Andersson et al[9] methanol respectively.

Gas chromatography. Fatty acid pyrrolidides have longer retention times than corresponding methyl esters under the same working conditions. Therefore higher temperatures are required and as stationary phase cyanosilicones are recommended (upper limit 275°C). The pyrrolidide derivative is thermostabile and good separation is reported for 16:0, 18:0, 18:1, and 18:2 fatty acid pyrrolidides at a working temperature of 240°C[4].

RESULTS AND DISCUSSION.

Fig. 1 shows the mass spectrum of N-octadecanoyl pyrrolidine in which the most prominent peaks in the high mass region are due to ions containing nitrogen. The base peak is a McLafferty rearrangement ion, m/e 113, it is of high abundance and therefore peaks above this mass number have to be enlarged in order to be vizualised in the normalized mass spectrum[3]. The spectrum shows a simple fragmentation pattern in the high mass region with peaks 14 atomic mass units apart, derived from fragmentations at each carbon-carbon bond of the chain. This fragmentation is present in **all** mass spectra of fatty acid pyrrolidides, more or less dominant depending on the presence of functional groups in the fatty acid chain.

One or more methyl branches in a fatty acid chain are indicated by peaks of lower intensities compared to the peaks in the spectra of normal chain acids[6]. In Fig. 3 m/e 126 is of very low abundance and includes carbon 3 of the chain without its methyl group. In the same way m/e 182 and m/e 294 are of low abundance and show methyl branches at carbons 6 and 13[12]. For GC separation of mixtures before MS-anlysis Fig. 2 gives the linear relationship between homologous series of normal, iso and anteiso fatty acid pyrrolidides on a cyanosilicone phase at 230°C[7].

Double bonds in fatty acid pyrrolidides disturb to a certain extent the typical fragmentation pattern shown in Fig.1. However, all isomers of monoenes of 18:1 and dienes of 18:2 gave different mass spectra[3,5]. A rule for their interpretation is also formulated and for most of the isomers it can be formulated as follows: If an interval of 12 a.m.u., instead of the regular 14, is observed between

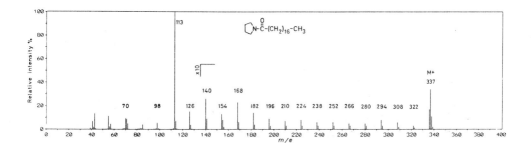

Fig. 1. Mass spectrum of N-octadecanoyl pyrrolidine.

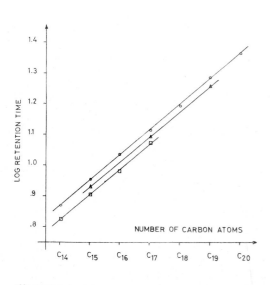

Fig. 2 Logarithmic retention time diagram of homologous series of saturated normal- (o), anteiso- (△) and iso- (☐) fatty acid pyrrolidides.

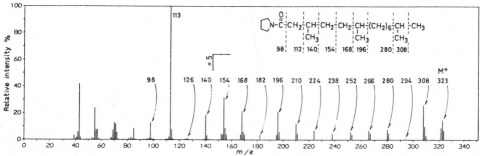

Fig. 3. Mass spectrum of N-3,6,13-trimethyltetradecanoyl pyrrolidine.

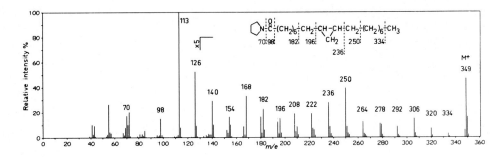

Fig. 4. Mass spectrum of N-octadec-9,12-dienoylpyrrolidine.

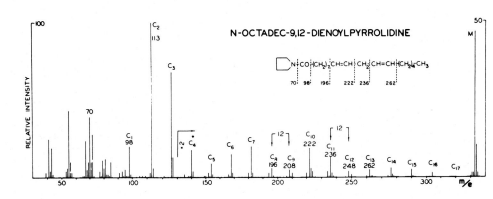

Fig. 5. Mass spectrum of N-9,10-methylene-octadecanoylpyrrolidine.

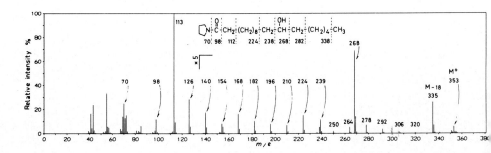

Fig. 6. Mass spectrum of N-12-hydroxy-octadecanoylpyrrolidine.

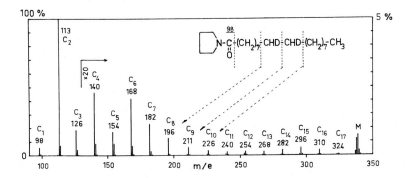

Fig. 7. Mass spectrum of N-9,10-dideuterio-octadecanoylpyrrolidine.

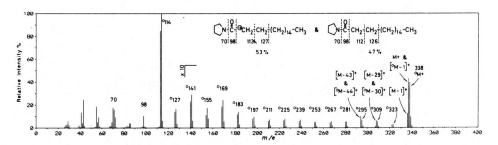

Fig. 8. Mass spectrum of 53% N-2-^{13}C-octadecanoylpyrrolidine and 47% N-2-^{13}C-octadecanoylpyrrolidine in a mixture.

the most intense peaks of clusters of fragments containing n and n+1 carbon atoms in the acidmoiety, a double bond occurs between carbons n and n+1. Fig. 4 illustrates this rule with the interval of 12 a.m.u. indicated by arrows.

Monoynoic[19] and monocyclopropanoic[12] fatty acid pyrrolidides show mass spectra thatsomewhat resemble those of the monoenoic. Fig. 5 shows a pyrrolidide with a cyclopropane ring at carbons 9 and 10 indicated by 12 a.m.u. between peaks m/e 196 and m/e 208.

Long-chain hydroxy acids show several competing fragmentation pathways and this is illustrated in Fig. 6 where a strong peak at m/e 268 indicates cleavage at the carbon carrying the hydroxyl group and m/e 335 where water is eliminated from the molecular ion[12].

Stabile isotopes such as deuterium and carbon-13 are also easily located in a fatty acid chain using pyrrolidides. Fig. 7 shows presence of two deuterium atoms at carbons 9 and 10 indicated by m/e 211 and m/e 226 respectively. Fig. 8 illustrates 53% carbon-13 incorporation at carbon 2 of the fatty acid chain. The possibility to locate isotopes, as shown above, opens new ways in tracing incorporation of isotopes in methabolic pathways of living organisms.

REFERENCES

1. Ryhage, R. and Stenhagen, E. (1961) J. Lipid Res., 1, 361-390.
2. Vetter, W., Walter, W. and Vecchi, M. (1971) Helv. Chim. Acta, 54, 1599-1605.
3. Andersson, B.Å. and Holman, R.T. (1974) Lipids, 9, 185-190.
4. Andersson, B.Å., Heimermann, W.H. and Holman, R.T. (1974) Lipids, 9, 443-449.
5. Andersson, B.Å., Christie, W.W. and Holman, R.T. (1975) Lipids, 10, 215-219.
6. Andersson, B.Å. and Holman, R.T. (1975) Lipids, 10, 716-718.
7. Andersson, B.Å. and Bertelsen, O. (1975) Chem. Scr., 8, 91-94.
8. Andersson, B.Å. and Bertelsen, O. (1975) Chem. Scr., 8, 135-139.
9. Andersson, B.Å., Dinger, F. and Dinh-Nguyen, Ng. (1975) Chem. Scr., 8, 200-203.
10. Andersson, B.Å., Dinger, F. and Dinh-Nguyen, Ng. (1976) Chem. Scr., 9, 155-157.
11. Andersson, B.Å., Dinger, F., Dinh-Nguyen, Ng. and Raal, A. (1976) Chem. Scr., 10, 114-116.
12. Andersson, B.Å. (1978) in Prog. Chem. Fats other Lipids, Holman, R.T. ed., Pergamon Press, London, pp. 279-308.
13. Gerson, T., Patel, J.J. and Nixon, L.N. (1975) Lipids, 10, 134-139.
14. Joseph, J. (1975) Lipids, 10, 395-403.
15. Gensler, W.J. and Marshall, J.P. (1977) J. Org. Chem., 42, 126-129.
16. Mayzaud, P. and Ackman, R.G. (1978) Lipids, 13, 24-28.
17. Valicenti, A.J., Chapman, C.J., Holman, R.T. and Chipault, J.R. (1978) Lipids, 13, 190-194.
18. Jordan, E.F. and Port, W.S. (1961) J. Am. Oil Chem. Soc., 38, 600-605.
19. Valicenti, A.J., Heimermann, W.H. and Holman, R.T. (1977) J. Am. Oil Chem. Soc., 54, 147A (Abstract).

A PRACTICAL PROCEDURE FOR THE ANALYSIS OF PLANT VOLATILES

BENGT Å. ANDERSSON, LENNART LUNDGREN AND GUNNAR STENHAGEN
Department of Biochemical Ecology, University of Göteborg,
Kärragatan 6, S-431 33 Mölndal (Sweden)

ABSTRACT

A practical procedure to isolate, concentrate and identify plant volatiles is described. The procedure involves use of an adsorption/desorption technique with subsequent chemical identification by a specially designed capillary gas chromatograph-mass spectrometer. Wallflower, *Cheiranthus cheiri*, and thyme, *Thymus vulgaris*, are used as examples and analytical results obtained from volatile substances of these flowers are discussed.

INTRODUCTION

The chemical plant defence against herbivores can be described as a triple defence system, namely chemicals spontanously emitted from the plant to the environment, a surface layer of taste substances, and chemicals released when the plant is wounded[1]. The released defence substances, allelochemics[2], are often volatiles and rapidly diluted into the environment of the plant. As they often occur in minute amounts it is necessary to isolate and concentrate these before the chemical identification. The most sensitive technique for this identification is the combination capillary gas chromatography- mass spectrometry (GC-MS) which both separates and identifies a mixture of volatiles.

This communication describes a practical procedure to isolate, concentrate and identify volatile substances emitted from plants. A survey of literature on this topic[1] shows that a direct chemical analysis of a volatile mixture naturally gives the best result if the components are present in measurable quantities. If an isolation and concentration step has to be introduced, adsorption on high-boiling liquids and porous polymers are to be recommended in order to get reproducible results and to avoid artefacts. Besides the adsorption technique is easy to integrate with a GC-MS system.

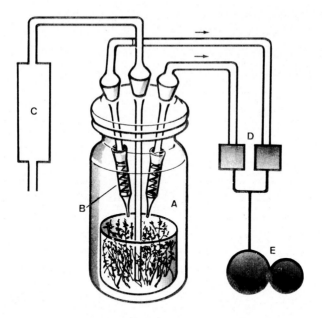

Fig. 1. Adsorption apparatus.

MATERIALS AND METHODS

Adsorption apparatus. Figure 1 shows the adsorption apparatus for isolation and concentration of plant volatiles. The plant material is placed in the chamber A which is kept under conditions similar to the plant habitat. Air is drawn through glass tubes B which are filled with appropriate adsorbent. Background impurities of incoming air are removed by a tube of activated charcoal C. The air flow through the adsorption tubes is kept constant by flow regulators D at a rate of 20 ml/min and the flow is produced by a vacuum pump E separated from the adsorption apparatus. The amount of plant material required for sample tubes containing 0.2 gram adsorbent varies between 1 and 5 gram, depending on plant material, and normal time for adsorption is 30-60 minutes. When adsorption is ready the sample tubes is placed either in the GC desorption inlet for immediate analysis or stored in screwcapped glass tubes at $-30°$ C under nitrogen for future analysis. This makes the identification step independent of time and space.

Adsorbents. The adsorption material used in our experiments are: Santovac-5 (polyphenyl ether), Lion S (alkylnaphtalene oil), SE-30 (methyl silicone), OV-17 (50% phenyl methyl silicone), XE-60 (25% cyanoethyl methyl silicone), Chromosorb 102, 100/120 mesh (styrene divinylbensene polymer), Tenax-GC (2,6-diphenyl-p-phenylene oxide).

The first two adsorbents are diffusion pump oils with low vapour pressure and high thermostability. The next three are GC stationary phases of varying polarity. These adsorbents are liquids and have to be applied as a film on a solid support. For this purpose 2cm x 5cm strips of glass fiber filter paper are coated by the liquids and inserted in the adsorption tubes. Before use the tubes are baked under a flow of inert gas at 200° C for 10 hours.

The GC and GC-MS instrumentations. A Carlo Erba Fractovap 2900 capillary gas chromatograph (Italy) is used. The injection port is modified to a desorption owen. After desorption the sample vapours are swept away by the carrier gas and collected in a U-shaped cold

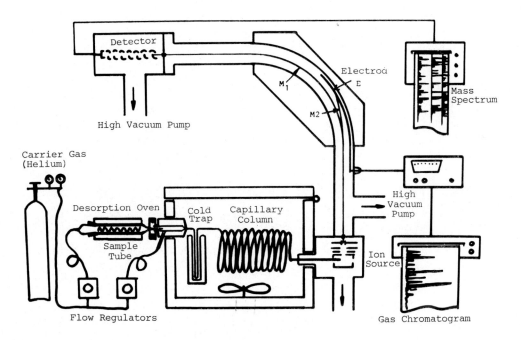

Fig. 2. Schematic drawing of the combination capillary gas chromatograph - mass spectrometer instrument.

trap immersed in a liquid nitrogen containing Dewar flask inside the GC owen. The cold trap consists of a glass lined metal tube placed between the desorption owen and the glass capillary column. Injection on column is effected by an electric current shock through the glass lined metal tube. The column is kept cool until injection whereafter temperature programming of owen starts.

Figure 2 shows a schematic drawing of the combination capillary GC-MS. The desorption owen and the cold trap are similar as described above. The outlet flow from the capillary column is allowed direct into the electron impact ion source of the MS without use of separator. This is possible as the carrier gas flow is low (2-5 ml He per minute) and as the vacuum system is constructed for this flow. In addition to the described GC part, the MS part of the combination instrument is also constructed at our laboratory by one of us (G.S.) and can be described as a 90° magnetic sector instrument with large radius (300 mm). Notice the total ion current electrode E placed inside the analyzer tube. The magnetic field can be adjusted so that low mass ions do not hit the electrode. This construction makes it possible to discriminate carrier gas, water and air ions (masses less than M_1) against ions from the sample (M_2). The measurement of the gas chromatogram can therefore be performed at high ion production (70 eV) together with total ion registration of all sample ions above a certain low mass number. This makes the described MS very sensitive as a GC detector. Less than 10 pg is needed for measureable GC peaks. Registration of mass spectra are performed through an electron multiplying detector and are recorded in linear form (figure 4) by use of a linear mass scale device[1].

RESULTS AND DISCUSSION

Figure 3 shows a capillary gas chromatogram obtained from volatiles of wallflower, *Cheiranthus cheiri*. The plant was kept deepfrozen before adsorption. One fraction A dominates the entire gas chromatogram. Its mass spectrum (figure 4) gives a molecular ion at m/z 149 and its isotope ions indicate presence of sulfur. Comparison of the mass spectrum with reference spectra revealed the structure as 3-methyl thiopropyl isothiocyanate[3]. This compound can be derived from a class of substances in Cruciferae, glucosinolates.

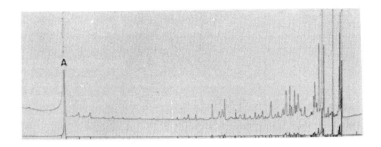

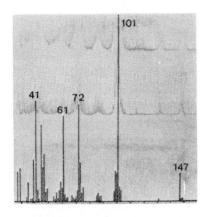

Fig. 3. Capillary gas chromatogram of volatiles from *Cheiranthus cheiri*.
Adsorbent: Tenax-GC
Column: SCOT, 53 meter, coated with FFAP, linearly temperature programmed from 40 to 200°C at 2° per min.

Fig. 4. Mass spectrum of component A in figure 3, 3-methylthiopropyl isothiocyanate, $CH_3S(CH_2)_3NCS$.

Further investigations of plant volatiles using thyme, *Thymus vulgaris*, as test object shows that there is no allround adsorbent. Polar adsorbents prefer polar volatiles and nonpolar adsorbents prefer nonpolar volatiles. Besides, the porous polymers have a preference for small molecules with high volatility, and the high-boiling liquids prefer less volatile and larger molecules. The latter is illustrated by the gas chromatograms in figure 5 representing thyme volatiles adsorbed on Tenax-GC and SE-30 respectively. Compound 1 (cis-3-hexenol-1) in the chromatograms is more dominant in figure 5A while components 2 (linalool) and 3 (thymol) are more dominant in figure 5B. Therefore a proper chemical analysis requires an adsorbent (or several) chosen according to the properties of the sample, e.g. a combination of Tenax-GC, SE-30 and XE-60.

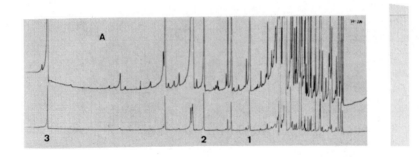

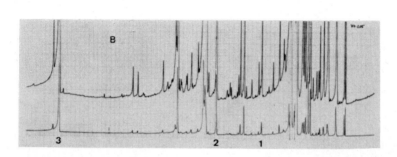

Fig. 5. Capillary gas chromatogram of volatiles from *Thymus vulgaris*. WCOT column, 130 meter, Carbowax 20M, linearly temperature programmed from 50 to 200°C at 2° per minute. Adsorbent upper curve (A): Tenax-GC and curve below (B): Glass fiber paper coated with SE-30. Adsorption time 45 minutes in both cases.

ACKNOWLEDGEMENTS

Grants in support of this Department were obtained from the National Swedish Environment Protection Board and the Wallenberg Fondation.

REFERENCES

1. Lundgren, L., Andersson, B.Å. and Stenhagen, G. (1978) National Swedish Environment Protection Board PM 1036, in Swedish (summary in English).
2. Whittaker, R.H., (1969) in Chemical Ecology, Sondheimer, E. and Simeone, J.S. eds., Academic Press, New York, chapter 3.
3. Kjaer, A., Ohashi, M., Wilson, J.M. and Djerassi, C. (1963) Acta Chem. Scand., 17, 2143-2154.

WATER STRESS, EPICUTICULAR WAX AND CUTICULAR TRANSPIRATION RATE

CURT BENGTSON, STIG LARSSON AND CONNY LILJENBERG
Department of Plant Physiology, University of Göteborg, Carl Skottsbergs Gata 22, S-413 19 Göteborg (Sweden)

ABSTRACT

Seven-day-old seedlings of oat (*Avena sativa* L. cv. Stormogul II, Risto, Sol II, Selma, Sang, and Pendek, arranged according to decreasing drought resistance) were exposed to water stress on four consecutive days. The stress treatment was mild and of short duration.

The stress treatment affected both the amount and the chemical composition of epicuticular wax in all varieties. Furthermore, due to the water stress the cuticular transpiration rate was reduced in all varieties.

INTRODUCTION

The resistance against diffusion of water through the cuticle is completely determined by the wax components (i.e. epicuticular and intracuticular wax) of the cuticle[1].

There are indications that in addition to the amount of epicuticular wax some properties of the wax might affect cuticular transpiration as well. The wax conformation is probably important[2] as well as the chemical composition[3]. On one hand, the chemical composition is crucial for the wax conformation[4,5,6,7] and on the other hand different wax components are not equally efficient in reducing evaporation[3].

The importance of the cuticle for the survival of plants during severe drought through its ability to reduce the water loss has been pointed out[8,9]. During longer stress periods (up to weeks) the deposition of epicuticular wax increases[10,11,12], which leads to a reduction in cuticular transpiration rate[11].

The aims of the present investigation are to study the initial changes in (a) amount, (b) chemical composition of epicuticular wax, and (c) cuticular transpiration rate in young oat seedlings exposed to mild water stress of short duration.

MATERIALS AND METHODS

Plant material. Oat seedlings (*Avena sativa* L.) of six varieties differing in drought resistance were cultivated under controlled conditions[13]. From field observations of growth and harvest yield during dry seasons Stormogul II has been found to be the most drought resistant of these varieties, whereas Pendek is the least resistant. The other varieties have been found to be intermediate in drought resistance and the order is assumed to be Risto, Sol II, Selma and Sang with a falling degree of drought resistance (V. Stoy, pers. comm.).

<u>Stress treatment</u>. Seven-day-old seedlings were water stressed on 4 consecutive days by cooling their roots for 3 h to $1.0 \pm 1.0^{\circ}C$. During this treatment the leaf water potential decreased from -7 to -12 bars[13].

<u>Extraction and determination of epicuticular wax</u>. The primary leaves were cut at their bases. The epicuticular wax was extracted by dipping 10 leaves for 10 s in chloroform (cf.[14]) while holding the cut ends out of the solvent. The chloroform was evaporated to dryness with nitrogen, and the wax samples were weighed.

After weighing, the samples of the respective variety were combined, and for a quantitative determination of the individual lipids the total wax was dissolved in a small volume of chloroform and applied on 0.5 mm Silica gel H plates. The plates were developed in carbontetrachloride-ethanol 99:1 (by volume). Authentic reference substances were traced by blowing with iodine vapour. The individual lipids were scraped off the plates and extracted from the gel with chloroform-methanol 2:1 (by volume).

The hydrocarbons and the free alcohols were analyzed quantitatively by using gas chromatography. Acetylation of the alcohols was done in bezene with acetic anhydride and pyridine[15]. The conditions during gas chromatography are described in detail by Bengtson et al.[16].

Esterified and free fatty acids (total fatty acids) were converted to the corresponding methyl esters with BF_3 in methanol[17] and analyzed using gas chromatography.

Further confirmation of the individual lipids was obtained from combined gas chromatography-mass spectrometry[16].

The β-diketones were roughly estimated from the UV-absorption at 273 nm[18].

<u>Determination of cuticular transpiration rate</u>. In previous investigations of cuticular transpiration from excised leaves it has been difficult to see from the "drying out curves" when the stomata are closed and the "pure" cuticular phase begins[19,20]. In the present investigation the method was improved by making sure that the stomata were closed during the measurements, i.e. that only the cuticular transpiration rate was measured.

The stomata were closed by drying excised leaves in air on a wire-netting for 3 h. Immediately before the determination of cuticular transpiration rate the leaves were rewatered, i.e. they were allowed to be water saturated.

Cuticular transpiration rate was determined by weighing samples of 5 leaves every 5 min during 35 min. As the stomata did not reopen during the determination (cf.[13]) there was only a slight reduction in leaf water content (on the average 4 % relative water content).

RESULTS

The stress treatment, although mild and of short duration, strongly affected the tested varieties. However, the impact varied greatly among the varieties.

Changes, due to water stress, in the chemical composition of the epicuticular wax, were examined by quantitative determinations of 4 wax components (free alcohols, alkanes, and fatty acids, see Table 1 and 2, and β-diketones). Determinations of β-diketones were unsuccessful due to interference from other substances absorbing in that region of the UV-spectrum. The main primary alcohol in all varieties was hexadocosanol and only minor amounts of other alcohols were present (cf.[21]).

TABLE 1

INDIVIDUAL LIPIDS OF THE EPICUTICULAR WAX IN SIX OAT VARIETIES EXPOSED TO WATER STRESS

The mean of two determinations is given in $\mu g \cdot cm^{-2}$. The varieties are arranged according to decreasing drought resistance (see MATERIAL AND METHODS).

Variety	Hexadocosanol		Total alkanes		Fatty acids	
	Control	Stress	Control	Stress	Control	Stress
Stormogul II	8.1	8.8	0.96	0.03	0.50	0.59
Risto	8.4	8.6	0.33	0.04	0.60	0.75
Sol II	6.9	7.0	0.06	0.09	0.79	1.09
Selma	9.2	9.3	0.10	0.17	0.50	0.54
Sang	9.2	10.1	0.68	0.33	0.71	0.60
Pendek	5.3	5.8	0.29	0.50	0.46	0.44

In all varieties the primary alcohol content of the wax was slightly higher in stressed seedlings than in controls.

Further, in Stormogul II and Risto the content of predominant alkanes was much lower in stressed seedlings than in controls. On the contrary, in Pendek the stressed seedlings showed a higher alkane content.

The total amount of fatty acids (free and esterified) was higher in stressed seedlings than in controls in the varieties which are most drought resistant (i.e. Stormogul II, Risto, and Sol II). There was no difference in Selma and Pendek, and in Sang the fatty acid content was lower in stressed seedlings.

Of the individual alkanes, nonacosane (C_{29}) and hentriacontane (C_{31}) content showed strong differences between unstressed and stressed seedlings in some varieties. However, the heptacosane (C_{27}) content was about the same in all varieties, and for unstressed and stressed seedlings as well (Table 2).

TABLE 2

ALKANES OF THE EPICUTICULAR WAX IN SIX OAT VARIETIES EXPOSED TO WATER STRESS (cf. Table 1).

Variety	C_{27}, heptacosane		C_{29}, nonacosane		C_{31}, hentriacontane	
	Control	Stress	Control	Stress	Control	Stress
Stormogul II	0.01	0.01	0.91	0.02	0.04	0.00
Risto	0.01	0.01	0.03	0.03	0.29	0.00
Sol II	0.01	0.02	0.05	0.07	0.00	0.00
Selma	0.02	0.02	0.08	0.09	0.00	0.06
Sang	0.01	0.00	0.20	0.07	0.47	0.26
Pendek	0.02	0.01	0.04	0.11	0.23	0.38

Unstressed seedlings of the most drought resistant variety, Stormogul II, showed the highest cuticular transpiration rate (Figure 1). After stress treat-

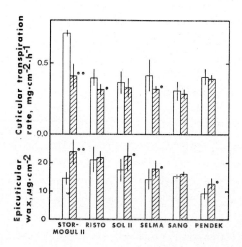

Fig. 1. Cuticular transpiration rate and total amount of epicuticular wax in six oat varieties exposed to water stress. Each value is the mean of five determinations. Unstressed seedlings (□) and stressed (▨). The vertical bars give the standard deviations. •• denotes significant difference between control and stressed seedlings at the 1 % probability level (P<1 %) and • at the 5 % level (P<5 %).

ment the cuticular transpiration rate was most strongly reduced in this variety, and at the same time it showed the largest increase in amount of epicuticular wax of the tested varieties. In Pendek and Sang, showing the slightest increase in epicuticular wax, the cuticular transpiration rate was only 5 % lower after stress treatment than before.

SUMMARY

It has been suggested that the cuticular transpiration rate is linearly related to the inverse of the amount of epicuticular wax[11]. However, if the cuticular transpiration rate is plotted against the inverse of the amount of epicuticular wax, using the data in Figure 1, it is obvious that no simple relation can be applied to all tested varieties (Figure 2).

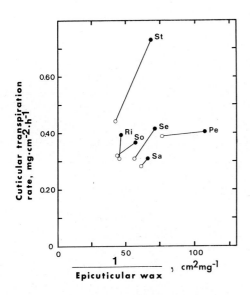

Fig. 2. The relation between cuticular transpiration rate and the inverse of amount of epicuticular wax in unstressed (●) and stressed (○) leaves of six oat varieties. St, Stormogul II; Ri, Risto; So, Sol II; Se, Selma; Sa, Sang; Pe, Pendek.

For instance, the difference in cuticular transpiration rate between Pendek and Stormogul II cannot be explained by the difference in the amount of epicuticular wax, which shows that some additional factor(s) affecting cuticular transpiration was (were) involved. One possible 'factor' (explanation) might be differences in chemical composition of the epicuticular wax in the tested varieties (cf. Introduction). Such differences are, in fact, indicated in the results (Table 1 and 2). Moreover, the composition of especially the alkanes was affected by the water stress and there were differences among the varieties as well. The mechanism for how these changes affect the properties of the epicuticular wax are at present under investigation.

REFERENCES

1. Schönherr, J. (1976) Planta (Berl.), 131, 159-164.
2. Chambers, T.C. and Possingham, J.V. (1963) Aust. J. Biol. Sci., 16, 818-825.
3. Grncarevic, M. and Radler, F. (1967) Planta (Berl.), 75, 23-27.
4. Hallam, N.D. (1970) Planta (Berl.), 93, 257-268.
5. von Wettstein-Knowles, P. (1972) Planta (Berl.), 106, 113-130.
6. Jeffree, C.E., Baker, E.A. and Holloway, P.J. (1975) New Phytol., 75, 539-549.
7. Chambers, T.C., Ritchie, I.M. and Booth, M.A. (1976) New Phytol., 77, 43-49.
8. Pisek, A. and Berger, E. (1938) Planta (Berl.), 28, 124-155.
9. Levitt, J. (1972) Responses of Plants to Environmental Stresses, Academic Press, New York, pp. 1-697.
10. Skoss, J.D. (1955) Bot. Gaz., 117, 55-72.
11. Clark, J.A. and Levitt, J. (1956) Physiol. Plant., 9, 598-606.
12. Daly, G.T. (1964) J. Exp. Bot., 15, 160-165.
13. Bengtson, C., Falk, S.O. and Larsson, S. (1977) Physiol. Plant., 41, 149-154.
14. Martin, J.T. and Juniper, B.E. (1970) The Cuticles of Plants, Edward Arnold (Publishers) Ltd, printed by R. and R. Clark, Ltd, Edinburgh, pp. 1-347.
15. Wellburn, A.R. and Hemming, F.W. (1966) J. Chromatogr., 23, 51-60.
16. Bengtson, C., Larsson, S. and Liljenberg, C. (1978, in press) Physiol. Plant.
17. Morrison, W.R. and Smith, L.M. (1964) J. Lipid Res., 5, 600-608.
18. Horn, D.H.S., Kranz, Z.H. and Lamberton, J.A. (1964) Aust. J. Chem., 17, 464-476.
19. Hygen, G. and Midgaard, E. (1954) Physiol. Plant., 7, 128-140.
20. Stålfelt, M.G. (1956) Handbuch der Pflanzenphysiologie Bd III, Ruhland, W. ed. Springer, Berlin, Göttingen, Heidelberg, pp. 342-350.
21. Tulloch, A.P. and Hoffman, L.L. (1973) Lipids, 8, 617-622.

DISTRIBUTION OF LONG-CHAIN FATTY ACIDS IN THE CUTICULAR LIPIDS
OF HEMP DURING PLANT DEVELOPMENT

PÉTER ÁKOS BIACS
Institute of Agricultural Chemical Technology, University of
Technical Sciences, Budapest, Gellért tér 4, 1111 Budapest
/Hungary/

ABSTRACT

Two different varieties of industrial cultivated hemp in Hungary:
Cannabis sativa L. var. Kompolti and its chlorophyll-deficient
variety were studied by lipid analysis of cuticular waxes. Samples
were taken in 8 different phases during plant development. Comparing the plant growth rate and cuticular lipids on leaves and stems
a relatively slow development and increased accumulation of lipids
was assigned to the chlorophyll-deficient "yellow" mutant. The
cuticular lipids of the "yellow" mutant were less enriched with
very long-chain acids from 20 to 30 carbon atom. The elongation of
wax acids of the mutant cultivar proceeded later, thus the protective function of cuticular waxes was less ensured.

INTRODUCTION

Lipids and lipid polymers are widely distributed in different
plant tissues according to their role. Presumably a protective
function is assigned to cuticular waxes covering the outside layer
of leaves and stems in many plants[1].

This phenomenon was studied on two different varieties of hemp:
Cannabis sativa L. var. Kompolti and its mutant, a chlorophyll-
deficient variety containing only 77% of chlorophyll and 67% of
carotenoids of the original hemp cultivar.

From technological point of view the "yellow" mutant showed an
intensive chlorophyll degradation at flowering stage of plant development giving a colourless stem for industrial processing of
fibres.

Samples were taken in 8 different phases after the stadium of
3-4 pairs of leaves from 18 May up the end of July. The rate of
plant development was followed by measuring the height of the plant
and the number of leaves. /Figure 1./

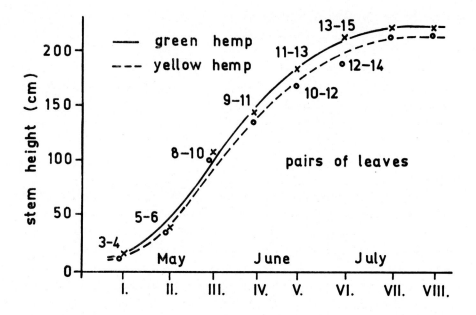

Fig. 1. Growth curves of two hemp varieties. Plotted values are the average data measured on 25 plants each.

Both cultivars were similarly growing up to the III. phase /8-10 pairs of leaves/, then "green" hemp quickly surpassed its "yellow" mutant. Last two phases were characteristic for the generative stages: blossoming and flowering.

MATERIALS AND METHODS

50-100 g fresh samples of leaves and stem particles were dried in room temperature by ventilling 50 °C warm air, then filled in Soxhlett extraction tubes. Neutral lipids were extracted in a mild procedure using light petroleum ether, removing the outside lipid layer. The extraction procedure was repeated by the method of Folch /chloroform-methanol 2:1 v/v / extracting polar lipids and some of the lipid polymers. Both extracts were subjected to evaporate solvent and the lipid dry weight was measured.

The distribution of lipid groups in both extracts was investigated by thin-layer chromatography. Lipid extracts have been saponified to acidic and neutral fraction using KOH/methanol for two hours at boiling temperature. The neutral, non-saponifiable fraction of hydrocarbons, wax alcohols, sterols were separated by petroleum ether - diethylether 1:1 v/v mixture. After removing neutral fraction a thin layer of diethylether was added to methanol-water phase and fatty acids were liberated by adding 10% HCl to the soap solution. Fatty acids were separated with 2 x 10 ml diethylether, then converted into methyl esters by Stahl's method[2].

The distribution of fatty acid methyl esters was carried out by different gas chromatographic measurements. Distribution according to the number of unsaturated bonds for components containing 14-20 carbon atoms was measured at a 10% DEGS /diethylene glycol succinate/ stationary phase under isothermic conditions at 185 $^\circ$C, while distribution according to chain lenght for long.chain fatty acids with 14-30 carbon atoms was determined on 1% OV-101 /methyl silicone/ as the stationary phase in the range of 110-315 $^\circ$C with temperature programmed instrument. Chromatogram of long-chain fatty acid methyl esters of "green" hemp cultivar from samples taken in the phase III. is presented in Fig. 2.

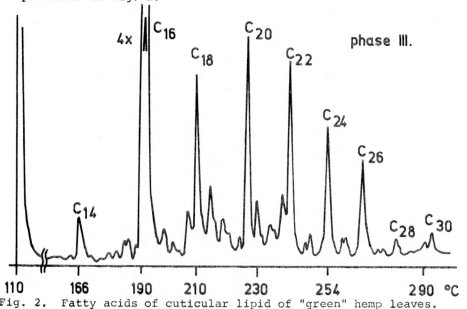

Fig. 2. Fatty acids of cuticular lipid of "green" hemp leaves.

Qualitative evaluation was made using standard fatty acid methyl ester mixtures and quantitative determination was performed by measuring the peak areas of the main components of the chromatogram. Long-chain fatty acids with even numbers were calculated.

RESULTS

Change in the cuticular lipid content. The change in the neutral lipid content of the two hemp varieties in the period of I.-VIII. phases is shown in Table 1., its fatty acid content in Table 2.

TABLE 1
CUTICULAR LIPID CONTENT OF HEMP LEAVES AND STEMS RELATED TO AIR-DRY MATERIAL /%/ IN THE PERIOD OF I.-VIII. PHASES OF GROWTH[a]

Sample	I.	II.	III.	IV.	V.	VI.	VII.	VIII.
Leaves								
"green"	1,8	0,7	1,2	1,6	2,3	1,0	2,1	1,9
"yellow"	2,5	0,4	2,4	2,1	1,6	2,4	4,3	3,6
Stems								
"green"	1,2	0,2	0,5	0,2	0,3	0,2	0,1	0,2
"yellow"	0,9	0,8	0,6	0,2	0,3	0,6	0,3	0,3

[a]Values obtained from 3 parallel measurements.

TABLE 2
FATTY ACID CONTENT OF HEMP LEAVES AND STEMS RELATED TO AIR-DRY MATERIAL /%/ IN THE PERIOD OF I.-VIII. PHASES OF GROWTH[a]

Sample	I.	II.	III.	IV.	V.	VI.	VII.	VIII.
Leaves								
"green"	1,2	0,3	0,6	0,8	1,4	0,4	0,6	0,1
"yellow"	1,5	0,2	1,8	1,6	0,6	1,2	1,1	1,1
Stems								
"green"	0,25	0,02	0,06	0,06	0,08	0,08	0,02	0,05
"yellow"	0,68	0,40	0,15	0,13	0,09	0,06	0,01	0,06

[a]Values obtained from 3 parallel saponification and measurements.

Qualitative investigation of the fatty acid fraction. The values of long-chain fatty acid composition of the petroleum ether extract of hemp leaves and stems are given in Table 3 and 4.

TABLE 3

ELONGATION OF LONG-CHAIN FATTY ACIDS OF CUTICULA LIPIDS ON HEMP LEAVES IN THE PERIOD OF I.-VIII. PHASES OF GROWTH

Distribution of C_{14}-C_{30} fatty acids in the petroleum ether extract[a]

Fatty acid	I.	II.	III.	IV.	V.	VI.	VII.	VIII.
"green" hemp leaves								
C_{14}	3,8	6,3	2,5	3,8	3,4	2,5	1,1	-
C_{16}	70,0	41,5	52,5	40,0	60,0	32,8	14,3	9,7
C_{18}	9,2	14,8	9,8	7,2	8,2	4,9	63,5	47,7
C_{20}	7,7	11,6	11,4	15,3	13,3	9,4	3,1	4,0
C_{22}	5,7	9,2	10,1	13,6	8,9	10,1	2,7	4,0
C_{24}	2,1	6,7	6,5	10,3	3,8	7,9	1,6	3,2
C_{26}	0,9	5,7	4,4	5,2	1,7	24,0	10,3	22,6
C_{28}	0,3	2,1	1,1	1,4	0,3	4,4	1,8	6,4
C_{30}	0,3	2,1	1,7	3,2	0,4	4,0	1,6	2,4
"yellow" hemp leaves								
C_{14}	1,0	9,5	3,5	5,7	9,7	0,7	13,8	8,9
C_{16}	19,3	39,6	63,5	53,0	33,3	17,4	35,2	36,3
C_{18}	74,0	13,2	8,6	10,5	22,2	69,0	17,2	13,0
C_{20}	2,8	10,7	13,0	14,0	10,3	2,5	9,4	11,2
C_{22}	1,8	8,4	7,8	9,1	9,4	2,2	7,2	8,1
C_{24}	0,7	8,4	2,2	4,5	3,0	0,7	4,0	5,6
C_{26}	0,2	6,4	0,8	1,7	5,6	4,7	6,3	10,0
C_{28}	0,1	2,0	0,3	0,5	1,9	1,0	1,8	2,8
C_{30}	0,1	1,8	0,3	1,0	4,6	1,8	5,1	4,1

[a] Values obtained from 3 parallel chromatographic measurements.

The lipids ot the "yellow" mutant are less enriched with long-chain fatty acids and the elongation of those wax acids proceeds later. A marked enrichment in wax acids of 26 carbon atom can be observed in the VI.-VIII. phase of the "green" hemp cultivar.

TABLE 4

ELONGATION OF LONG-CHAIN FATTY ACIDS OF CUTICULA LIPIDS ON HEMP STEMS IN THE PERIOD OF I.-VIII. PHASES OF GROWTH

Distribution of C_{14}-C_{30} fatty acids in the petroleum ether extract[a]

Fatty acid	I.	II.	III.	IV.	V.	VI.	VII.	VIII.
"green" hemp stems								
C_{14}	3,0	2,8	4,0	3,3	16,0	1,3	-	2,5
C_{16}	40,5	48,7	12,8	3,7	3,2	3,4	10,0	37,5
C_{18}	15,9	18,9	1,5	1,6	1,4	0,9	4,0	7,9
C_{20}	12,1	10,2	1,7	0,7	0,5	0,8	0,5	4,9
C_{22}	11,1	9,0	6,8	2,5	1,2	2,5	0,5	9,0
C_{24}	4,6	3,7	27,8	13,7	9,4	6,7	5,0	4,5
C_{26}	7,2	3,4	28,8	51,0	47,5	59,0	58,5	18,4
C_{28}	3,1	1,9	9,7	11,4	11,4	13,4	12,5	6,3
C_{30}	2,5	1,4	6,9	12,0	9,4	12,0	7,0	9,0
"yellow" hemp stems								
C_{14}	-	1,4	3,6	-	-	-	3,3	2,4
C_{16}	86,5	79,3	35,0	41,6	48,5	3,1	39,0	8,1
C_{18}	7,3	8,8	12,2	12,4	12,5	0,7	6,6	3,2
C_{20}	3,1	5,3	7,8	14,4	7,6	1,9	5,0	1,0
C_{22}	1,6	3,5	9,6	10,2	13,2	3,8	9,3	0,7
C_{24}	1,0	0,7	8,6	7,9	5,0	10,8	4,6	6,2
C_{26}	0,5	0,7	13,8	11,3	13,2	55,0	16,2	58,6
C_{28}	-	0,3	3,6	0,8	-	12,5	6,6	12,8
C_{30}	-	-	5,8	1,4	-	12,2	9,4	7,0

[a] Values obtained from 3 parallel chromatographic measurements.

SUMMARY

According to the chromatographic data the elongation of fatty acids from 20 to 30 carbon chain is demonstrated.

REFERENCES

1. Tulloch, A.P. /1976/ in Chemistry and biochemistry of natural waxes, Kollattukudy, P.E. ed., Elsevier, Amsterdam, pp. 266-279.
2. Stahl, E. /1969/ Thin-layer chromatography, Springer Verlag, Berlin-New York, pp. 175.
3. Biacs, P.A. Gruiz, K. and Holló, J. /1976/ Acta Alimentaria, 5, 403-423.

MASS SPECTROMETRY OF GLYCOLIPIDS

M.E. BREIMER, G.C. HANSSON, K.-A. KARLSSON, G. LARSON, H. LEFFLER, I. PASCHER, W. PIMLOTT and B.E. SAMUELSSON
Department of Medical Biochemistry, University of Göteborg, Göteborg (Sweden)

Cell surface carbohydrates exist in two forms, glycoproteins and glycolipids. Both have been implicated as receptors in cell surface recognition phenomena (i.e. differentiation, tumour development, blood group systems, host-patogen interaction) in animals[1,2,3], bacteria and plants[4]. Glycosphingolipids are plasma membrane lipids. The receptor function of their carbohydrate chain is well established (i.e. blood group antigens, cholera toxin receptor), and their ceramide moieties make up part of the membrane lipid bilayer[3]. A function for sulphatide (ceramide galactose-3-sulphate) in sodium-potassium transport has been postulated[3,5].

For the investigation of the chemical structure of glycosphingolipids mass spectrometrical techniques have been developed. These include analysis of ceramide structure (type of long-chain base and fatty acid[6], methyl branch position[7], double bond position[8]) and carbohydrate chain structure (type and sequence of sugars[9,10], sugar residue ring size[10], sulphate group substitution[11]).

Mass spectrometry of intact glycosphingolipids as methylated or methylated and reduced derivatives is of special value[9]. In mass spectra of methylated and reduced glycosphingolipids the abundant fragments containing the fatty acid and the complete carbohydrate chain give specific information on the number (at present up to 9) and type of sugars in the chain, see fig. 1. In mass spectra of methylated glycolipids abundant fragments containing terminal mono- and oligosaccharides are useful for detecting different immunological determinants of antigenic glycolipids, see fig. 2. The combined interpretation of these and other fragment types in mass spectra of the two derivatives allows a safe conclusion concerning not only type and number of sugars, but also their sequence including branching in the carbohydrate chain. Mass spectra also give information on the structure of the ceramide part (type of long-chain base and fatty acid, number of carbon atoms and degree of unsaturation).

This technique has successfully been tested for analysis of protein linked oligosaccharide (submitted). It has also been used for analysis of a glycerylether-based sulphoglycolipid from Halobacterium Salinarum (unpublished), see fig. 3. The mass spectrum shows that the molecule contains a chain of three hexoses attached to a saturated glycerylether containing 40 carbon atoms.

The sulphate group (replaced by a trimethylsilyl group in the derivative used) is shown to be attached to the terminal hexose.

The specificity of fragment masses for different carbohydrate structures allows the successful analysis of mixtures of complex glycolipids including the identification of previously unknown compounds. This should be of special value for the analysis of very small samples, see fig. 5.

A programmed temperature dependent evaporation of the sample in the mass spectrometer, combined with computerized monitoring of specific fragments, has been used for the separation and analysis of mixtures of glycolipids with 1-8 sugars[12], see fig. 4.

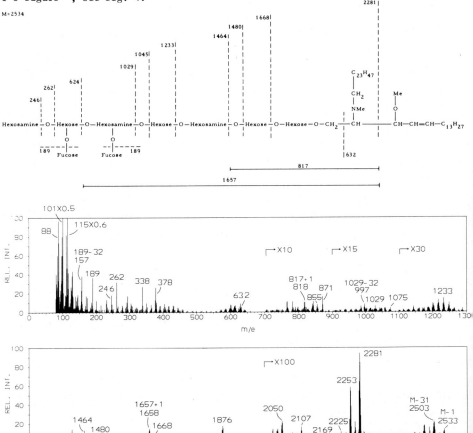

Figure 1. Mass spectrum of a blood-group A like nonaglycosylceramide isolated from human erythrocyte membranes. The glycolipid was permethylated and reduced ($LiAlH_4$). The spectrum was recorded at a probe temperature of $320°C$.

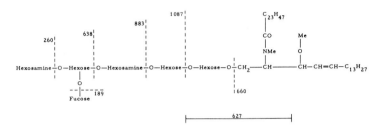
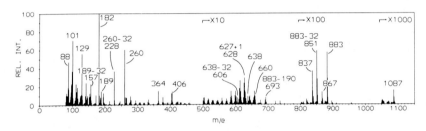
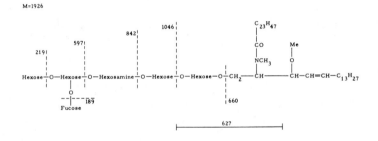
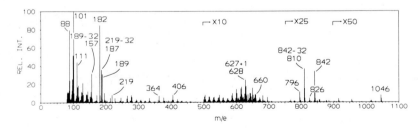

Figure 2. Mass spectra and simplified formulae of the methylated derivative of the major blood-group A active (upper) and blood-group B active (lower) glycosphingolipid from human A and B erythrocyte membranes, respectively. The spectra were recorded at a 290–330°C probe temperature.

284

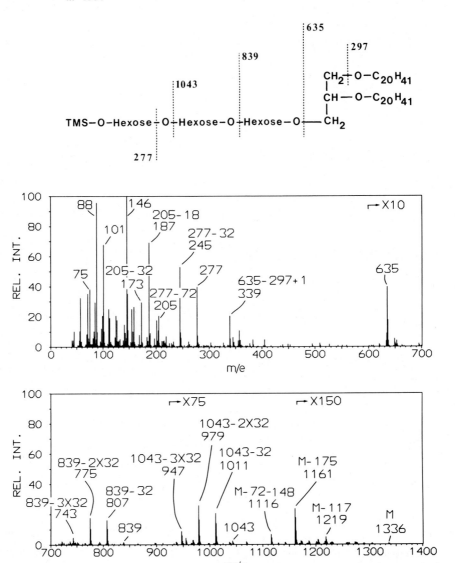

Figure 3. Mass spectrum of the sulphoglycolipid (described by M. Kates et. al. in ref. 13) isolated from Halobacterium Salinarium. The lipid was permethylated and the sulphate group was replaced by a trimethylsilyl substituent. A simplified formula is shown for interpretation.

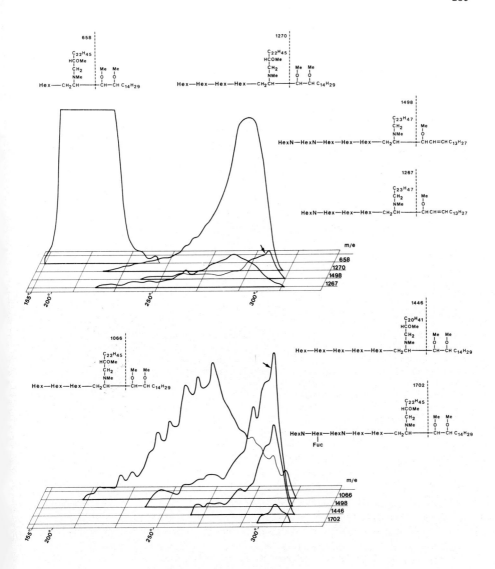

Figure 4. Mass fragmentograms and simplified formulae for interpretation concerning methylated-reduced non-acid glycolipids of cat small intestine (200 μg). The sample temperature was raised 5°C/min in the region 150-200°C and 1°C/min in the region 200-325°C. There were 10 scans within each section of the bottom frame. Electron energy 48 eV, trap current 500 μA, ion source temperature 290°C, acceleration voltage 4 kV. The arrow indicates the same curve in two reproductions with different intensity scales. Hex means hexose, HexN, hexosamine and Fuc, fucose.

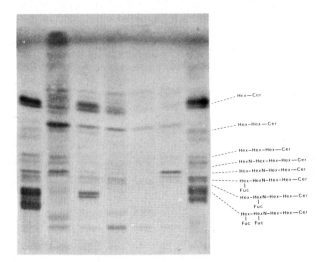

Figure 5. Thin-layer chromatogram of total non-acid glycosphingolipids of human embryonic tissues. References of an intestine of a week 17 fetus to the left and a meconium (the first feces of newborns) from a blood-group 0 individual to the right (see ref. 14). The major components of the latter portion, as identified by mass spectrometry, are shown on far right (for abbreviations, see fig. 4). The samples in between the references are, from the left, kidneys and intestine of a week 10 fetus, intestine and fetal membranes of a week 11 fetus and fetal membranes of another week 11 fetus, all with 20 µg out of about 200 µg mounted on the TLC-plate. Chloroform/Metanol/Water, 60:35:8 (by vol.) was used as solvent and the anisaldehyde reagent for detection.

REFERENCES

1. Hughes, R.C. (1975) in: Essays in Biochemistry (Campbell, P.N., Aldridge, W.N. eds.) pp 1-36, vol. 11, Academic Press, London.
2. Hughes, R.C., Sharon, N. (1978) Nature 274, 637-638.
3. Yamakawa, T., Nagai, Y. (1978) Trends Biochem. Sci. 3, 128-131.
4. Albersheim, P., Andersen-Prouty, A.J. (1975) Ann-Rev. Plant Physiol. 26, 31-52.
5. Karlsson, K.-A. (1977) in: Structure of Biological Membranes (Abrahamsson, S., Pascher, I., eds.) pp 245-274, Plenum Press, New York.
6. Karlsson, K.-A., Samuelsson, B.E., Steen, G.O. (1973) Biochim. Biophys Acta 316, 336-362.
7. Karlsson, K.-A., Samuelsson, B.E., Steen, G.O. (1973) Chem. Phys. Lip. 11, 17-38.
8. Karlsson, K.-A. (1970) Chem. Phys. Lip. 5, 6-43.
9. Karlsson, K.-A. (1976) in: Glycolipid Methodology (Witting, L. ed) pp 97-122, Am. Oil Chem. Soc., Champaign, Illinois.
10. Karlsson, K.-A., Samuelsson, B.E., Steen, G.O. (1972) J. Lip. Res. 13, 169-176.
11. Karlsson, K.-A., Samuelsson, B.E., Steen, G.O. (1969) Biochem. Biophys. Res. Comm.37, 22-
12. Breimer, M.E., Hansson, C.G., Karlsson, K.-A., Leffler, H., Pimlott, W., Samuelsson, B.E. (1978) FEBS Letters 89, 42-46.
13. Kates, M., Deroo, P.W. (1973) J. Lip. Res. 14, 438-445.
14. Karlsson, K.-A., Larson, G. (1978) FEBS Letters 87, 283-287.

TRITERPENES, STEROIDS, ALKANES AND WAXES FROM THE PEEL OF GRAPES

CARL HEINZ BRIESKORN AND GÜNTER BLOSCZYK
Institut für Pharmazie und Lebensmittelchemie der Universität,
D 8700 Würzburg (Federal Republik of Germany)

ABSTRACT

In the peel of grapes, besides oleanolic acid, we isolated little quantities of oleanonic acid, 29-hydroxyoleanolic acid, 3-hydroxyolean-12-en-28-aldehyd and two till now not identified oleanen derivatives. The main-compound of steroids is ß-sitosterol. The alkanes consist of odd- and evennumbered compounds. The main alkanes have the chainlength C_{25} and C_{27}. The waxes are identical in the evennumbered alcohols from C_{22} to C_{28}. They distinguish themselves in their fatty acids.

INTRODUCTION

The chemical composition of the surface wax of grapes has received only little attention. This is caused especially by the trouble to remove the peel of a little fruit and to take care that no grape seeds get in the investigation material. Until now is pointed out the presence of oleanolic acid[1], of ß-sitosterol[1], of wax[2] and free alcohols[3].

MATERIALS AND METHODS

Material. 100 - 150 kg of husks of Roedelseer Kuechenmeister, Mueller-Thurgau, were the starting material. We separated the fruit rests of the peels by a solution of ammonium oxalate (5 %) and oxalic acid (2,5 %) at the temperature of 80 - 90 °C. By this process the seeds, because specific lighter, ascented and could be separated. The rest of seeds was separated by hand after drying the peels. By the mechanical stirring in the ammonium oxalate-oxalic acid solution the original peels were tended in pieces of 1 cm². We received 1815 g of pure peels.

Method. After grinding the peels in a mill, the powder was extracted exhaustive by light petroleum (Bp. 50 - 70 °C) then by ether. We received 4,97 % of light Petroleum- and 2,6 % of ether-extract, related to dried peels, free of seeds and rests of the fruit. By column chromatography (SiO_2, Al_2O_3 and Al_2O_3 mixed with $AgNO_3$) the compounds of the two extracts were separated. The elution began with light petroleum (50 - 70°C). Its polarity was increased very slowly by addition of ether. The isolated fractions of alkanes, waxes, free alcohols, triterpenoic acids and sterols were identified by gas liquid chromatography, mass spectrometry, nmr and thinlayer chromatography.

RESULTS

TABLE 1

MAIN COMPOUNDS FROM GRAPE PEEL

Compounds	% of total extract impure	pure	% of peels
Alkanes	---	0,9	0,07
Waxes	---	15,8	1,2
Free alcohols	9,7	2,4	∼0,74
Sterols	16,8	3,8	∼1,23
Triterpenes	---	40,5	3,1

Table 1 shows the groups of compounds and their quantities, we have isolated.

TABLE 2

COMPOSITION OF ALKANES

C-number	%	C-number	%
15	7,1	24	3,7
16	0,1	25	20,6
17	5,8	26	3,6
18	1,3	27	15,1
19	0,6	28	2,5
20	1,7	29	9,2
21	5,2	30	1,8
22	3,4	31	8,9
23	9,5		

Table 2 informs about the exact composition of the alkanes,

TABLE 3

COMPOSITION OF WAXES (ester of high alcohols with high fatty acids)

C-number	I %	II %	III %	IV %	V %
Alcohols					
21	trace	trace	trace	trace	trace
22	12,0	5,7	7,0	6,0	6,2
23	trace	trace	trace	trace	1,5
24	40,7	28,1	32,1	30,3	26,2
25	trace	trace	trace	trace	3,7
26	36,4	49,1	42,7	44,1	43,4
27	trace	trace	trace	trace	1,5
28	10,9	17,1	18,1	19,6	17,5
Fatty acids					
15	---	---	---	trace	trace
16	5,5	15,4	3,8	1,7	3,0
17	---	---	---	trace	trace
18	14,7	15,5	1,7	4,2	6,7
18:1	---	56,9	4,4	---	---
18:2	---	5,2	38,3	---	---
18:3	---	---	15,6	---	---
19	---	---	---	0,1	trace
20	53,0	2,8	14,7	24,0	41,5
21	---	---	---	1,5	1,2
22	20,8	1,3	9,9	16,8	20,8
23	---	---	---	1,2	0,8
24	6,0	1,7	7,4	13,5	10,4
25	---	---	---	1,5	1,4
26	---	1,2	4,0	8,1	6,7
27	---	---	---	0,1	0,4
28	---	---	---	6,2	2,8
29	---	---	---	1,0	0,2
30	---	---	---	2,8	1,0
31	---	---	---	---	0,1
32	---	---	---	---	1,0

separated by gas-liquid-chromatography. There are eight even- and nine oddnumbered alkanes. Among the odd alkanes the hydrocarbons C_{23}, C_{25} and C_{27} are the main compounds. Among the even-numbered alkanes C_{24} is with 3,7 % the quantitative utmost.

Table 3 shows the five different fractions of waxes. Every fraction contains after saponification the same four even-numbered alcoholes with the carbon numbers C_{22}, C_{24}, C_{26} and C_{28}. Their quantity is nearly the same in every fraction. The different elution by column-chromatography should be influenced by the higher fatty acids of each of the waxes. The acids differ in the C-number, the quantity and in their degree of saturation.

TABLE 4
COMPOSITION OF FREE ALCOHOLS

	Fraction		
C-number	A	B	C
	%	%	%
22	2,4	17,5	3,9
23	---	---	0,8
24	41,2	45,7	28,6
25	---	---	2,5
26	42,8	30,7	36,5
27	---	---	1,4
28	8,2	---	17,8

Table 4 shows the composition of the free alcohols. They could be separated in three fractions. Fraction A and B consist only of even-numbered alcohols. Fraction C contains also little quantities of odd-numbered alcohols.

TABLE 5
COMPOSITION OF STEROLS

	%
ß-Sitosterol	87,4
Stigmasterol	4,6
Campesterol	7,5

Table 5 shows the composition of the steroid fraction. Until now only ß-sitosterol was known. We have also found campesterol and stigmasterol. ß-sitosterol with 87,5 % is the main compound.

TABLE 6
COMPOSITION OF TRITERPENES

Triterpene	Light petroleum-extract	Ether-	%
3ß-hydroxyolean-12-en-28-oic acid	+	+	95,6
3-oxoolean-12-en-28-oic acid	+	+	2,5
3ß-hydroxyolean-12-en-28-aldehyde	+	−	0,4
3ß.29α-dihydroxyolean-12-en-28-oic acid	−	+	together 0,4
T_5	−	+	
T_6	−	+	

Table 6 shows the triterpenes of grape peel. We also found oleanolic acid (1) as the main triterpene. By careful column chromatography of the ether- and light petroleum-extract and after separation of oleanolic acid we could isolate five further triterpenes. All belong to the oleanen-type. We identified its product of oxidation at C-3, oleanonic acid (2), of oxidation at C-28, 29α-hydroxyoleanolic acid (3) and of reduction at C-28, 3-hydroxy-olean-12-en-28-aldehyde (4). The two last mentioned oleanen-derivatives belong to the group of triterpenes very rarely found in plants. 3 was for a long time only known as a glycosid in cactus[4]. Some years later we isolated the same compound from the leaf of Melissa off. L.[5]. 4 has been found in plants two times[6,7]. The two other until now not identified triterpenes (T_5, T_6) are also oleanen-derivatives. Both contain a further oxogroup, T_6 contains another double bond.

A minor part of 1 we found esterified about its C-3 OH group or about its C-28-carboxylic group. The chainlength of the fatty acids differ between C_{15} to C_{28}, of the alcohols between C_{21} to C_{28}.

	R_1	R_2	R_3
1	OH	COOH	CH_3
2	=O	COOH	CH_3
3	OH	COOH	CH_2OH
4	OH	CHO	CH_3

SUMMARY

The quantity of triterpenoids and biogenetical related steroids is about the half greater than the quantity of lipids. Nevertheless Radler[8] has pointed out that only lipids are responsible for the low water permeability of the grapes. Within the lipids alkanes and higher alcohols are the most efficient. They reduce the water permeability from one third to one seventh of the efficiency of oleanolic acid. Derivatives of ursen group we did not found in the cuticular substance of the peel.

ACKNOWLEDGEMENTS

We gratefully acknowledge the Julius-Spital, Wuerzburg, for giving the husks.

REFERENCES

1. Markley, K.S. et. al. (1938) J. biol. Chem. 123, 641
2. Radler, F. (1965) Aust. J. biol. Sci 18, 1045
3. Dudman, W.F., and Grncarevic, M. (1962) J. Sci. Food Agric. 13, 221
4. Tursch, B. et. al. (1967) Tetrahedron Letters 539
5. Brieskorn, C.H., and Krause, W. (1974) Arch. Pharm. 307, 603
6. Shamma, M., and Rosenstock, P. (1959) J. org. Chem. 24, 726
7. Corsano, S., and Mincione, E. (1967) Ric. Sci. 37(4), 370
8. Radler, F. (1965) Nature, Lond. 207, 1002.

ON PHOSPHOLIPIDS AND FATTY ACIDS IN CELL-FREE FRACTIONS FROM THE RHODOSPIRILLACEAE Rhodopseudomonas palustris AND Rps. spheroides.

KARL KNOBLOCH and FREDI GEMEINHARDT
Institute of Botany and Pharmaceutical Biology, University of Erlangen-Nürnberg, Schlossgarten 4, D-8520 Erlangen, Germany

INTRODUCTION

Photosynthetic bacteria contain a membrane system which separates the cytoplasm into two cell compartments. Under anaerobic conditions, the cytoplasma membrane grows by invagination to form the intracytoplasmic membrane system where bacteriochlorophyll is inserted. The various species differ in the arrangement of the membrane structure[1]. While Rhodopseudomonas spheroides forms membrane vesicles (Fig. 1 and 2), the Rps. palustris species develops a continuous intracytoplasmic membrane (Fig. 1). In dark aerobic environment, the Rhodospirillaceae do not synthezise bacteriochlorophyll and/or the intracytoplasmic membrane systems. The cells bleach out performing life by respiratory energy metabolism catalyzed by the cytoplasma membrane.

After breaking the cells by use of high pressure, the chromatophore membranes (P-144) can be separated from the "soluble" supernatant fraction (S-144) by ultracentrifugation (144,000 x g for 60 min) (Fig. 1). However, the supernatant still catalyzes electron transfers[2-4] which, due to their characteristics, ought to be integrated into membranes. For this reason, phospholipids as building stones of biological membranes are of interest.

MATERIALS AND METHODS

Rps. palustris (ATCC 17001) was grown photolithotrophically on thiosulfate[3]. Aerobic dark culture conditions with pyruvate as the substrate were applied as well using the thiosulfate-less medium given under ref.[3]. Rps. spheroides (strain 2.4.1) was cultured either photoheterotrophically on succinate[5] or under dark aerobic conditions in the similar medium where succinate had been replaced by pyruvate.

The extraction of lipids was performed by a varied procedure described by Bligh and Dyer[6]. - Fatty acids were determined after mild alkaline deacylation using the method given by Kates[7].

RESULTS

As Fig. 1 illustrates, the two Rhodospirillaceae studied create a different intracytoplasmic membrane system in the light. As a result, upon disruption, the Rps. spheroides cell yields a chromatophore fraction which represents uniform rounded vesicles (Fig.2). Their interior holds contents of one of the cell compartments. On the other hand, the Rps. palustris cell yields chromatophore membranes which are leaking[8,9] (Fig. 3).

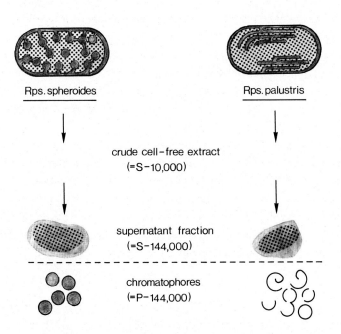

Fig. 1. Schematic arrangements of intracytoplasmic membrane systems of Rps. spheroides, forming closed chromatophore vesicles, and of Rps. palustris, producing connected thylakoid sheets. The cell-free extracts are separated by ultracentrifugation to yield the supernatant fraction (S-144) holding, in both cases, contents of both the cell compartments, and into the pellet (P-144) consisting out of the chromatophore structures.

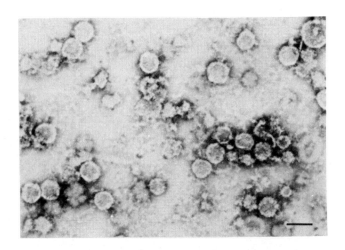

Fig. 2. Chromatophore fraction (P-144) obtained from Rps. spheroides. (The bar represents 0.1 um)

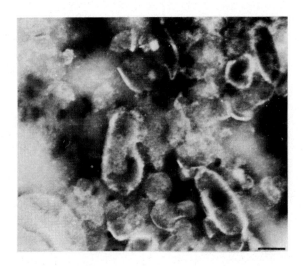

Fig. 3. Negative staining of chromatophore membranes from Rps. palustris. (Electron micrographs by Dr. G. Acker)

The supernatant fractions from both the bacteria have to keep the contents of both the cell compartments (Fig. 1). The supernatants catalyze electron transfers such as a succinate- or thiosulfate-linked energy-dependent reversed electron flow resulting in the reduction of NAD^+ [3]. The chromatophore membrane pieces from Rps. palustris, empty through the centrifugation process, do not perform the electron transfer. The closed up chromatophores from Rps. spheroides do catalyze the energy-dependent reaction (in the presence of reduced mammalian cytochrome c).

After aerobic heterotrophic growth in the dark, the similar supernatant fractions obtained from the bacteria also reveal enzymatic activities with respiratory electron transport chains involved. Electron transport inhibitors are active.

The amounts of phospholipids in the supernatant fractions from Rps. palustris after photosynthetic and dark heterotrophic growth conditions (Table 1) and the corresponding data from Rps. spheroides cultures (data not given) indicate the presence of slender membrane fragments within the supernatants.

TABLE 1
LIPID AND PHOSPHOLIPID CONTENT OF CELL-FREE FRACTIONS FROM PHOTOSYNTHETICALLY AND DARK GROWN Rps. palustris

cell-free fraction	growth on			
	light plus thiosulfate		pyruvate plus O_2	
	lipid/protein	P/lipid	lipid/protein	P/lipid
S- 10	0.09	0.0095	0.12	0.0158
P-144	0.12	0.0095	0.20	0.0156
S-144	0.03	0.0080	0.08	0.0141

The phospholipids identified within the several fractions from both the organisms after growth in the light and/or in the dark are phosphatidylethanolamine, cardiolipin, phosphatidylglycerol, and phosphatidylcholine (Fig. 4). The proportional composition of the different phospholipids, however, seems not to be the same under light or under dark culture conditions.

The pattern of fatty acids produced by both the bacteria is basically similar. However, again it varies in the relative amounts of the several components produced under the different conditions of growth. The fatty acids identified are tetradecanoic, lipoic (LipS$_2$), hexadecenoic, hexadecanoic, C$_{17}$-cyclopropanoic (17cy), octadecenoic, and octadecanoic acid.

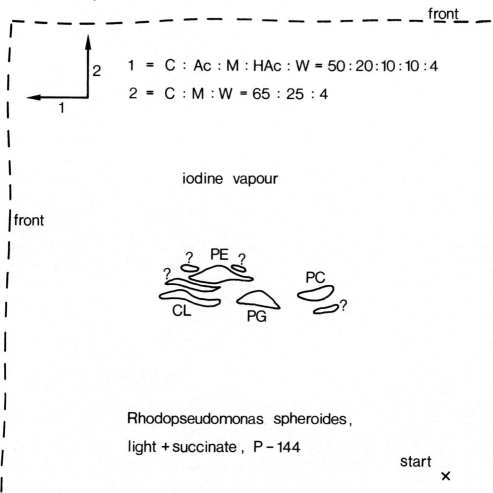

Fig. 4. Thin layer chromatogram of phospholipids from Rps. spheroides grown photosynthetically on succinate, chromatophore fraction P-144. Silica gel plate (MERCK 60). The first dimension was developed in chloroform : aceton : acetic acid : methanol : water = 50 : 20 : 10 : 10 : 4, the second dimension in chloroform : methanol : water = 65 : 25 : 4. Visualization and identification by successive staining with iodine vapour, phosphate and ninhydrin reagent.

DISCUSSION

From previous results on inhibitor sensitive electron transport systems[2-4], we had to conclude that the supernatants do contain membranes. Electron micrographs support the presence of membrane fragments which are 100 to 500 Å in size with subunits of about 50 Å in diameter[10].

The data presented here prove the presence of phospholipids in the supernatant fractions, hereby indicating directly the membranous character.

The phospholipids identified are not unusual. Among the fatty acids, the C_{17}-cyclopropanoic acid is of interest. Of importance, however, seems to be the identification of oleic and lipoic acid ($LipS_2$). In a reaction with activated acetaldehyde (Ac-TPP), $LipSH_2$ forms the energy-rich compound S-acetyl-dihydro-lipoic acid. Lipoic acid has been found to be required in a system to substitute for ATP in the supernatant fraction S-144 from Rps. palustris in order to drive energy-dependent electron transfers[11]. A dihydrolipoate- and oleate-dependent ATP synthesis in yeast promitochondria has been reported[12].

As far as the appearance of membrane (sub-)units in the supernatant fractions is concerned, we think about the possibility that during the process of cell disruption slender membrane parts, are disconnected which naturally are attached to or embedded into the heavier chromatophore containing membrane layer.

ACKNOWLEDGMENT

This work was supported by grants from the Deutsche Forschungsgemeinschaft.

REFERENCES

1. Oelze, J. and Drews, G. (1972) Biochim. Biophys. Acta, 265, 209.
2. Knobloch, K. (1975) Z. Naturforsch., 30c, 342.
3. Knobloch, K., Eley, J.H. and Aleem, M.I.H. (1971) Arch. Mikrobiol., 80, 97.
4. Eley, J.H., Knobloch, K. and Aleem, M.I.H. (1971) Arch. Biochem. Biophys., 147, 419.
5. Sistrom, W.R. (1960) J. Gen. Microbiol., 22, 778.
6. Bligh, E.G. and Dyer, W.J. (1959) Can. J. Biochem. Physiol., 37, 911.

7. Kates, M. (1975) Techniques of lipidology: isolation, analysis and identification of lipids, North-Holland, Amsterdam, pp. 558-565.
8. Knobloch, K. and Acker, G. (1977) Hoppe-Seyler Z. Physiol. Chem., 358, 1232.
9. Knobloch, K. (1978) Hoppe-Seyler Z. Physiol. Chem., 359, 286.
10. Knobloch, K. (1974) Ber. Deut. Bot. Ges., 87, 119.
11. Knobloch, K. (1977) Hoppe-Seyler Z. Physiol. Chem., 358, 262.
12. Griffiths, D.E., Hyams, R.L. and Bertoli, E. (1977) FEBS Letters, 74, 38.

APPLICATION OF TLC/FID IN LIPID ANALYSIS

BENGT HERSLÖF
AB Karlshamns Oljefabriker, Research Laboratory, S-292 00 Karlshamn (Sweden)

INTRODUCTION

Natural lipids are generally a complex mixture of chemical species. This complexity has presented considerable difficulties to deal with for the analytical lipid chemist. The possibility of studying a single chemical species came with the advent of chromatographic techniques, which were introduced in the 1940-60's.

Many different forms of chromatography exist in the modern laboratory. In principle we can talk about two main types: gas chromatography and liquid chromatography. The liquid systems are capable of being used for all substances, whereas gas chromatography has certain limitations depending on the volatility of the material.

Similar to GC, several of the LC techniques (e.g. HPLC) can be characterised as column chromatography. In these cases the chromatogram is recorded after the column by the aid of a detector or by collecting fractions which are then subjected to further identification. Unlike the above procedures, thin layer and paper chromatography are techniques which can be regarded as open column chromatography. The pattern of separated compounds is permanently available for examination when the actual chromatographic procedure is finished.

The most universal and sensitive detectors exist for gas chromatography; the flame ionisation detector (FID) is probably the commonest used. In LC column systems many different detectors are used, e.g. UV, RI, IR and flourescence detectors. The open column systems have the advantage that visual inspection can be made (UV-light, spray reagent, charring) but the disadvantage of difficult quantification.

During recent years attempts have been made to combine the open column system with a universal detector[1,2,3]. The simple operation of the TLC technique and the fact that all of the spotted sample remains after development, makes it ideal for flame ionisation detection. A very successful device in this respect is the commercially available chromarods combined with the Iatroscan TH-10 Analyser.

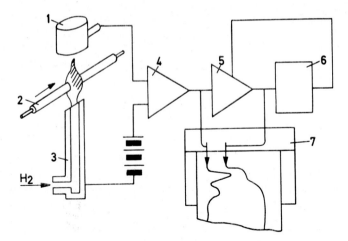

Fig. 1. Principle of method of detection: 1:Collector, 2:Chromarod, 3:Burner, 4:Current Amplifier, 5:Integral amplifier, 6:Integral Reset Circuit and 7:Recorder.

DESCRIPTION OF THE EQUIPMENT

The separation is performed on the so-called chromarods[4]. These rods, which are made of quartz (0.9 mm in diameter), have silica gel or alumina powder fused onto the support using fine glass powder as a binding agent. The sintered rods can be used repeatedly (100 determinations are guaranteed by the manufacturer).

The rods are scanned directly in the hydrogen flame of the detector (Figure 1). 10 rods can be mounted at the same time and recorded automatically.

ANALYTICAL PROCEDURE

The first part of the procedure is practically identical to the ordinary TLC procedure. Ten rods in a metal or glass holder are spotted and developed together in a tank with a suitable solvent mixture. The developing time is usually 20-45 minutes, depending on the nature of the solvent. The solvent is dried off and the rods are transferred to the scanning frame, which is part of the detector. The second part of the procedure is the scanning which takes about 30 sec./rod (this time is adjustable). After scanning, the rods are replaced in the holder and can be immediately spotted with new samples.

RESULTS

The result of a separation on a chromarod is a chromatogram which is very similar to a gas chromatogram. The applications are as numerous as ordinary TLC. Experience in separations performed on silica gel or alumina can successfully be translated to the TLC/FID method. However, the solvent mixtures generally have to be slightly modified, in most cases by lowering the polarity.

APPLICATION TO LIPID ANALYSIS

So far only a few reports have appeared in litterature concerning the use of TLC/FID. However, it is interesting to note that most of these are concerned with lipids.

Kawai et al.[5], Nakano et al.[6], Ueda et al.[7], and Vandamme et al.[8] have reported analysis of plasma lipids. Phospholipids have been analysed by Tokunaga et al.[9]. Gantois et al.[10] reported on the analysis of glycerides and Tanaka et al.[11] described the technique for triglyceride analysis depending on the degree of unsaturation.

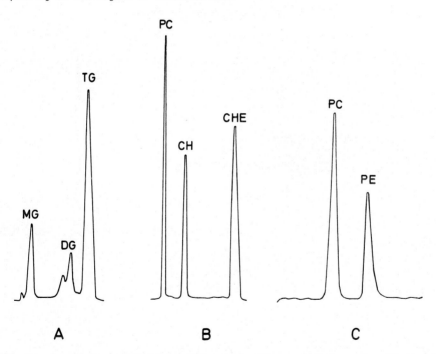

Fig. 2. Separation of lipids on chromarods. MG:monoacylglycerols; DG:diacylglycerols; TG:triacylglycerols; PC:phosphatidylcholine; CH:cholesterol; CHE:cholesterol esters; PE: phosphatidyl ethanolamine. A and B: hexane - diethylether - acetic acid 80:20:1. C:chloroform - methanol - ammonium hydroxide 60:30:5.

In our laboratory we have been able to use the TLC/FID technique for a few months. We are concerned with lipid analyses in general, for routine as well as research applications. Our aim has hitherto been to find out the general applicability of the new technique, and to obtain practical experience.

The most useful types of analyses are the separation and quantification of different lipid classes. Figure 2 shows examples of chromatograms of lipids with different polarity.

In collaboration with other laboratories we are working on a number of - different projects. Figure 3 shows chromatograms from a study of dietary lipids[12], and Figure 4 chromatograms from a study of the minor components in vegetable oils[13].

In order to separate different components within the same lipid class GLC or HPLC are generally the preferred techniques; however a reverse phase type of separation, is in principle, possible to perform on chromarods. The stationary hydrocarbon phase being removed with vacuum. Of more practical importance is the $AgNO_3$ impregnated rods which are capable of separating a lipid class somponents according to the degree of unsaturation.

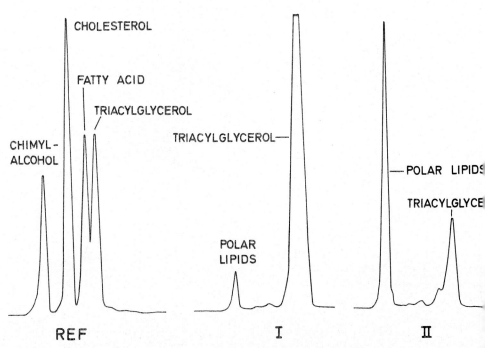

Fig. 3. Dietary lipids studied by the duplicate portion technique. I 1:st extraction (chloroform), II 2:nd extraction (chloroform - methanol 1:1).

GENERAL COMMENTS

We have found the technique to be of considerable value. The sensitivity is very high (10^{-7} g) and our experience is that 2-5 µg of a lipid mixture normally gives a well-defined chromatogram. The rods normally have a long life and are quite resistant with regard to mechanical wear. Cleaning the rods after 15-20 determinations, in hydrochloric acid, prolongs their life time.

The response of the detector can vary with different types of lipids, but we have not found this in the case of glycerides, fatty acids or phosphoglycerides, but sterols do seem to give a higher response. This point must be taken into account and calibration curves have to be made.

One of the techniques main advantages is the rapid scanning procedure. This allows a very fast screening of a large number of samples. The flexibility of the developing part of the analyses is another advantage. The separate detector unit permits any type of rod to be scanned anytime. This is different for GLC or HPLC where the separational procedure is linked with the detector in a single system.

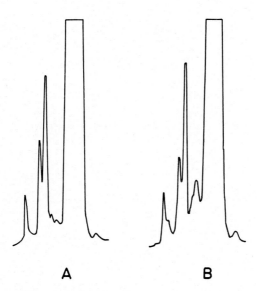

Fig. 4. Minor components in Shea oil. A:concentrated at -40°C; B:concentrated at -50°c. Solvent:Hexane - diethylether - acetic acid 80:20:1.

REFERENCES

1. Cotgreave, T. and Lynes A. J. Chromatography 30(1967)117.
2. Padley, F.B., J. Chromatography 39(1969)37.
3. Szakasits, J.J., et al., Anal. Chem. 42(1970)251.
4. Okumura, T. and Kadono, T., Japan Analyst 22(1973)980.
5. Kawai, T. et al., Jap. J. Clin. Pathol. 19(1971)293.
6. Nakano, E. et al., Jap. J. Clin. Pathol. 20(1972)186.
7. Ueda, H. et al., Jap. J. Med. Techn. 19(1975)639.
8. Vandamme, E., Blaton V. and Peeters, H., J. Chromatography 145(1978)151.
9. Tokunaga, M. et al., Jap. Conf. Biochem. Lip. 15(1973)195.
10. Gantois, E., Mordret, F., and Le Barbanchon, N., Rev. Franç. Corps Gras 24(1977)167.
11. Tanaka, M. et al., 15th Ann. Meet. Jap. Oil Chem. Soc. 1976.
12. Åkesson, B., unpublished data.
13. Appelquist, L.-Å., Johansson A. and Herslöf, B., unpublished data.

EFFECTS OF LIGHT ON LIPIDS IN POTATO TUBERS

CONNY LILJENBERG, ANNA STINA SANDELIUS and EVA SELSTAM
Department of Plant Physiology, Botanical Institute, University of Göteborg,
Carl Skottsbergs Gata 22, S-413 19 Göteborg (Sweden)

ABSTRACT

The contents and the fatty acid composition of monogalactosyl diglycerides, digalactosyl diglycerides and triglycerides in potato tubers were investigated during post harvest storage in darkness and in light.

Storage in total darkness resulted in only small changes in the amounts of the above mentioned lipid classes. Light treatment caused greening of the peripheral parts of the tubers, together with an increase in the digalactosyl diglyceride contents in these parts. At the same time the fatty acid composition of the galactolipids changed towards higher unsaturation. Simultaneously, the fatty acid composition of the triglycerides changed to higher saturation.

INTRODUCTION

In potato tubers, lipids constitute only about 0.5 % of the dry weight[1,2]. Approximately 85 % of the total lipids are polar lipids[1] and the total content of linoleic acid + linolenic acid is approximately 75 % of the total fatty acids[1,3,4]. The main part of the lipids in potato tubers is considered to be membrane lipids[2]. During post harvest storage, the total lipid content slowly decreases[5,6,7], while their fatty acid composition remains fairly constant[1,3,5].

Irradiation of potato tubers results in biosynthesis of chlorophylls[8] and glycoalkaloids[9]. The former process might reflect a formation of plastids similar to the chloroplast.

In this study we have investigated whether irradiation of potato tubers, apart from stimulating chlorophyll biosynthesis, also influences their lipids in respect to content and composition, as compared with non-irradiated tubers.

Abbreviations: MGDG, monogalactosyl diglyceride, DGDG, digalactosyl diglyceride, TG, triglyceride, 12:0, dodecanoic acid, 14:0 tetradecanoic acid, 16:0, hexadecanoic acid, 18:0 octadecanoic acid, 18:1, cis-9-octadecenoic acid, 18:2, cis,cis-9,12-octadecadienoic acid, 18:3, all cis-9,12,15-octadecatrienoic acid.

MATERIALS AND METHODS

The varieties Bintje and Desirée of potato (Solanum tuberosum L.) were grown under standard conditions at Svalöv in southern Sweden and harvested in September. The tubers were stored in darkness at 6°C. Tubers used for analysis were randomly chosen. For estimation of light effects on the tuber lipids, some tubers were irradiated with white light for 14 h per day in a chamber at 12°C with an irradiance of 12 W·m^{-2}.

Before lipid extraction, halves of potato tubers were vacuum dried. The dry potato-half was weighed and placed in chloroform-methanol 2:1 (v/v). for 24 hours at 2°C. This soaked potato half was then homogenized and the homogenate was stirred for 1 hour at 2°C in darkness[5]. The homogenate was filtered and the lipid extract washed with one volume of 0,73 % NaCl in distilled water[10].

A part of the total lipid extract was taken to dryness on a rotary evaporator and dissolved in chloroform and separated into lipid classes on a silicic acid column[11,12]. The triglycerides and the galactosyl diglycerides were further purified by thin layer chromatography. Silica gel H plates (0,5 mm thick, Merck) were developed in light petrol (b.p. 40-60°C)-diethyl ether-formic acid 85:15:1 (v/v/v) for separation of the neutral lipids and in chloroform-methanol-acetic acid-water 85:15:10:3.5 (v/v/v/v) for separation of the galactolipids[13]. Reference substances were run on both sides of the sample and located with iodine vapour. TG, DGDG and MGDG were scraped off the plates and extracted from the gel with chloroform:methanol 2:1 (v/v).

The fatty acids of the lipids were analysed as methyl esters[14] by gas chromatography. The column was 0.2 x 250 cm and packed with 5 % butane-1,4-diol-succinate on Chromosorb W (80-100 mesh). The chromatograph was run isothermally at 185°C. When not analysed, lipid extracts, at all stages of analysis, were stored under N_2 at -18°C.

All values are the mean of three experiments.

RESULTS

Content of DGDG, MGDG and TG (Table 1 and 2)

Storage in darkness. In the potato variety Desirée, the contents of DGDG, MGDG and TG remained fairly constant. In Bintje, the content of DGDG, and to a lesser degree also that of TG, increased initially during storage in darkness. The content of MGDG in Bintje was fairly constant.

Storage in darkness + light treatment. Light caused a marked increase in the DGDG content in Bintje. DGDG in Desirée, as well as MGDG in both varieties, remained fairly constant compared with tubers stored in darkness. TG content in both varieties decreased slightly during light treatment.

TABLE 1

CONTENTS OF DGDG, MGDG AND TG IN THE POTATO VARIETY BINTJE

Time of storage (days)	Content of lipid (μmol lipid $\cdot$ (mg dry weight of potato tuber)$^{-1}$)		
	DGDG	MGDG	TG
18 d[a]	0.16	0.37	0.19
41 d	0.75	0.37	0.36
69 d	0.60	0.30	0.41
69 d + 12 l[b]	1.05	0.38	0.33

[a]18 d = storage in darkness for 18 days.
[b]69 d + 12 l = storage in darkness for 69 days + light treatment for 12 days.

TABLE 2

CONTENTS OF DGDG, MGDG AND TG IN THE POTATO VARIETY DESIREE

Time of storage (days)	Content of lipid (μmol lipid $\cdot$ (mg dry weight of potato tuber)$^{-1}$)		
	DGDG	MGDG	TG
14 d[a]	0.49	0.34	0.20
36 d	0.49	0.28	0.29
63 d	0.56	0.38	0.21
63 d + 12 l[b]	0.62	0.40	0.10

[a,b]for explanation, see footnotes Table 1.

Fatty acid composition of DGDG, MGDG and TG (Figures 1, 2 and 3)

The changes in fatty acid composition of DGDG and MGDG, during storage in darkness and in darkness + light, were similar in both potato varieties.

Storage in darkness. The fatty acid compositions of DGDG and TG, respectively, remained fairly constant during storage in darkness. In MGDG, however, the percentage of 18:2 increased, and that of 18:3 decreased in the beginning of the storage period.

Storage in darkness + light treatment. The percentages of 16:0 and 18:0 + 18:1 in DGDG and MGDG remained on the same levels as during storage in darkness. In both DGDG and MGDG the percentage of 18:2 decreased while that of 18:3

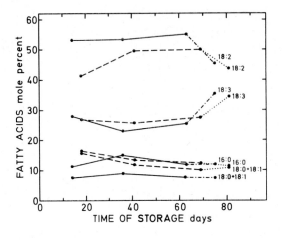

Fig. 1. Fatty acid composition of DGDG in Bintje and Desirée versus storage in darkness and in light.
– – – – Bintje, stored in darkness,
••••••• Bintje, stored under white light,
——— Desirée, stored in darkness,
•–•–•– Desirée, stored under white light.

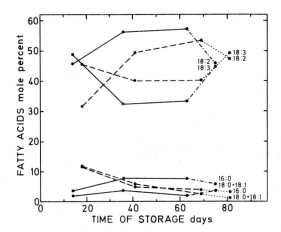

Fig. 2. Fatty acid composition of MGDG in Bintje and Desirée versus storage in darkness and in light.
Otherwise as in Fig. 1.

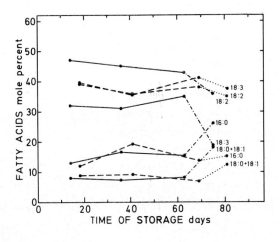

Fig. 3. Fatty acid composition of triglycerides in Bintje and Desirée versus storage in darkness and in light.
Otherwise as in Fig. 1.

increased. The percentage of 18:2 + 18:3 remained on approximately the same level as during storage in darkness.

In TG, light treatment caused quite different changes in fatty acid composition. The percentages of 16:0 and 18:0 + 18:1, respectively, increased, while those of 18:2 and 18:3 decreased. All light-induced changes in fatty acid composition of TG were more pronounced in the potato variety Desirée than in the variety Bintje. In Desirée the percentage of 18:2 + 18:3 in TG decreased from c. 78 % during storage in darkness to c. 55 % in the irradiated potato tubers. In Bintje, the corresponding values were c. 80 % and c. 70 %, respectively.

DISCUSSION

The potato tuber contains only a small amount of lipids. For a storage organ, it has a rather unusual lipid composition, with relatively low contents of neutral lipids[1,15]. Polar lipids, mainly galactolipids and phospholipids, make up the major part of the total lipids[1]. Galactolipids are generally known as membrane components in photosynthetic tissues. Assuming that the mitochondria do not contain higher amounts of galactolipids[16,17,18,19,20], in the potato tubers, these lipids are therefore probably membrane constituents of the amyloplasts and the proplastids.

Irradiation of the tubers caused an increase in the amounts of galactolipids as well as a marked change in the fatty acid composition of these lipids (Table 1 and 2, Fig 1 and 2). At the same time, greening of the tubers was observed.

It has been shown that irradiation of potato tubers causes ultrastructural changes in the amyloplasts of the tubers[21,22]. Formation of thylakoid-like membranes has been demonstrated[21,22].

The results of the present investigation might be interpreted as a de novo formation of thylakoid membranes containing galactolipids with high contents of polyunsaturated fatty acids. The observed light-induced change in the fatty acid composition of the triglycerides (Fig 3) might then reflect a utilization of the fatty acids of the triglycerides for thylakoid membrane formation.

REFERENCES

1. Lepage, M. (1968) Lipids, 3, 477-481.
2. Galliard, T. (1973) J. Sci. Food Agric., 24, 617-622.
3. Galliard, T. (1972) Phytochemistry, 11, 1899-1903.
4. Bolling, H. and El Bayà, A.W. (1973) Z. Pflanzenphysiol., 69, 402-408.
5. Liljenberg, C., Sandelius, A.S. and Selstam, E. (1978) Physiol. Plant., 43, 154-159.

6. Mondy, N.I., Mattich, L.R. and Owens, E. (1963) J. Agric. Food Chem., 11, 328-329.
7. Berkeley, H.D. and Galliard, T. (1974) J. Sci. Food Agric., 25, 861-867.
8. Vecher, A.S., Nenadovich, R.A., Mas˘ko, A.A. and Reshetnikov, V.N. (1978) Fiziologija i biochimija kul˘turnych rastenij, 10, 269-275.
9. Jadhav, S.J. and Salunkhe, D.K. (1975) in Advances in Food Research, Chichester, C.O. ed., 21, pp. 307-353.
10. Folch, J., Lees, M. and Stanley, G.H.S. (1957) J. Biol. Chem., 226, 497-509.
11. Rouser, G., Kritchevsky, G., Simon, G. and Nelson, G.J. (1967) Lipids, 2, 37-40.
12. Appelqvist, L.Å. (1972) J. Am. Oil Chem. Soc., 49, 151-152.
13. Nichols, B.W., Harris, R.V. and James, A.T. (1965) Biochem. Biophys. Res. Commun., 20, 256-262.
14. Morrison, W.R. and Smith, L.M. (1964) J. Lipid Res., 5, 600-608.
15. Galliard, T. (1968) Phytochemistry, 7, 1907-1914.
16. McCarty, R.E., Douce, R. and Benson, A.A. (1973) Biochim. Biophys. Acta, 316, 266-270.
17. Schwertner, H.A. and Biale, J.B. (1973) J. Lipid Res., 14, 235-242.
18. Mackender, R.O. and Leech, R.M. (1974) Plant Physiol., 53, 496-502.
19. Ohmori, M. and Yamada, M. (1974) Plant Cell Physiol., 15, 1129-1132.
20. Mazliak, P. (1977) in Lipids and Lipid Polymers in Higher Plants, Tevini, M. and Lichtenthaler, H.K. eds., Springer-Verlag, Berlin, pp. 48-74.
21. Rezende-Pinto, M.C. de (1962) Port. Acta Biol. Ser. A 6, 239-242.
22. Lobov, V.P., Bondar, P.I., Egorova, L.D., Ostapljuk, A.N., Sakalo, V.D. and Dashaljuk, A.P. (1977) Fiziologija i biochimija kul˘turnych rastenij, 9, 40-47.

POLYPRENYLPHOSPHATE DERIVATIVES AS INTERMEDIATES IN THE BIOSYN-
THESIS OF CELLULOSE PRECURSORS. SUBCELLULAR LOCALIZATION.

H. E. HOPP, P. A. ROMERO, G. R. DALEO AND R. PONT LEZICA
Departamento de Biología, Fundación Bariloche (Argentina) and
Fachbereich Biologie, Universität Kaiserslautern, Postfach 3049,
D-6750 Kaiserslautern (Federal Republic of Germany)

ABSTRACT

Cellular membranes from the green alga Protheca zopfii were isolated by isopycnic centrifugation. The enzymes responsible for the glucosylation of dolichol derivatives and protein were found in the endoplasmic reticulum rich fraction. On the other hand, cellulose synthetase was found in the dictyosome rich fraction.

INTRODUCTION

In spite of cellulose being the most abundant substance on earth, very little is known about the mechanism of its biosynthesis. It is generally assumed that cellulose is a very simple polymer formed by linear β-1,4-linked glucoses. Recent work with particulate preparations of the Chlorophyta Prototheca zopfii had shown that lipid intermediates are involved in the synthesis of a "cellulosic protein". Glucolipids and oligosaccharide-linked lipids are formed from UDP-Glc and endogenous lipids[1]. The lipid moiety was identified as dolichyl phosphate, with a chain length ranging from C_{90} to C_{105}[2]. The oligosaccharides could be transferred from the lipid carrier to a protein acceptor, this reaction being inhibited by coumarin[3]. The glucoprotein serves as primer for the formation of an alkali insoluble polymer with the properties of cellulose[1].

We present here evidence for the subcellular localization of different steps involved in cellulose biosynthesis in green algae.

MATERIAL AND METHODS

Cell disruption

Prototheca zopfii strain PR-5 (ATCC 16533) was grown as described[2] and harvested from the log-phase cultures at 50% of maximum growth. Disruption by grinding was achieved by mixing packed cells with an equal volume of aluminium oxide to obtain a semi-solid paste, in a mortar at $0°C$; considerable pressure was employed during the 5 min grinding period. The paste was mixed thoroughly with its own volume of buffer: 50 mM Tris-HCl, pH 7.5, 2 mM EDTA, 10 mM KCl, 0.1 mM $MgCl_2$ and 8% sucrose for the high EDTA preparations; and 50 mM Tris-HCl, pH 7.5, 0.1 mM EDTA, 5 mM $MgCl_2$ and 8% sucrose for the high Mg^{++} ones. When the alumina had settled, the supernatant was decanted. The alumina was washed with more buffer and the supernatant combined. The homogenate was centrifuged at 300 x g for 10 min to remove cell wall debris, intact cells, chloroplast and nuclei. For the data in Table 1, the supernatant was successively centrifuged at 1 000 x g for 10 min, 10 000 x g for 10 min, 38 000 x g for 30 min and 105 000 x g for 120 min. The pellets from these centrifugations were suspended in 50 mM Tris-HCl, pH 7.4.

Linear sucrose gradients

For linear sucrose gradient analysis, a preliminar centrifugation at 10 000 x g was made in order to separate most mitochondria, plastid and large particles. Particles in the supernatant from this centrifugation were concentrated by layering it over a double spread composed of 10 ml 45% (w/w) sucrose and 10 ml 12% sucrose in buffer (10 mM Tris-HCl, pH 7, 0.1 mM $MgCl_2$ and 1 mM EDTA for the high EDTA experiments, and 10 mM Tris-HCl, pH 7, 3 mM $MgCl_2$ and 0.1 mM EDTA for the high Mg^{++} ones), centrifuged at 40 000 x g for 90 min. About 6 ml corresponding to the 10%/45% sucrose interphase were diluted until their concentration was below 15% (w/w), and layered over a 15 to 45% (w/w) sucrose linear gradient. Runs were made for 14 hr at 23 000 rpm in a Beckman SW 25.2 rotor. The gradient was fractionated into 4.5 aliquots, which were used for enzyme assay.

Enzyme assay

Marker enzymes were used for the determination of purity of subcellular fractions. NADH-cytochrome c reductase (antimicyn A insensible) was assayed as described[4] for endoplasmic reticulum.

5'nucleotidase was used as plasma membrane marker and assayed according to Morré[5]. Mitochondria were associated with cytochrome c oxidase activity and measured as described[6]. UDP-Glc: sterol glucosyltransferase was measured as described and used as Golgi apparatus marker[7].

UDP-Glc:polyprenyl phosphate glucosyltransferase was assayed as following: 10 µg dolichyl-P on chloroform-methanol (2:1, v/v) were mixed with 1 µmol EDTA and 2.5 µmol $MgCl_2$ and dried in test tube under N_2. Then 15 µmol Tris-HCl, pH 7.4, 0.5% Triton X-100 and 1 nmol UDP-(^{14}C)Glc (269 Ci/mol) were added, plus 0.1 ml of different fractions to a total volume of 0.25 ml. Incubations were carried out at 30°C for 30 min. The reaction was stopped with 0.5 ml butanol and glucolipids were separated as described[8]. Oligosaccharide-linked lipids and glucoprotein formation were assayed according to Hopp et al[1]. Cellulose synthetase was measured as reported[9] using UDP-(^{14}C)Glc (200 000 cpm) or UDP-(^{14}C)Glc (250 000 cpm) plus non radioactive GDP-Glc as precursors.

RESULTS

Previous work indicated that the enzymes involved in cellulose initiation were associated with particulate fractions[1]. Differential centrifugation were performed to determine the centrifugal

TABLE 1
FORMATION OF GLUCOLIPIDS AND POLYMERS IN DIFFERENT FRACTIONS

Subcellular membrane fractions obtained by differential centrifugation were assayed for the activity of enzymes involved in cellulose synthesis. Specific activity was expressed as (^{14}C) glucose incorporated from sugar nucleotides (cpm x 10^{-3}/mg of protein) in different fractions. Dol-P: dolichyl phosphate. Dol-PP: dolichyl pyrophosphate.

Fraction	Dol-P-Glc[a]	Dol-PP-(Glc)$_n$[a]	Gluco-protein[a]	Cellulose synthetase[b]
300–1 000 x g	47.29	8.49	14.10	30.14
1 000–10 000 x g	37.20	11.00	16.25	42.40
10 000–38 000 x g	73.89	19.93	39.67	78.13
38 000–105 000 x g	223.87	51.09	72.86	74.79

[a]UDP-(^{14}C)Glc as precursor. [b]GDP-(^{14}C)Glc as precursor.

force necessary to pellet these enzymes. Table 1 indicates that the highest specific activities of all the enzymes were found in the pellet sedimenting between 10 000 and 105 000 x g. For more precise analysis by linear sucrose gradients, the fraction sedimenting under 10 000 g was discarded.

Figure 1 shows that UDP-Glc:Dol-P glucosyltransferase, the enzymes responsible for the formation of oligosaccharide-lipids and the subsequent transfer from lipid to protein were associated to antimicyn A insensible NADH-cytochrome c reductase activity.

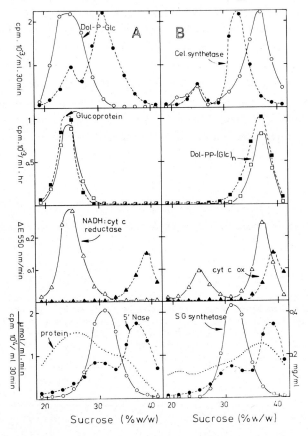

Fig. 1. Sucrose density gradient distribution of particulate enzymes from the cellulose pathway and the marker enzymes. SG synt: steryl glucoside synthetase (UDP-Glc:sterol glucosyltransferase); 5'Nase:5'nucleotidase; A:plus EDTA; B:plus Mg .

This is an indication that these activities were located in the endoplasmic reticulum. In addition, all these enzyme activities were shifted to a higher density in the gradient when the membranes were prepared in the presence of Mg^{++}. This change is caused by the release of ribosomal subunits from the rough endoplasmic reticulum by quelating agents[10, 11]. These results are in agreement with those obtained by other authors for the system involved in protein glycosylation in Phaseolus aureus[12] and Pisum sativum[13]. On the other hand, cellulose synthetase was consistently found associated with the Golgi rich fractions and non with the plasma membrane. Similar results were obtained by Shore and Maclachlan[11] in peas.

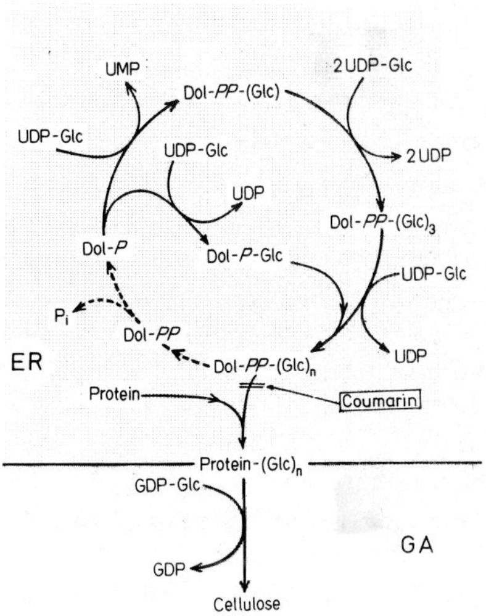

Fig. 2. General scheme with the reactions involved in the synthesis of cellulose and the localization within the cell. Dol-P: dolichyl phosphate; Dol-PP: dolichyl pyrophosphate; ER: endoplasmic reticulum; GA: Golgi apparatus.

Figure 2 shows the proposed scheme of cellulose biosynthesis in Prototheca zopfii with the localization of the enzymes involved. However, remarks are necessary: in our experiments only the UDP-Glc: dolichyl-P glucosyltransferase and cellulose synthetase were measured with external substrates. The other enzymes were tested with internal substrates and the result might reflect the actual amount of glucosyl accepting molecules and not different amounts of enzymes.

ACKNOWLEDGEMENTS

H.E.H. holds a fellowship from the Centro de Investigaciones Agronómicas, INTA (Argentina). R.P.L. is a career investigator from the Consejo Nacional de Investigaciones Científicas y Técnicas (Argentina) and holds an Alexander von Humboldt research fellowship. We are indebted to Prof. H. Kauss for continuous help and laboratory facilities.

REFERENCES

1. Hopp, H.E., Romero, P.A., Daleo, G.R. and Pont Lezica, R. (1978) Eur. J. Biochem. 84, 561-571.
2. Hopp, H.E., Daleo, G.R., Romero, P.A. and Pont Lezica, R. (1978) Plant Physiol. 61, 248-251.
3. Hopp, H.E., Romero, P.A. and Pont Lezica, R. (1978) FEBS Letters 86, 259-262.
4. Donaldson, R.P., Tolbert, T.E. and Schnarrenberger, C. (1972) Arch. Biochem. Biophys. 152, 199-215.
5. Morré, D.J. (1971) Methods Enzymol. 22, 130-148.
6. Simon, E.W. (1958) Biochem. J. 69, 67-74.
7. Hopp, H.E., Romero, P.A., Daleo, G.R. and Pont Lezica, R. (1978) Phytochemistry 17, 1049-1052.
8. Pont Lezica, R., Romero, P.A. and Dankert, M.A. (1976) Plant Physiol. 58, 675-680.
9. Barber, G., Elbein, A.D. and Hassid, W.Z. (1964) J. Biol. Chem. 239, 4056-4061.
10. Lord, J.M., Kagawa, T., Moore, T.S. and Beevers, H. (1973) J. Cell Biol. 57, 659-667.
11. Shore, G. and Maclachlan, G.A. (1975) J. Cell Biol. 64, 557-571.
12. Lehle, L., Bowles, D.J. and Tanner, W. (1978) Plant Science Letters 11, 27-34.
13. Nagahashi, J. and Beevers, L. (1978) Plant Physiol. 61, 451-459.

SEPARATION OF PRENYLLIPIDS BY HIGH PERFORMANCE LIQUID CHROMATOGRAPHY

URSULA PRENZEL and HARTMUT K. LICHTENTHALER
Botanical Institute (Plant Physiology), University of Karlsruhe,
D-7500 Karlsruhe, Kaiserstraße 12 (FRG)

ABSTRACT

The separation of plant pigments (chlorophylls, carotenoids), prenylquinones and prenylvitamins by high performance liquid chromatography (HPLC) can be achieved by using either adsorption or reversed phase chromatography. Generally, those prenyllipids which appear as one band in the adsorption HPLC system, are well resolved by reversed phase HPLC.

INTRODUCTION

Plant pigments (carotenoids and chlorophylls), prenylquinones (Fig. 1), prenylvitamins (Fig. 1 and 2) and prenols (Fig. 1 and 2) were separated in a recently developed chromatographic system: the high performance liquid chromatography. The adsorption HPLC is comparable to the separation sequence found in adsorption thin layer chromatography[1]. The most polar compounds have long retention times while less polar compounds appear earlier.
Reversed phase HPLC gives a good resolution for samples with different chainlength. Polar prenyllipids are resolved first.

MATERIALS

Instrument: Chromatograph S 100 and column switching valve, Siemens, FRG.
Detector: UV spectralphotometer, Zeiss, FRG.
Column: V 4 A steel, 250 x 3 mm, 125 x 3 mm.
Packing: adsorption HPLC = LiChrosorb SI 60, 5 μm, Merck, FRG
         reversed phase HPLC = LiChrosorb RP 8, 5 μm, Merck, FRG.
Solvents: adsorption HPLC = n-hexane + dioxan, Merck, FRG
         reversed phase HPLC = methanol + water.

RESULTS

Adsorption HPLC: Figure 3 shows an example of the separation of non polar (β-carotene) and more polar compounds (α-tocoquinone) on silicagel with a non polar solvent (hexane + 1 % dioxan). Decreasing the polarity of the solvent results in longer retention times but provides good separation of K_1 and PQ-9

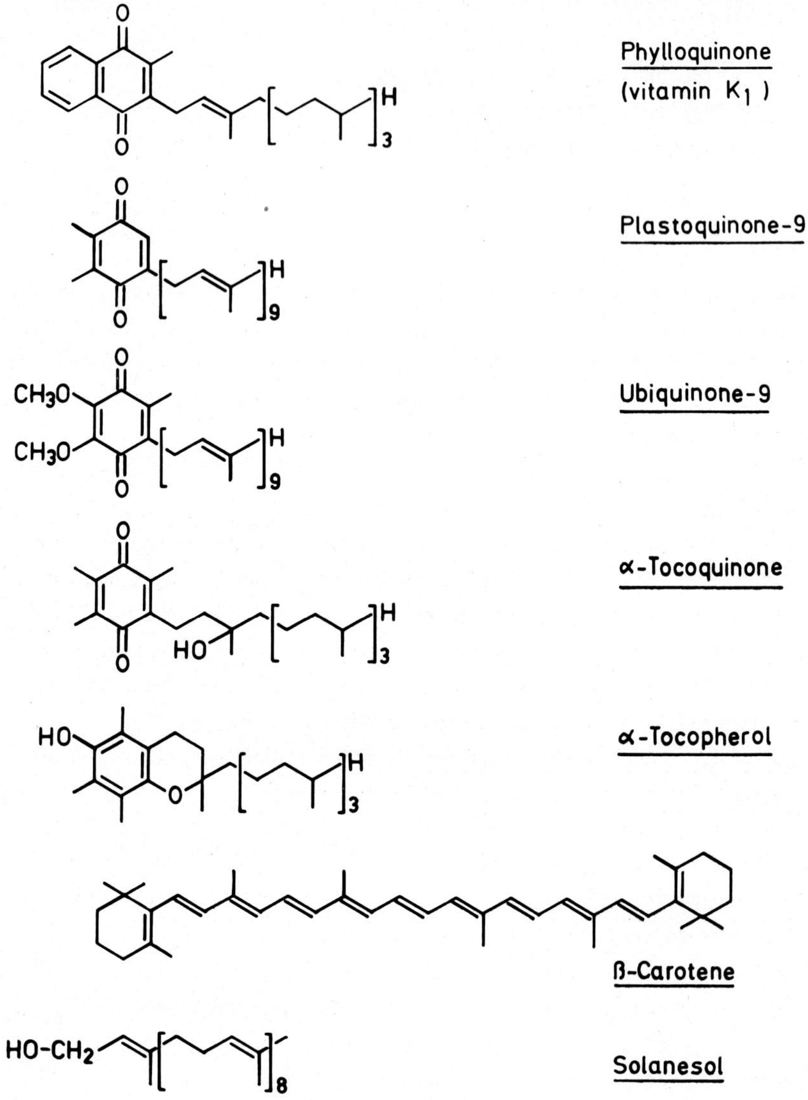

Fig. 1. Chemical structure of prenylquinones and prenols.

(Fig. 4 a).

By increasing the polarity of the solvent one obtains shorter retention times for α-tocoquinone, the resolution of phylloquinone K_1 and plastoquinone-9 are, however, poor (Fig. 4 b).

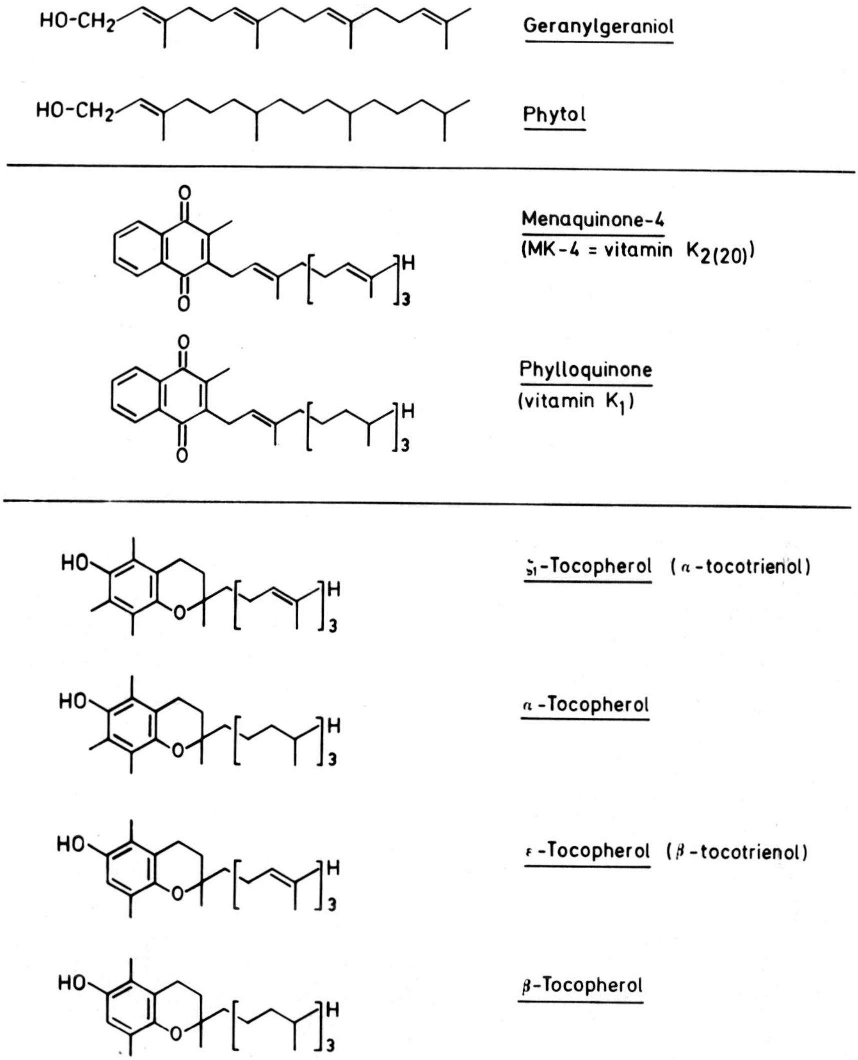

Fig. 2. Chemical structure of prenols and prenylvitamins which differ in the number of double bonds.

Prenyllipids differing in the number of double bonds are completly separated e.g. phytol and geranylgeraniol (Fig. 3), vitamin K_1 and MK-4 (K_2) (Fig. 6), α-tocopherol and ζ_1-tocopherol, β-tocopherol and ε-tocopherol (Fig. 5).

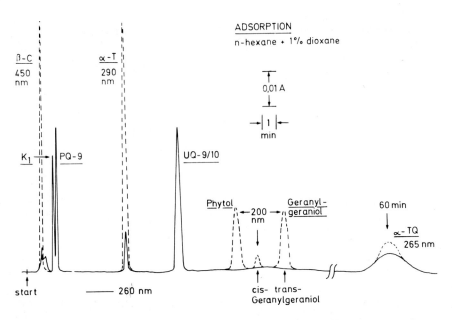

Fig. 3. <u>Adsorption HPLC</u>. Column: LiChrosorb SI 60, 5 μm, 250 x 3 mm. Solvent: 1 % dioxan in hexane. ß-C = ß-carotene, K_1 = vitamin K_1, PQ-9 = plastoquinone-9, α-T = α-tocopherol, UQ-9/10 = ubiquinone-9 or 10, α-TQ = α-tocoquinone.

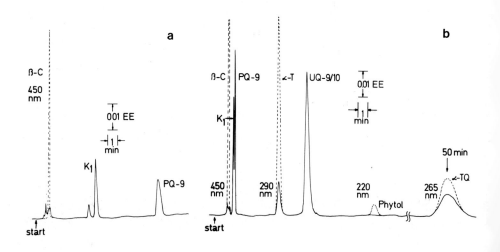

Fig. 4. <u>Adsorption HPLC</u>. Column: LiChrosorb SI 60, 5 μm, 250 x 3 mm. Solvent: a) 0.3 % dioxan in hexane, b) 1.25 % dioxan in hexane.

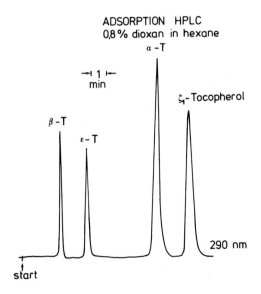

Fig. 5. <u>Adsorption HPLC</u>. Column: LiChrosorb SI 60, 5 μm, 250 x 3 mm. Solvent: 0.8 % dioxan in hexane. β-T = β-tocopherol, ε-T = β-tocotrienol, α-T = α-tocopherol, ζ₁-Tocopherol = α-tocotrienol.

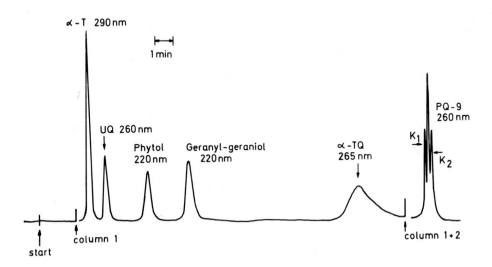

Fig. 6. <u>Adsorption HPLC</u>. Column: LiChrosorb SI 60, 5 μm, 125 x 3 mm + 250 x 3 mm Solvent: 1 % dioxan in hexane. α-T = α-tocopherol, UQ-9 = ubiquinone-9, α-TQ = α-tocoquinone, K_1 = vitamin K_1, PQ-9 = plastoquinone-9, K_2 = menaquinone-4.

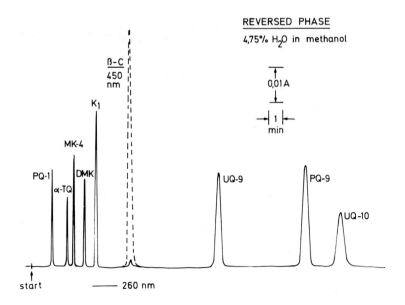

Fig. 7. <u>Reversed phase HPLC</u>. Column: LiChrosorb RP 8, 5 μm, 250 x 3 mm. Solvent: 4.75 % water in methanol. PQ-1 = plastoquinone-1, MK-4 = menaquinone-4, DMK = desmethyl vitamin K_1.

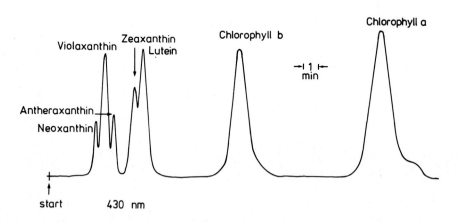

Fig. 8. <u>Reversed phase HPLC</u> of plant prenyl pigments. Column: LiChrosorb RP 8, 5 μm, 250 x 3 mm. Solvent: 10 % water in methanol.

A column switching valve is used, instead of a gradient elution, if the separation time for polar compounds is too long. This provides simultaneous separation of polar and unpolar prenyllipids in short retention times (Fig. 6)[2].

Reversed phase HPLC[3]: Figure 7 is an example for the separation of prenyl-lipids of different polarity, different chainlength and different number of double bonds. Chromatography takes place on reversed phase column packing with methanol and water as a solvent. Polar prenyllipids appear first, ubiquinone-9 before plastoquinone-9 (Fig. 7). Homologues of different chainlength are separated , opposite to the adsorption HPLC, e.g. UQ-9 and UQ-10 (Fig. 7). Compounds with a different number of double bonds appear as clearly separated peaks, e.g. vitamin K_1 and MK-4 (Fig. 7), phytol and geranylgeraniol, tocopherols and tocotrienols.

The retention times of prenyllipids decrease with a decreasing number of CH_3-groups. The increase of water in the elution solvent results in an increased retention time.

The main prenyl pigments of green plants (chlorophylls and carotenoids) are resolved by reversed phase HPLC (Fig. 8). The separation of minor carotenoid components and of secondary carotenoids is under investigation.

The lowest amounts of prenyllipids detectable in uv-light using HPLC are in the range of 0.1 μg or less, depending on the individual absorbance coefficient.

SUMMARY

The tested prenyllipids can be resolved as individual peaks in less than 30 minutes. By using the column switching valve even non polar compounds can be separated extremely fast in an isocratic system.

REFERENCES
1. Lichtenthaler, H.K. (1977) in Lipid and Lipid Polymers in Higher Plants, Tevini, M. and Lichtenthaler, H.K. eds., Springer Berlin.
2. Lichtenthaler, H.K. and Prenzel, U. (1977) J. Chromatogr. 135, 493 - 498.
3. Lichtenthaler, H.K. (1979) in Lipids and Technical Lipid Derivates, CRC Handbook of Chromatography.

ACKNOWLEDGEMENTS

This work was sponsored by a grant from the Deutsche Forschungsgemeinschaft.

LIPID BIOSYNTHESIS

LIPID BIOSYNTHESIS IN PLANT CELL CULTURES

*M. KATES, **A.C. WILSON and **A.I. DE LA ROCHE, *Department of Biochemistry, University of Ottawa, Ottawa, Canada and **Research Branch, Agriculture Canada, Ottawa, Canada.

ABSTRACT

A review covering recent studies of the lipid composition of plant cell cultures, including neutral and polar lipids and their fatty acid moieties, is presented. The use of plant cell cultures in studies of the biosynthesis of individual lipid classes and of the C_{18}-polyunsaturated fatty acids is also reviewed.

INTRODUCTION

Plant cell cultures provide useful models for studying lipid biosynthesis. They can be grown reproducibly and uniformly in defined media, providing control over cell type and growth stage (e.g., synchronous culture)[1,2]. In contrast to the situation with whole plants or plant tissues, adminstration of labeled precursors to plant cell cultures is relatively easy to control and subsequent processing is greatly facilitated. Thus, pulse-chase and other kinetic studies may be readily carried out[3]. Furthermore, the effect of environmental conditions, such as temperature, illumination, aeration, and of inhibitors and growth factors, etc., can readily be studied.

In this paper we will review recent studies on the overall lipid composition of plant cell cultures, as well as investigations of fatty acid and glycerolipid biosynthesis using labeled precursors.

Lipid Composition

A considerable amount of information is now available concerning the lipid composition of plant cell cultures from different species in comparison with that of photosynthetic and non-photosynthetic plant tissues and organelles. Most studies have been done with heterotrophic, non-green plant cell cultures, but in a few studies, photosynthetically active cultures have been examined in comparison with the corresponding non-green cultures (Table 1).

TABLE 1

LIPID COMPOSITION OF VARIOUS PLANT CELL CULTURES[a]

(Values calculated as % of total lipids)

Major Lipid Components	Kalanchoë crenata[4] non-green	Kalanchoë crenata[4] green	Ipomoea sp.[5] (maximum values)	Daucus carota Ref. 6	Daucus carota Ref. 7	Nicotiana tabacum[8] Non-green	Nicotiana tabacum[8] green	Glycine max[9]
Total lipids, % dry wt.	0.3	1.2	6	0.5	–	1.7	6.2	6
Phospholipids								
PC	43	11	16	28	30	33	23	23
PE	12	4	13	25	17	12	14	24
PG	3	2	9	2	1	2	5	7
PI	10	2	–	8	9	8	5	4
PS	–	–	–		tr	–	–	tr
PA	11	4	–	–	0.6	–	–	0.2
Glycolipids								
MGD	15	52	3	–	3	10	24	4
DGD	6	21	3	–	4	33	24	8
SQD	N.D.	4	–	–	tr	2	3	N.D.
SG	–	–	–	–	0.6	–	–	3
ESG	–	–	–	–	0.8	2	3	4
Neutral lipids								
TG	–	–	10	16	30	–	–	13
MG + DG	–	–	2	11	tr	–	–	0.4
FFA	–	–	–	9	tr	–	–	0.1
Sterols	–	–	5	–	3	–	–	7

[a] Abbreviations: PC, phosphatidylcholine; PE, phosphatidyl ethanolamine; PG, phosphatidyl glycerol; PI, phosphatidyl inositol; PS, phosphatidyl serine; PA, phosphatidic acid; MGD, monogalactosyldiglyceride; DGD, digalactosyldiglyceride; SQD, sulfoquinovosyldiglyceride; SG, sterol glycoside; ESG, esterified sterol glycoside; TG, triacyglycerol; DG, diacylglycerol, MG, monoacylglycerol; FFA, free fatty acids. N.D., not detected; tr, trace; –, not determined.

Results of earlier studies, reviewed previously[1], in which extraction procedures were inadquate to inactivate phospholipases and degradative enzymes are not included in the present survey. The low values of free fatty acids, phosphatidic acid (PA) and lyso-compounds for most of the cultures[5,7,8,9] listed in Table 1 confirm the absence of significant phospholipase activity during the lipid extraction procedures. All cultures except the photosynthetic ones had lipid compositions similar to those of non-photosynthetic tissues[10]. Phospholipids accounted for 50-80% of total lipids and consisted mainly of phosphatidyl choline (PC), phosphatidyl ethanolamine (PE), phosphatidyl inositol (PI) and phosphatidyl glycerol (PG), in decreasing order (Table 1). Glycolipids accounted for about 10-40% of total lipids and included digalactosyl diglyceride (DGD), monogalactosyl diglyceride (MGD), sterol glycoside (SG) and esterified sterol glycoside (ESG), but only traces or small amounts of sulfoquinovosyl diglyceride (SQD). Neutral lipids accounted for 15-35% of total lipids including mono-, di- and triacyl glycerols (MG, DG and TG), sterols, free fatty acids and sterol esters, the latter component being present only in traces.

The photosynthetic cultures (green)[4,8] had much higher proportions of PG, MGD, DGD and SQD, and lower proportions of PC, PE, PI and ESG than the corresponding heterotrophic, non-photosynthetic cultures (Table 1; also refs. 11 & 12). Furthermore, the total lipid content in non-photosynthetic cultures is considerably lower than in photosynthetic cultures[4,8], with exception of those of red goosefoot (Chenopodium rubrum) which are reported to have similar lipid contents for both heterotrophic and mixotrophic cultures[11,12].

Fatty Acid Composition

The major fatty acids in the heterotrophic cell cultures listed in Table 2 were found to be 18:2, 16:0, 18:3 and 18:0. The 18:2 acid was present in highest proportion in all cultures except soybean[9,13], nasturtium[14] and lettuce[13] in which 18:3, 18:1 and 16:0, respectively, were the predominant acids. In photosynthetic (green) cultures[8,12] the proportions of 18:3 are much higher than those in heterotrophic (non-green) cultures of the same species (Table 2).

TABLE 2

TOTAL FATTY ACID COMPOSITION OF PLANT CELL CULTURES FROM VARIOUS CELL LINES.[a]

Plant Species	Reference	Fatty Acid Composition, weight %						
		16:0	16:1	18:0	18:1	18:2	18:3	>18:0
Ipomonea Sp.	13	23.0	-	6.7	3.0	29.5	37.7	-
Ipomonea Sp. var. Heavenly Blue	13	26.2	-	11.5	9.9	38.5	13.9	-
Daucus carota	13	19.1	-	6.2	8.3	55.9	10.5	-
Solanum pseudo capsicum	13	25.0	-	7.0	5.4	55.0	7.6	-
Solanum dulcamara	13	20.0	-	5.1	3.6	60.9	10.4	-
Atropa belladonna	13	26.1	-	10.8	8.7	41.8	12.6	-
Cannabis sativa 95-D	13	30.8	-	3.9	7.2	40.6	17.5	-
Cannabis sativa 95-C	13	30.0	-	6.1	9.2	42.2	12.5	-
Phaseolus vulgaris	13	34.0	-	5.7	6.6	33.8	19.9	-
Lactuca sativa	13	54.8	-	14.6	10.2	16.8	3.6	-
Glycine max	13	21.9	-	4.0	6.6	16.3	51.2	-
	9	22.0	tr	3.1	5.5	18.9	50.9	-
Nicotiana tabacum (non green)	8	24.6	-	4.8	3.6	54.7	11.6	1.3
Nicotiana tabacum (green)	8	20.3	-	4.3	5.9	34.6	32.9	-
Brassica napus 30°C	14	27.6	tr	7.2	19.0	22.6	17.6	6.0
Brassica napus 5°C	14	21.6	tr	5.6	12.8	24.0	32.8	3.2
Tropaeolum majus 30°C	14	27.1	tr	tr	35.3	14.6	23.0	tr
Tropaeolum majus 5°C	14	25.6	tr	tr	24.0	24.1	26.3	tr
Chenopodium rubrum (non green)	12	32.7	1.5	1.5	19.4	28.2	4.5	10.8
Chenopodium rubrum (green)	12	24.9	1.5	1.5	2.8	45.5	16.0	5.9

[a] Abbreviations: tr, trace; -, not determined.

The fatty acid composition of heterotrophic cultures is of course influenced by environmental conditions such as aeration and temperature[11]. For example, callus cultures of Brassica napus or nasturtium grown at 5° showed an almost two-fold increase in 18:3 or 18:2, respectively (Table 2).

The individual phospholipid components in non-photosynthetic cultures had a characteristic fatty acid pattern in which 16:0 and 18:2 acids were predominant (Table 3). Soybean cultures appeared to be exceptional in that 18:3 was also predominant, in addition to 16:0 and 18:2, in the phospholipids (Table 3). In the photosynthetic tobacco culture[8], the phospholipids generally showed an increase in 18:3 and 18:1 and a decrease in 18:2 and 16:0. The $16:1-\Delta^3$-trans acid, present only in PG, showed an increase in the photosynthetic cultures, consistent with the idea that the $16:1-\Delta^3$-trans acid containing species of PG is present in chloroplast lamellae[10].

The galactolipids (MGD, DGD) had a characteristically high 18:3 content in the photosynthetic tobacco culture (and also small amounts of 16:3), but had lower contents of 18:3 in the non-photosynthetic tobacco culture (Table 3). However, high contents of 18:3 are also found in MGD and DGD of the heterotrophic soybean culture used in our studies[9], which was shown by electron microscopy to contain proplastids but no developed chloroplasts (cf. ref. 33). The absence of SQD in this culture is thus consistent with the absence of both chlorophyll and chloroplasts, since synthesis of SQD is directly associated with chlorophyll synthesis[4,10]. SQD however was reported present in non-green tobacco cultures[8], but had a lower 18:3 content than that in the green tobacco cultures (Table 3). Esterified sterol glycoside appears to have a characteristically high 16:0 content in the cultures examined (Table 3).

In regard to positional distribution of fatty acids between C-1 and C-2 of the glycerol moiety in phospholipids and glycolipids, Siebertz et al.[8] have shown that green cultures of tobacco have essentially the same fatty acid distribution as in green tobacco leaves, e.g., 16:0 in SQD is present both at C-1 and C-2, $16:1-\Delta^3$-trans and 16:3 are found at C-2 in PG and MGD, respectively. Non-green cultures of tobacco[8] showed a more

TABLE 3

FATTY ACID COMPOSITION OF MAJOR INDIVIDUAL LIPIDS IN VARIOUS PLANT CELL CULTURES

Plant Culture	Reference	Lipid Component[a]	Fatty Acid Composition, %					
			16:0	16:1-Δ^3-t	18:0	18:1	18:2	18:3
Daucus carota	6	PC	22.7	-	-	7.0	61.5	8.8
		PE	25.1	-	-	6.0	65.6	3.3
		PG	29.1	-	-	5.4	60.2	5.3
		PS	33.2	-	-	6.4	55.9	4.7
		TG	14.2	-	0.6	6.3	68.9	10.0
		FFA	31.9	-	1.8	5.8	52.8	5.4
Glycine max	9	PC	18.4	-	4.1	16.0	26.5	35.1
		PE	24.1	-	2.9	14.4	30.7	28.0
		PG	35.0	-	4.0	10.0	22.5	28.2
		MGD	3.6	-	1.5	5.0	7.0	82.2
		DGD	12.3	-	3.2	6.0	8.5	70.0
		ESG	37.3	-	8.3	16.1	11.7	26.6
		TG	2.9	-	1.0	11.3	24.6	60.2
		FFA	21.6	-	5.7	14.8	17.5	36.8
Nicotania tobacum (non green)	8	PC	23.3	-	3.8	1.6	62.6	8.5
		PE	24.6	-	2.5	0.4	66.1	5.8
		PG	64.6	4.0	2.8	0.9	22.0	4.0
		MGD	11.9	-	1.4	0.4	27.8	58.5
		DGD	33.0	-	6.8	2.3	39.7	17.1
		SQD	56.0	-	3.4	1.8	17.3	19.7
		ESG	43.4	-	10.5	7.9	27.0	6.1
Nicotania tobacum (green)	8	PC	19.6	-	4.6	8.9	50.4	16.0
		PE	21.7	-	4.0	3.9	58.2	11.6
		PG	46.7	7.3	5.2	4.9	19.9	16.1
		MGD	3.6	-	0.9	1.4	8.5	82.8
		DGD	16.9	-	4.3	2.7	10.3	65.5
		SQD	40.4	-	4.7	4.4	16.7	32.9
		ESG	40.8	-	11.5	8.9	28.8	6.4

[a] Abbreviations as in Table 1

uniform distribution, characterized by a general preference of 16:0 for C-1 in all lipid components, as is usually found in extra-plastidic phospholipids[8].

Lipid Biosynthesis

a) <u>Glycerolipid Synthesis</u>. Radiolabel incorporation studies of lipid biosynthesis in plant cell cultures have been reported only relatively recently. Thomas and Stobart[4] used ^{32}P- and ^{35}SO$_4$-incorporation in callus tissue of <u>Kalanchoë crenata</u> to characterize the lipids associated with greening of the originally dark-grown callus. These studies established that only phosphatidyl glycerol and sulfoquinovosyl diglyceride, in addition to the galactolipids, showed an increase on exposure to light that was associated with chloroplast development, confirming previous findings obtained with photosynthetic tissues[10].

Efficient incorporation of labelled ethanolamine, choline and serine into phospholipids has been reported for soybean[15] and carrot[7] cultures, but inositol was incorporated less well. The results indicated that the methylation pathway of synthesis of PC from PE was operating in both cultures[7,15]; the synthesis of PE by decarboxylation of PS was also demonstrated in the soybean cultures[15]. Incorporation of labelled glycerol into the glycerol moiety of phospholipids was demonstrated in soybean[15] and carrot[6,7] cultures; all of the major phospholipids were labelled but PC and PG had highest specific activities in the carrot culture cells[6].

Glycerol labelling has also been used in pulse-chase experiments to determine phospholipid turnover rates in soybean[15] and carrot cell[7] cultures. The half-lives of PC and PE in soybean cells[15] were 67 and 66 h respectively, while in carrot[7] the values were 45 h and 40 h. These values compare reasonably well with estimates of degradation rates obtained for labelling the polar head groups of the phospholipids in soybean cultures and there appeared to be pools of individual phospholipids turning over at different rates[15]. The turnover rate is also sensitive to the growth stage of the cells tested being almost three times greater in the stationary phase than in exponential phase[15].

The most widely used labelled precursor for studies on lipid synthesis has been [^{3}H] or [^{14}C]acetate. Fatty acids are predominantly labelled by this precursor but a significant labelling of sterols and other non-polar lipids does occur[7,9].

In our studies of lipid synthesis in 4-day-old soybean suspension cultures[9], rapid incorporation of [1-^{14}C]acetate was observed into the total lipids up to 16 min, an essentially constant incorporation level of about 25% of the added ^{14}C being reached between 45 min and 2.6 h. Most of the radioactivity (60-70%) was associated with phospholipids of which PC was the most rapidly labelled, accounting for a maximum of 40-45% of the total ^{14}C in the first 16 min; thereafter ^{14}C in PC decreased rapidly to low values at 22 h. PE and PI + PG were labelled much more slowly to maxima of 15-20% at 2.6 h and 85 min, respectively, the labelling decreasing thereafter to low values at 22 h. In contrast, an unidentified phospholipid (PX)* which may be identical with bis-phosphatidic acid (bis-PA) previously detected in soybean suspension cultures[16], showed little labelling up to 2.6 h but at 22 h was found to contain most of the ^{14}C present in the total lipids. Galactolipids (MGD + DGD) were also labelled very slowly (1-2%) up to 2.6 h but then showed an increase to 10% of the total ^{14}C at 22 h. Tri- and diacyl glycerols accounted for about 5-10% of the ^{14}C at all times. In general, the overall pattern of [^{14}C]acetate labelling of glycerolipids in the soybean suspension culture was similar to that reported previously for leaf tissue[18-20].

b) <u>Fatty Acid Synthesis</u>. In general, callus cultures have not been found to synthesize fatty acids characteristic of differentiated plant tissues, e.g., erucic acid (22:1) in rapeseed[11,14]. However, Yano et al.[21] demonstrated the synthesis of cyclopropane and cyclopropene fatty acids from both [^{14}C]oleic acid and [^{14}C]acetate by <u>Malva parviflora</u> callus tissue, and Seibertz et al.[8] have shown that green cultures of tobacco contain 16:1-Δ^3-<u>trans</u> and 16:3 acids in their PG and MGD components, respectively,

---

*The unidentified phospholipid had TLC mobilities close to those reported for bis-PA[16] but distinctly different from those of phosphatidylmethanol, shown to be formed by the action of solvent-activated phospholipase D phosphatidyl transferase[17]. The absence of phosphatidyl methanol is thus indicative of the absence of phosphatidyl transferase activity during the extraction procedure used in our studies[9].

which are associated specifically with chloroplast lamellae[10].

Labelled acetate was found to be readily incorporated into fatty acids by callus tissue culture of several species of Malvaceae[21], and by cell suspension cultures of soybean[15,22,23] and carrot[7]. Kinetic studies of acetate incorporation showed that 18:1 and 16:0 are the most rapidly labelled fatty acids, and analysis of the subsequent variation with time in the distribution of radioactivity suggested a sequential desaturation pathway of 18:1 → 18:2 → 18:3[15,21-23]. This pathway was further supported by the results obtained with ^{14}C labelled long chain fatty acids as precursors[23], such as 18:1 or 18:2, which are very rapidly taken up by soybean suspension cells and incorporated into membrane lipids, particularly PC, followed after a 30 min lag period by the desaturation of 18:1 → 18:2 and 18:2 → 18:3. Soybean cell cultures were unable to desaturate added saturated fatty acids[15,23], confirming the presence of an ACP-specific desaturase as found in other plant tissues[24] and the presumptive absence of acyl CoA - ACP transacylase activity in plant cells[23].

In our studies[9] on [^{14}C]acetate labelling of soybean culture cells, a high proportion (>95%) of the ^{14}C incorporated into the glycerolipids from [^{14}C]acetate was associated with the fatty acid moieties, but the most striking feature was the distribution of the ^{14}C among the fatty acids of each of the individual glycerolipids (Table 4).

In PC, oleic acid accounted for about 75% of the radioactivity at 16 min, thereafter decreasing to 12% at 22 h. The proportion of ^{14}C in the 18:2 component of PC was very low at 16 min but increased ten-fold to a maximum at 2.6 h, thereafter decreasing up to 22 h (Table 4). Only 1% of the ^{14}C in PC was present in the 18:3 component at 16 min, this value increasing slowly to 6.0% at 2.6 h, and then to 50% at 22 h (Table 4); as a percentage of the total fatty acid ^{14}C, the proportion at 22 h was, however, only about 1%. The decrease in the absolute amounts of ^{14}C in 18:1, 18:2 and 18:3 of PC during the period from 2.6 h to 22 h was compensated for by increases in these acids, particularly 18:3, in the unidentified phospholipid (bis-PA) and in the glycolipid fraction; the di- and triacylglycerol fraction, however, showed an increase only in the 18:3 component (Table 4).

TABLE 4

INCORPORATION OF [^{14}C]ACETATE INTO FATTY ACIDS OF MAJOR LIPIDS OF SOYBEAN SUSPENSION CELLS AS A FUNCTION OF TIME [a].

Lipid[b]	Time after addition of [^{14}C]acetate min	Total ^{14}C in fatty acids 10^{-3} x dpm per ml culture	Distribution of [^{14}C]fatty acids in individual lipid component (%)			
			16:0	18:1	18:2	18:3
Total	16	490	29	60	9	2
	45	535	29	53	17	2
	85	546	26	46	23	6
	155	548	26	34	28	11
	22.0 h	522	14	8	25	53
PC	16	289	20	76	3	1
	45	227	22	61	16	2
	85	215	22	51	23	3
	155	182	23	43	28	6
	22.0 h	13	20	12	18	51
PE	16	89	37	57	5	1
	45	97	37	52	9	2
	85	121	38	45	14	3
	155	169	37	39	19	6
	22.0 h	16	31	15	24	30
PX	16	5	54	41	4	1
	45	16	42	48	8	1
	85	18	40	44	13	3
	155	17	36	45	15	4
	22.0 h	277	28	8	25	40
TG + DG	16	59	26	61	10	3
	45	105	23	56	18	4
	85	92	19	48	26	7
	155	95	16	37	34	14
	22.0 h	77	15	6	19	60
MDG+ DGD+ ESG	16	18	38	38	11	13
	45	21	37	39	17	9
	85	25	27	34	24	14
	155	31	22	27	27	24
	22.0 h	109	27	11	15	47

[a]Data taken from ref. 9. [b]Abbrev: As in Table 1. PX, unidentified phospholipid.

These results, while consistent with the overall sequential desaturation of 18:1 → 18:2 → 18:3[22,23] discussed above raise questions concerning the precise pathways involved. The pattern of labelling of the fatty acids in the individual glycerolipids which we observed (Table 4) is suggestive of a more complex desaturation sequence, summarized in Fig. 1.

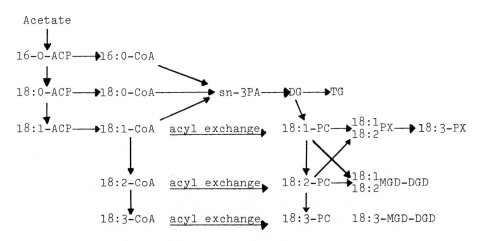

Fig. 1. Scheme for proposed mechanism of desaturation sequence of fatty acids in soybean cell cultures [adapted in part from Stumpf and Weber[23]].

In this sequence, labelled 18:1-PC rapidly accumulates by synthesis de novo from DG or by "acyl exchange" of unlabelled PC with labelled 18:1-CoA[25]. The 18:1 of PC is then rapidly converted to 18:2 of PC during the first 2 h of incubation, either by direct desaturation of 18:1-PC, as demonstrated previously with Chlorella chloroplasts[26], with pea-leaf microsomes[27], and recently with microsomes of developing safflower seeds[28]; alternatively, "acyl exchange" with 18:2-CoA formed by desaturation of 18:1-CoA[23,29,30] could also occur. Subsequently, both 18:1 and 18:2 of PC are then transferred to other glycerolipids, in particular the unidentified phospholipid PX (bis-PA) and the glycolipids (MGD + DGD). Linoleic acid accumulates in these lipids during the same period and also to a limited extent in PC and PE (Table 4).

The 18:3-PC could be formed by desaturation of 18:2-PC or by "acyl exchange" with 18:3-CoA, but the low amounts of labelled 18:3 detected in PC would tend to eliminate PC as a significant source of 18:3 for transfer to PX and MGD + DGD. The formation of 18:3 in PC and the galactolipids may perhaps be explained by desaturation of 18:2 → 18:3 in situ in these components or by "acyl exchange" with 18:3-CoA formed by desaturation of 18:2-CoA[23]. Pulse-chase experiments using [^{3}H]glycerol and [^{14}C]acetate in developing maize leaves[31] have confirmed that the 18:1 of PC is a major precursor of the 18:3 of galactolipids. The conversion of 18:1 → 18:2 has been associated with the endoplasmic reticulum of maize and pea-leaves[18], whereas desaturation of 18:2 → 18:3 has been proposed to occur in the chloroplasts[18,32].

The mechanism of these desaturations and of the interorganelle lipid transfer still remains to be elucidated. However, knowledge of the cellular localization of the fatty acid synthetase, the desaturases and the phospholipid and glycolipid synthetases is required.

So far, fatty acid synthesis in soybean cell cultures has been shown to be localized in proplastids[33]. Intact proplastid preparations incorporated labelled acetate, malonate and pyruvate into saturated (16:0, 18:0) and unsaturated (mainly 18:1 + traces in 18:2 and 18:3) fatty acids by a combination of de novo synthesis, elongation and desaturation[33]. With disrupted proplastids, similar products were formed from labelled malonyl-CoA, which was a better precursor than acetyl-CoA, and a slightly higher incorporation into 18:2 and 18:3 occurred. However, compared to whole soybean culture cells which make high amounts of 18:2 and 18:3 (Table 2), the synthesis of these acids by whole or broken protoplasts is very low. This would be consistent with the conversion of 18:1 → 18:2 occurring by action of a phospholipid desaturase on 18:1 esterified to PC[26-28], as discussed above. Presumably the proplastids lack the phospholipid desaturase system which may be localized in another organelle (most likely in the endoplasmic reticulum).

Thus, soybean and other plant cell suspension cultures should prove to be very useful systems with which to pursue these problems of the mechanism of fatty acid desaturation, particularly the formation of the important polyunsaturated fatty acids, linoleic and linolenic acids.

ACKNOWLEDGEMENTS

This work was supported by Grant A-5324(MK) from the National Research Council of Canada, and Grant EMR-7603(MK) from Agriculture Canada. We thank N. Long for growing the cultures.

REFERENCES

1. Radwan, S.S. and Mangold, H.K. (1976) Adv. Lipid Res., 14, 171-211.
2. Street, H.E. (1973) in Annual Proceedings of the Phytochemical Society, No. 9, Milborrow, B.V. ed., Academic Press, London and New York, pp. 93-125.
3. Jones, L.H. (1974) in Industrial Aspects of Biochemistry, Spencer, B. ed., Federation of European Biochemical Societies, North Holland Publishing Co., Vol. XXX, pp. 813-834.
4. Thomas, D.R. and Stobart, A.K. (1970) J. Exp. Bot., 21, 274-285.
5. Song, M. and Tattrie, N. (1973) Can. J. Bot., 51, 1893-1897.
6. Gregor, H.D. (1977) Chem. Phys. Lipids, 20, 77-85.
7. Kleinig, H. and Kopp, C. (1978) Planta, 139, 61-65.
8. Siebertz, H.P., Heinz, E. and Bergmann, L. (1978) Plant Science Lett. 12, 119-126.
9. Wilson, A.C., Kates, M. and de la Roche, A.I. (1978) Lipids, 13, 504-510.
10. Kates, M. (1970) Adv. Lipid Res. 8, 225-265.
11. Mangold, H.K. (1977) in Plant Tissue Culture and its Biotechnological Application, Barz, W., Reinhard, E. and Zenk, M.H. eds.,Springer Verlag, Berlin-Heidelberg, New York, pp. 55-65.
12. Radwan, S.S., Mangold, H.K., Hüsemann, W. and Barz, W. Chem. Phys. Lipids, in press.
13. Tattrie, N.H. and Velicky, I.A. (1973) Can. J. Bot., 51, 513-516.
14. Radwan, S.S., Grosse-Oetringhaus, S. and Mangold, H.K. (1978) Chem. Phys. Lipids, 22, 177-184.
15. Moore, T.S. (1977) Plant Physiol., 60, 754-758.
16. Stearns, E.M. Jr. and Morton, W.T. (1977) Lipids, 12, 451-453.
17. Roughan, P.G., Slack, C.R. and Holland, R. (1978) Lipids, 13, 497-503.

18. Slack, C.R. and Roughan, P.G. (1975) Biochem. J., 152, 217-228.
19. Roughan, P.G. (1975) Lipids, 10, 609-614.
20. Heinz, E. and Harwood, J.L. (1977) Hoppe-Seyler's Z. Physiol. Chem., 358, 897-908.
21. Yano, I., Morris, L.J., Nichols, B.W. and James, A.T. (1972) Lipids, 7, 35-45.
22. Stearns, E.M. Jr. and Morton, W.T. (1975) Lipids, 10, 597-601.
23. Stumpf, P.K. and Weber, N. (1977) Lipids, 12, 120-124.
24. Jaworski, J.G. and Stumpf, P.K. (1974) Arch. Biochem. Biophys. 162, 158-164.
25. Shine, W.E., Mancha, M. and Stumpf, P.K. (1976) Arch. Biochem. Biophys., 173, 472-479.
26. Gurr, M.I., Robinson, M.P. and James, A.T. (1969) Eur. J. Biochem. 9, 70-78.
27. Slack, C.R., Roughan, P.G. and Terpstra, J. (1976) Biochem. J., 155, 71-80.
28. Stymne, S. and Appelqvist, L.A. (1978) Eur. J. Biochem. 90, 223-229.
29. Vijay, I.K. and Stumpf, P.K. (1972) J. Biol. Chem., 247, 360-366.
30. Abdelkader, A.B., Cherif, A., Demandra, C. and Mazliak, P. (1973) Eur. J. Biochem., 32, 155-165.
31. Slack, C.R., Roughan, P.G. and Balsingham, N. (1977) Biochem. J., 162, 289-296.
32. Tremolieres, A. and Mazliak, P. (1974) Plant Sci. Lett. 2, 193-201.
33. Nothelfer, H.G., Barckhaus, R.H. and Spener, F. (1977) Biochim. Biophys. Acta, 489, 370-380.

BIOSYNTHESIS OF LINOLEIC AND LINOLENIC ACIDS - SUBSTRATES AND SITES

LARS-ÅKE APPELQVIST AND STEN STYMNE
Department of Food Hygiene, Swedish University of Agricultural Sciences, Roslagsvägen 101, S-104 05 Stockholm (Sweden)

ABSTRACT

Recently published findings and hitherto unpublished results on linoleic and linolenic acid biosynthesis are reviewed in the light of earlier literature data. Emphasis is placed on the type of substrate activation (thio-ester or oxygen-ester) and the subcellular location of the enzymes involved. Special consideration is given to the conditions prevailing in developing oilseed tissue, which accumulates large amounts of triacylglycerols.

INTRODUCTION

Extensive reviews on fatty acid biosynthesis in plant tissues each with numerous references have appeared in later years (Gurr[1], Harwood[2], Stumpf[3]). A recent review by Stumpf[4] on the biosynthesis of lipids in developing oilseed tissue deliberately deleted a discussion of the biosynthesis of polyunsaturated fatty acids. Hence, it is timely to discuss at this conference the biosynthesis of linoleic and linolenic acid, notably in relation to the conditions in developing oilseed tissue. For a more general discussion on the accumulation of lipids in developing oilseed tissue, a review by Appelqvist[5] may be consulted.

In view of the importance of the production of oils in oil-rich seeds to provide necessary calorie additions to the diets in developing countries and to provide essential fatty acids both in industrialized and developing countries, it is amazing how little we know of the pathways and control mechanisms in the biosynthesis of triacylglycerols in developing oilseed tissue.

This paper will not give a full account of all recent papers relating to linoleic and linolenic acid biosynthesis but will discuss some of the recent literature and present new results from our own laboratory. Further, some slightly provocative suggestions will be put forward as a basis for future experiments and discussions.

SOME DATA FROM *IN VIVO* AND TISSUE SLICE LABELLING

The ability of leaf tissue to convert ^{14}C-oleic acid to linoleic acid was reported some 20 years ago by James and Webb[6] and since then many papers have made the sequence oleic → linoleic → linolenic acid in leaf tissue most likely, see e.g. James[7] and Harris and James[8]. From still earlier studies of developing soya bean plants, being supplied with ^{14}C-sucrose, it was concluded that the highest specific activity in the seed lipids was observed for oleic acid and that this acid at least to some extent was converted to linoleic and linolenic acids (Simmons and Quackenbush[9]). The results of such early *in vivo* or tissue slice work was carefully abstracted by Stearns Jr[10] in a very comprehensive review covering the literature up to 1970.

In vivo or tissue slice experiments are still being reported but the interpretation of the results is open to much uncertainty. It appears as results from pulse-chase type of experiments with tissue slices are more easy to interpret than results from long-time, continuous exposure to the substrate with analysis of labelling patterns at various time intervals. Recently, several *in vivo* and tissue slice studies have been published e.g. by Bolton and Harwood[11,12], by Cherif et al.[13], by Siebertz and Heinz[14], by Williams et al.[15]. The conclusions from these studies will be discussed after the presentation of some *in vitro* studies.

THE SUBCELLULAR LOCATION OF THE DESATURATION OF OLEIC ACID

In view of the predominance of linolenic acid among the total fatty acids of chloroplasts, one would tend to assume that the chloroplast also has the capacity to desaturate oleic and linoleic acids. The failure to observe any labelling in the linoleic and linolenic acids of chloroplasts after incubation *in vitro* with ^{14}C-acetate or other appropriate substrates from the time of an early report by Stumpf and James[16] up to the paper by Kannangara and Stumpf[17] has been a constant nuisance to researchers in this area.

It should however be noted that although Harris and James[8] were unsuccessful in isolating subcellular fractions from leaf tissue, which converted oleate to linoleate, they did report an active oleoyl-CoA desaturase activity in a "broken chloroplast" fraction from *Chlorella vulgaris* which yielded both linoleic and linolenic acids. Thus in 1965 they concluded that the result of the fractionation studies point to the chloroplast as the site of the desaturation (Harris and James[8]).

Probably the first report on *in vitro* desaturation of oleic acid in higher plants was the preliminary report by McMahon and Stumpf[18] that a crude 4.000 g pellet (called plastids) from developing safflower seeds was the most active fraction in converting oleoyl-CoA into linoleic acid. Later studies, authored by Vijay and Stumpf[19,20], define the microsomes as the primary, if not the sole, site. In other studies of oleoyl-CoA desaturase from "aged" potatoe slices, the predominant location was reported to be the microsomes (Abdelkader et al.[21]). Other investigations in Mazliak's laboratory have clearly pointed to the microsomes as the major, or sole, site of oleic acid desaturation (Dubacq et al.[22]). Also the oleate desaturase of pea leaves by Slack et al.[23] was mainly confined to the microsomal fraction.

Although this review has to cover only the biosynthesis of linoleic and linolenic acids, it is appropriate to mention very briefly data on the biosynthesis of oleic acid (or a derivative thereof) in various parts of the cell. The stearoyl-ACP desaturase, which yields free oleic acid as the main product *in vitro*, seems to be found in the stroma of spinach chloroplast (Jaworski and Stumpf[24]) as well as in non-chloroplast containing tissue, cauliflower inflorescences and extracts of developing safflower seeds (Porra and Stumpf[25]). Therefore it is likely that oleic acid is available by direct biosynthesis in chloroplasts as well as in extra-plastidic compartments, which is of interest to note for further detailed studies on the oleate desaturase in various cell compartments.

THE SUBCELLULAR LOCATION OF THE DESATURATION OF LINOLEIC ACID

As mentioned above, the "broken chloroplast" fraction (27.000 g pellet) from *Chlorella* was found to yield both ^{14}C-linoleic and linolenic acids from ^{14}C-oleoyl-CoA (Harris and James[8]). Work in Mazliak's laboratory clearly point to the chloroplast as the major, if not the sole, site of linoleic acid desaturation in pea leaves[26]. It is thus likely that chloroplasts are capable of synthesizing linolenic acid from linoleic acid. However in developing tissue of such seeds which accumulate large amounts of triacylglycerols rich in linolenic acid (flax, soya beans, rapeseed) it appears rather unlikely that the chloroplast is the sole site for linolenic acid biosynthesis; cf. the discussion by Appelqvist[5] on the high rate of tricylglycerol accumulation also during the stage of rapid decline in the content of typical chloroplast lipids, such as MGDG, indicating chloroplast breakdown.

Very recent work in our laboratory has shown that homogenates of highly immature soya bean cotyledons are capable to convert with good yield ^{14}C-oleoyl-CoA not only into ^{14}C-linoleic acid but also into ^{14}C-linolenic acid. It is interesting to note that Roughan[27] suggested that linolenic acid was synthesized from linoleic acid while attached to the phosphatidylcholine. Since the newly synthesized ^{14}C-linolenate in soya bean homogenates is preferentially in the phosphatidylcholine fraction (Stymne and Appelqvist, to be published) it appears highly likely that microsomes of developing soya beans can carry out a sequential desaturation of oleoyl-PC to linolenoyl-PC. If this can be verified, it could mean that linolenic acid can be synthesized from linoleic acid both in plastids and an extraplastidic compartment.

AN ALTERNATIVE PATHWAY FOR LINOLENIC ACID BIOSYNTHESIS

Although many authors have reported the biosynthesis of saturated fatty acids and oleic acid by isolated chloroplasts (see a review by Stumpf[3]) no substantial labelling of linoleic and linolenic acids had been reported until Kannangara and Stumpf[17] presented their studies on "immature" chloroplasts from spinach leaves. In these studies it was found that CN^- inhibited the synthesis of linoleic acid but not of linolenic acid, indicating the possibility of another mode of linolenic acid formation than by desaturation of linoleic acid. This pathway was revealed in a series of papers from P.K. Stumpf's laboratory (Jacobson et al.[28,29], Kannangara et al.[30]) where it was shown that linolenic acid could be synthesized under anaerobic conditions from 7,10,13-hexadecatrienoic acid (16:3) and ^{14}-C-acetate by a disrupted spinach chloroplast system. Further, it was suggested that the 16:3 acid probably was synthesized by elongation of 12:3 which was supposed to be formed by sequential desaturation of 12:0. These findings were summarized and discussed by Stumpf[3] shortly after the original reports.

Recently, additional work on this elongation system from a chloroplast stromal fraction demonstrated that various fatty acids added as acyl-CoAs could be elongated by acetyl-CoA in a reaction which required NADPH or NADH and was greatly stimulated by nucleoside triphosphates (Vance and Stumpf[31]). Molecular sieve chromatography indicated that the elongation was performed by individual enzymes and not by a multi-enzyme system.

The significance of this pathway for linolenic acid formation under *in vivo* conditions has not been established, and great difficulties in such attempts have been foreseen by Vance and Stumpf[31]. Recent studies by Brar and Thies[32] have however indicated that in rapeseed leaf and developing seed tissues incubated with various ^{14}C-labelled fatty acids, e.g. 18:2 and 16:3 the elongation pathway and the desaturation pathway operate simultaneously in both tissue types.

THE SUBSTRATES FOR OLEIC AND LINOLEIC ACID DESATURATION - COA DERIVATIVES OR GLYCEROLIPIDS?

Whereas the substrates for the synthesis of linolenic acid by elongation appear to be activated as thio-esters (Vance and Stumpf[31]) there is some controversy as regards the desaturation of oleic and linoleic acids. In their studies of ^{14}C-oleoyl-CoA metabolism by "broken chloroplasts" of *Chlorella*,

Gurr et al.[33] found a close association between newly formed ^{14}C-linoleic acid and phosphatidylcholine. One of the alternative interpretations of their data was that oleoyl-phosphatidylcholine was the actual substrate for the desaturation. From studies of safflower microsomes, Vijay and Stumpf[19] concluded that the true substrate was oleoyl-CoA and the immediate product was linoleoyl-CoA which was rapidly esterified to form linoleoyl-phosphatidylcholine. In the studies of linoleic acid biosynthesis by potatoe microsomes, Abdelkader et al.[21] assume that the true substrate is oleoyl-CoA but no figures are reported on the kinetics of appearance of labelled ^{14}C-linoleic acid in various lipid classes. Thereby their data could be interpreted either way.

Recent studies of the safflower microsomal desaturase in our laboratory (Stymne and Appelqvist[34]) tend to contradict the previous results of Vijay and Stumpf as reported briefly below. Microsomes of developing safflower seeds (*Carthamus tinctorius*) prepared according to Vijay and Stumpf[19], desaturated added 1-^{14}C-oleoyl-CoA into ^{14}C-linoleate in the presence of NADH. Studies of the kinetics revealed that linoleate formation continued after the disappearance of oleoyl-CoA (Figure 1). The first lipid to be labelled with linoleate was phosphatidylcholine (PC). Similar results with safflower microsomes have independently been obtained by Slack and Roughan (personal communication).

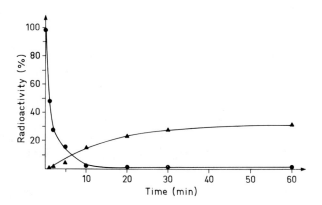

Fig. 1. Time course of disappearance of 1-^{14}C-oleoyl-CoA (●—●) and appearance of ^{14}C-linoleate (▲—▲) in incubations of 1-^{14}C-oleoyl-CoA with microsomes of developing safflower seeds. Radioactivity is expressed as a percentage of that recovered. From Stymne and Appelqvist (1978), Eur. J. Biochem., 90, 223-229.

In the absence of NADH no desaturation occurred. However, the oleate from oleoyl-CoA was still rapidly incorporated mainly into PC as well as some other lipids. After 10 minutes essentially no oleoyl-CoA remained in the incubation mixture. Upon subsequent addition of NADH, linoleate synthesis took place to about the same extent as when NADH was present at the beginning of the incubation (Table 1).

TABLE 1

INFLUENCE OF NADH ADDITION AND TIME OF INCUBATION ON THE METABOLISM OF ^{14}C-OLEOYL-COA IN MICROSOMES OF DEVELOPING SAFFLOWER SEEDS

9 nmol 1-^{14}C-oleoyl-CoA was added at the start of the incubation
(Stymne and Appelqvist (1978), Eur. J. Biochem., 90, 223-229)

Incubation time	NADH	Distribution of ^{14}C label, %					Overall 18:2 yield
		PC		Other lipids in CHCl$_3$ phase		Acyl-CoA (H$_2$O/MeOH phase)	
Min.	mM	18:1	18:2	18:1	18:2		%
10	1	55.9	13.5	26.9	2.9	0.8	16.4
10	0	73.7	0.0	25.4	0.0	0.9	0.0
10/40	0/1	42.3	25.1	25.1	6.8	0.7	31.9
30	1	45.6	24.6	23.3	5.8	0.7	30.4

These data should be compared to those from the work on safflower microsomes by Vijay and Stumpf[19] claiming that oleoyl-CoA is the immediate substrate and linoleoyl-CoA is the primary product. However it should be noted that their method for determination of thiol esters according to Barron and Mooney[35] has been reevaluated by Nichols and Safford[37] and shown not to be satisfactory for mixtures with polar lipids.

In view of the great difficulties we encountered in obtaining active microsomes from developing soya bean cotyledons, we consider it of interest to report data on the metabolism of ^{14}C-oleoyl-CoA by homogenates of immature soya beans: Freshly prepared homogenates of developing soya beans (30 days after flowering) (*Glycine max* var. Fiskeby V) desaturated added 1-^{14}C-oleoyl-CoA. Addition of NADH and BSA greatly stimulated the desaturation. A time course study of the incorporation of oleate from oleoyl-CoA and the appearance of linoleate in different lipid classes revealed that oleate was mainly incorporated into PC and that linoleate appeared in PC long before it was detected in any other lipid class. (Figure 2).

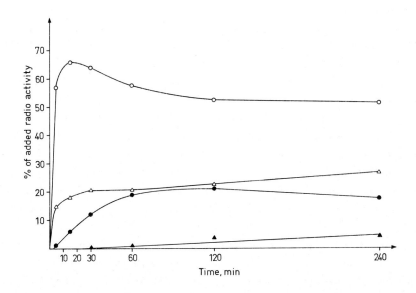

Fig. 2. Time course changes in the ^{14}C-activity of oleate and linoleate in various lipids after addition of ^{14}C-oleoyl-CoA to homogenates of developing soya beans. Each incubation mixture contained per ml: 7 nmol 1-^{14}C-oleoyl-CoA, 3 μmol NADH, 10 mg BSA, 0.1 mmol potassium phosphate buffer pH 7.2 and homogenate of developing soya bean cotyledons 30 days after flowering (20 mg of protein). Incubations were carried out with air as gas phase in a waterbath at 25°C with constant shaking and stopped at the times indicated. Lipids were analyzed as described by Stymne and Appelqvist[34].
Symbols used: oleate in PC (o—o); oleate in other lipids of the chloroform phase (△—△); linoleate in PC (●—●); linoleate in other lipids in the chloroform phase (▲—▲).

Small amounts of linoleate occurred in lipid classes other than PC after prolonged incubation time (Figure 2). Among the other lipid classes investigated, the first to be labelled were the diacylglycerols (DG) and after that phosphatidyl ethanolamine (PE) and the triacylglycerols (TG). This sequence of labelling corresponds to the results of ^{14}C-acetate and ^{3}H-glycerol feedings to whole tissues of developing soya beans reported by Slack et al.[23] and supports their suggestions that oleate is desaturated while esterified to PC and the newly formed linoleate then being transferred to TG via DG.

It thus appears as the major if not the sole pathway of linoleic acid biosynthesis in pea leaves, and in developing soya beans and safflower seeds is by desaturation of oleoyl-phosphatidylcholine. It would be interesting to carry out similar analyses of the products of the oleoyl-CoA desaturase of "aged" potatoe microsomes to find out whether the true substrate is oleoyl-CoA or oleoyl-phosphatidylcholine.

Although the indirect evidence for an oleoyl-phosphatidylcholine desaturase appears to be very strong, a separation of the two steps assumed, the acylation of lysophosphatidylcholine and the desaturation of the phosphatidylcholine are necessary for a final proof. Therefore it is regretable that Vijay and Stumpf[19,20] had to report that "attempts to solubilize the desaturase were unsuccessful". We have so far not made any such attempts.

As regards substrate, it should however be remarked that only oleoyl-CoA was desaturated and not oleoyl-ACP or other CoA-derivatives tested such as vaccenyl-CoA, elaidyl-CoA, palmitoleoyl-CoA and linoleoyl-CoA. Therefore, the desaturation enzyme appears to have very strict requirement for oleate (Vijay and Stumpf[20]).

Except for the work in Mazliak's laboratory, reported in this volume by Trémolierès et al. we are not aware of any reports on linoleoyl-CoA desaturase experiments. However it should be noted that ammonium-linoleate or free linoleic acid has been used as substrate in some *in vitro*-experiments (Harris and James[8]).

Thus, our very recent findings that ^{14}C-linoleoyl-CoA is metabolized by homogenates of highly immature soya bean cotyledons into ^{14}C-linolenate with good yield is of great interest. The mode of appearance of label in phosphatidylcholine versus that in other lipid classes strongly suggest that also the desaturation of linoleic acid is carried out while the acid is esterified to phosphatidylcholine (Stymne and Appelqvist, to be published). This is in line with the suggestions by Roughan[27] from results of ^{14}C-labelling of pumpkin leaf slices.

It thus appears as there is mounting evidence that a microsomal fraction from several plant species have a high capacity to carry out the conversion of oleoyl-CoA via oleoyl-PC to linoleoyl-PC and probably further to linolenoyl-PC. However, this does not exclude a desaturation also at the CoA level. Howling et al.[37] suggested that there are two mono-ene desaturases in plants, one recognizing the distance from the methyl end of the mono-ene to the double bond and using an oxygen ester (e.g. PC) as substrate, the other recognizing the distance from the carboxyl end and the double bond, and probably using a CoA or ACP derivative. Actually, Pugh and Kates[38] presented some evidence for the co-existence of an oleoyl-PC-desaturase and an oleoyl-CoA desaturase in *Candida*.

EVIDENCE FOR DESATURATION OF FATTY ACIDS ACYLATED TO VARIOUS GLYCEROLIPIDS

An interrelationship between fatty acid biosynthesis and acyl-lipid synthesis was suggested already in 1967 from the data which Nichols et al.[39] obtained in analysis of lipid labelling after ^{14}C-acetate feeding to *Chlorella vulgaris*. They proposed that lipids with high rates of turnover viz. phosphatidylcholine, phosphatidylglycerol, monogalactosyl diacylglycerol and "neutral lipid" might be involved in fatty acid synthesis.

From similar studies of two blue-green algae, Nichols[40] concluded that "the metabolic behaviour of chloroplast lipids may vary considerably according to the class of alga concerned". Detailed studies of the structure and metabolism of different monogalactosyl diacylglycerol "species" in *Chlorella vulgaris* being supplied with ^{14}C-acetate, gave Nichols and Moorhouse[41] reason to state that significant changes in the fatty acid composition of the monogalactosyl diacylglycerol fraction of *Chlorella* occurred after its *de novo* synthesis.

A direct involvement of monogalactosyl diacylglycerols as substrates for linoleate biosynthesis was inferred by Appleby et al.[42] from studies of the blue-green algae *Anabaena variabilis*, a species lacking phosphatidylcholine. Concomitant studies on *Chlorella vulgaris* pointed to two independent sites of linoleate synthesis in the latter species, the lipid intermediates being phosphatidylcholine and monogalactosyl diacylglycerol respectively. As mentioned previously the PC-pathway in *Chlorella* gained further support from the results of Gurr et al.[33], studying the metabolism of ^{14}C-oleoyl-CoA.

As regards higher plants, labelling studies with intact pumpkin leaves (*Cucurbita pepo*), by Roughan[27], using ^{14}CO$_2$ as well as ^{14}C-acetate, showed PC to have the highest specific activity at short incubation times of all lipids investigated. Whereas the specific activity of the galactolipids (MGDG and DGDG) did not change notably with time after incubation, that of PC was the only lipid to show an appreciable turnover of radiocarbon. From these and other data it was suggested that "the primary site of linoleic and linolenic acid biosynthesis in leaf cells is within the PC molecule".

This suggestion obtained further support by the results of Willemot and Verret[43] who studied the kinetics of labelling within the PC molecules of alfalfa leaves supplied with ^{14}C-choline. These authors found that the highest specific activities were shifted from the (16:0/18:2) and (18:2/18:3) species to the (16:0/18:3) species respectively as time of incubation increased.

In further studies by Roughan[44] of the metabolism of ^{14}C-acetate by pumpkin leaves, the role of PC in linoleic acid biosynthesis was verified but data were also reported which were in favour of a desaturation of palmitic acid to trans-

3-hexadecenoic acid while attached to the phosphatidylglycerol molecule. This role of PG fits nicely to suggestions by Bartels et al.[45] based on studies of the metabolism of ^{14}C-hexadecenoic acid by *Chlorella vulgaris* and lettuce leaves. Whereas Roughan[27] noted no change in the oleic acid labelling of PG and no appearance of labelled linoleic or linolenic acids in this lipid even after very long incubation times of "fully expanded" pumpkin leaves, Slack and Roughan[46] noted considerable labelling in oleic, linoleic and linolenic acids in PG in the different organelles studies after ^{14}C-acetate feeding to "expanding" spinach leaves. These authors also found a decrease in palmitic acid labelling and concomitant increase in trans-3-hexadecenoic acid labelling of PG with time of incubation, indicating a desaturation at the acyl lipid level.

Studies by Williams et al.[15] applying $^{14}CO_2$ to *Vicia faba* leaves followed by analysis of ^{14}C-activity of fatty acids in different acyl lipids are in favour of a desaturation of oleic and linoleic acids at the PC and PG level but not in the galactolipids, which are presumed to obtain their unsaturated fatty acids from the PC and PG molecules.

Bedord et al.[47] found that newly synthesized linolenic acid in the algae *Cyanidium caldarum* was mainly associated with phosphaticylcholine and a sulfolipid, whereas MGDG, PG and PE were slowly labelled in 18:3. These authors also reported that both the desaturation and the elongation pathways for 18:3 synthesis are operating in their system, but that the former contributes most to the 18:3 formed.

Labelling experiments with leaves of plant species which have 16:3 besides 18:3 in their monogalactosyl diacylglycerol (Siebertz and Heinz[14]) gave results which suggested a desaturation of 16:0 to 16:1 and further to 16:2 and 16:3 at the MGDG molecule. Their data were consistent with a PC-coupled desaturation of oleic acid, but not with PC as donator of polyunsaturated fatty acids. Rather their data pointed to a relation between fatty acid desaturation and many glycerolipids.

Analysis of lipid labelling by leaf tissue of barley, maize, rye grass and wheat being supplied with ^{14}C-acetate were in favour of linoleic acid desaturation at the MGDG level rather than at the PC level since the former accumulated considerable amount of ^{14}C-linolenic acid whereas PC only showed a trace of 18:3 (Bolton and Harwood[11,12]).

Quite recently, studies in our laboratory on lipid biosynthesis in very young barley shoots, using ^{14}C-oleic acid as a precursor, demonstrated that PC was the first lipid to become labelled with 18:3 and that after a short but definite, time lag linolenic acid appeared also in MGDG (Keith Stobart, personal communication). After long incubation times, a major part of 18:3 was found in MGDG, which is consistent with other data on leaf labelling.

At the present time it seems impossible to draw final conclusions from the different reports briefly abstracted under this subheading, since the results obtained with different species in various physiological conditions are partially contradictory. Hopefully ongoing *in vitro* studies will help clarifying the subject. The role of PC is further discussed below.

THE ROLE OF PHOSPHATIDYLCHOLINE AS DONOR OF LINOLEIC AND LINOLENIC ACIDS TO TRIACYLGLYCEROLS AND OTHER LIPIDS IN DEVELOPING OILSEED TISSUE

Considerable support for the proposal by Roughan[27] that PC is a donor of 18:2 and 18:3 to DG and further to other lipids, such as MGDG and triacylglycerols, has been presented by further work from their laboratory.

Most interesting is the fact that there is a close resemblance between the fatty acid patterns of the PC and the DG of developing oilseed tissue from species which accumulate triacylglycerols with different levels of polyunsaturated fatty acids (Slack et al.[48]). As seen from Table 2 there is a remarkable similarity in the content of different fatty acids in PC and DG of linseed, soya beans and three genetic lines of safflower which are high, intermediate and low in linoleic acid in the triacylglycerols of the fully mature seeds.

TABLE 2

FATTY ACID COMPOSITION OF 3-SN-PHOSPHATIDYLCHOLINE AND DIACYLGLYCEROL FROM DEVELOPING SEED COTYLEDONS OF VARIOUS SPECIES (From Slack et al.[48])

Abbreviations: PC, phosphatidylcholine; DG, diacylglycerol

Species	Lipid	Amounts of individual fatty acids in each lipid (g/mol)				
		Palmitate	Stearate	Oleate	Linoleate	Linolenate
Linseed	PC	50	41	193	90	175
	DG	35	35	215	77	188
Soya bean	PC	81	28	200	210	35
	DG	78	26	184	212	55
Safflower	PC	65	20	43	420	
Cultivar O-22	DG	69	20	51	405	
Cultivar UC-1	PC	33	4	371	152	
	DG	27	3	378	152	
Intermediate	PC	69	11	106	362	
linoleate	DG	53	10	115	370	

It thus appears as the DG-pool in these tissues at the stage of development characterized by rapid lipid accumulation is derived mainly from PC by the action of phospholipase C. However, this does not exclude a small pool of DGs derived directly from the glycerolphosphate pathway. Support for the operation of the glycerolphosphate pathway of triacylglycerol synthesis in developing seeds of *Crambe abyssinica* has been obtained by Gurr et al.[49] from studies of the incorporation of ^{14}C-glycerol in these seed lipids.

The real significance of the sequence PC → DG → TG in developing oilseed tissue is further supported by the observation (see Table 3) that there is a rapid and concomitant change in PC and DG fatty acid patterns of developing soya bean cotyledons as the plants are moved from an intermediate temperature to a higher or lower temperature (Slack and Roughan[50]).

TABLE 3

CHANGES IN THE FATTY ACID COMPOSITION OF 3-SN-PHOSPHATIDYLCHOLINE AND 1,2-DIACYL-GLYCEROLS OF DEVELOPING SEED COTYLEDONS OF SOYA BEAN PLANTS AFTER CHANGES IN AMBIENT TEMPERATURE

Uniform batches of developing soya bean pods were selected on four soya bean plants growing at a temperature of 22.5°C. Each batch contained six pods. At mid-photoperiod four batches of pods were harvested, then two plants transferred to 13°C and the others ot 32°C, and single batches harvested at 6 h and 24 h thereafter. The fatty acid compositions of the lipids from individual batches were analysed by GLC.

Lipid	Ambient temperature (°C)	Time after transfer from 22.5° C (h)	Fatty acid composition (mol/100 mol)				
			16:0	18:0	18:1	18:2	18:3
PC	22.5	0	14.3	4.8	31.7	39.9	9.3
	13	6	14.3	4.6	14.2	56.1	10.8
	13	24	14.8	4.1	10.0	60.0	10.6
	32	6	15.5	6.2	44.0	27.8	6.4
	32	24	13.2	5.4	55.3	22.0	4.2
DG	22.5	0	12.1	5.1	35.4	37.5	10.2
	13	6	11.1	5.2	20.2	52.2	11.3
	13	24	13.6	4.3	16.5	55.0	10.5
	32	6	12.4	6.5	46.1	28.5	6.5
	32	24	11.3	5.0	53.5	24.2	5.9

From Slack and Roughan (1978) Biochem J., 170, 437-439

Such compositional data and previously discussed tracer data from Roughan and Slack and their associates fit nicely to earlier observations by Dybing and Craig[51] These authors, in analysis of the labelling patterns in various lipid classes after feeding of ^{14}C-acetate to developing flaxseed, noted a rapid labelling of PC and DG. Further, when the tracer source was withdrawn, TG fatty acids increased in activity, apparently at the expense of P-lipid and DG fatty acids.

Also the data obtained so far in our laboratory on developing soya beans easily fit to the scheme proposed by Roughan.

Attempts to estimate the contribution of different phospholipids as precursors of DG to be used for TG synthesis were made by Wilson and Rinne[52]. Since the different ^{14}C-phospholipids were supplied in ethanol solution to the excised soya bean cotyledons, the metabolism observed may also reflect differences in uptake by the cells. However, it may be of interest that these authors obtained results in favour of a recycling into phospholipids of the diacylglycerols (once formed from the P-lipids), as well as a utilization for tricylglycerol synthesis.

Some reports on the metabolic behaviour of different phospholipids of developing soya beans must be evaluated cautiosly since Roughan et al.[53] recently reported the formation of rather large quantities of an artefact, phosphatidylmethanol, during extraction of developing soya beans with chloroform/methanol unless certain precautions are made. Activity ascribed to various natural phospholipids might thus refer to phosphatidylmethanol, which is probably derived from several natural phospholipids.

Stumpf[54] has suggested that desaturation of complex polar lipids could occur as a result of altered temperature in the environment in order to rapidly change membrane fluidity, whereas the production of polyunsaturated fatty acids which accumulate in the triacylglycerols could be a result of desaturation at the CoA level. Although this model would make the stereospecificity of the triacylglycerols easier to explain, data in the literature so far tend to support the idea of Roughan.

CONCLUDING REMARKS

Generally, in discussions on the subcellular location of fatty acid biosynthesis in plant tissue, it is assumed that there is a cooperation between the chloroplast and extra-plastidic compartments in carrying out the full biosynthetic sequence from activated acetate to linolenic acid (see e.g. Dubacq et al.[22], Slack et al.[48]). See also Weaire and Kekwick[55] for an account on the true site(s).

The senior author of this paper has now and then taken an alternative view: Until the opposite is finally proved, it is reasonable to assume that the chloroplast is autonomous in synthesizing the lipids necessary for its own need and the rest of the cell is cooperating in the synthesis of lipids located in the mitochondria, the microsomes, the oil bodies of developing seeds and other extraplastidic organelles. Many experiments in the past have only defined the "major" site for fatty acid synthesis or desaturation and many negative findings as regards stearic acid desaturation were reported before the disclosure of the stearoyl-ACP desaturase.

Also the developmental stage of the tissue seems to be of paramount importance for the success in isolation of highly active, in $vitro$, desaturases. The first positive reports on polyunsaturated fatty acids in chloroplasts was from studies by Kannangara and Stumpf[17], using "immature" spinach leaves. Slack and Roughan[46] found high oleate desaturase activity in "immature" maize laminae but little desaturation in "mature" leaves. In our laboratory all attempts to prepare active microsomes from developing soya bean cotyledons were unsuccessful until we used cotyledons which had barely started accumulating triacylglycerols. Since the wet weight of such cotyledons is very low compared to that of cotyledons which are in the peak of triacylglycerol accumulation, the most "logical" object for studies on seed tissue polyunsaturated acid biosynthesis, it is easy to understand that nobody deliberately chooses tiny cotyledons with very low oil content as the favorite object of study.

Hence it appears worthwhile to look seriously for an oleic acid desaturase in the chloroplast (thylakoids, stroma or envelope). So far published data and our own unpublished evidence point to the possibility for a complete synthetic pathway from acetyl-CoA to linolenic acid in extraplastidic locations. In view of the great interest in "designing" plants of soya beans and rapeseed very low in or lacking linolenic acid in their triacylglycerols, further attempts to disclose the independence of polyunsaturated fatty acid biosynthesis in the chloroplasts from that in extra-plastidic compartment are of great theoretical and practical interest.

ABBREVIATIONS

DG, diacylglycerol; TG, triacylglycerol; PC, phosphatidylcholine; PE, phosphatidylethanolamine; PG, phosphaticylglycerol; MGDG, monogalactocyldiacylglycerol.

ACKNOWLEDGEMENTS

The financial support by the Bank of Sweden Tercentenary Foundation and the use of the Phytotron at the Department of Ecophysiology, the Swedish University of Agricultural Sciences, are gratefully acknowledged.

REFERENCES

1. Gurr, M.I. (1974) in Biochemistry, Series One, vol. 4, Biochemistry of Lipids, Goodwin, T.W. ed., Butterworths, London, pp. 181-235.
2. Harwood, J.L. (1975) in Recent Advances in the Chemistry and Biochemistry of Plant Lipids, Galliard, T. and Mercer, E.I. eds., Academic Press, London, pp. 43-93.
3. Stumpf, P.K. (1975) in Recent Advances in the Chemistry and Biochemistry of Plant Lipids, Galliard, T. and Mercer, E.I. eds., Academic Press, London, pp. 95-113.
4. Stumpf, P.K. (1977) in Lipids and Lipid Polymers in Higher Plants, Tevini, M. and Lichtenthaler, H.K. eds., Springer-Verlag, Berlin, pp. 75-84.
5. Appelqvist, L-Å. (1975) in Recent Advances in the Chemistry and Biochemistry of Plant Lipids, Galliard, T. and Mercer, E.I. eds., Academic Press, London, pp. 248-286.
6. James, A.T. and Webb, J.P.W. (1957) Biochem. J., 66, 515.
7. James, A.T. (1963) Biochim. Biophys. Acta, 70, 9-19.
8. Harris, R.V., and James, A.T. (1965) Biochim. Biophys. Acta, 106, 456-464.
9. Simmons, R.O. and Quackenbush, F.W. (1954) J. Am. Oil Chemists' Soc., 31, 441-443.
10. Stearns, Jr, E.M. (1970) in Progress in the Chemistry of Fats and Other Lipids, vol. IX, Polyunsaturated Acids, Part 4, Holman, R.T. ed., Pergamon Press, Oxford, pp. 455-516.
11. Bolton, P. and Harwood, J.L. (1978) Planta, 138, 223-228.
12. Bolton, P. and Harwood, J.L. (1978) Planta, 139, 267-272.
13. Cherif, A., Dubacq, J.P., Mache, R., Oursel, A. and Trémolières, A. (1975) Phytochem., 14, 703-706.
14. Siebertz, H.P. and Heinz, E. (1977) Z. Naturforsch., 32 c, 193-205.
15. Williams, J.P., Watson, G.R. and Leung, S.P.K. (1976) Plant Physiol., 57, 179-184.
16. Stumpf, P.K. and James, A.T. (1963) Biochim. Biophys. Acta, 70, 260-270.
17. Kannangara, C.G. and Stumpf, P.K. (1972) Arch. Biochem. Biophys., 148, 414-424.
18. McMahon, V. and Stumpf, P.K. (1964) Biochim. Biophys. Acta, 84, 361-364.
19. Vijay, I.K. and Stumpf, P.K. (1971) J. Biol. Chem., 246, 2910-2917.
20. Vijay, I.K. and Stumpf, P.K. (1972) J. Biol. Chem., 247, 360-366.
21. Abdelkader, A.B., Cherif, A., Demandre, C. and Mazliak, P. (1973) Eur. J. Biochem., 32, 155-165.
22. Dubacq, J.P., Mazliak, P. and Trémolières, A. (1976) FEBS Letters, 66, 183-186.
23. Slack, C.R., Roughan, P.G. and Terpstra, J. (1976) Biochem. J., 155, 71-80.
24. Jaworski, J.G. and Stumpf, P.K. (1974) Arch. Biochem. Biophys., 162, 158-165.
25. Porra, R.J. and Stumpf, P.K. (1976) Arch. Biochem. Biophys., 176, 53-62.
26. Trémolières, A. and Mazliak, P. (1974) Plant Science Letters, 2, 193-201.

27. Roughan, P.G. (1970) Biochem. J., 117, 1-8.
28. Jacobson, B.S., Kannangara, C.G. and Stumpf, P.K. (1973) Biochem. Biophys. Res. Commun., 51, 487-493.
29. Jacobson, B.S., Kannangara, C.G. and Stumpf, P.K. (1973) Biochem. Biophys. Res. Commun., 52, 1190-1198.
30. Kannangara, C.G., Jacobson, B.S. and Stumpf, P.K. (1973) Biochem. Biophys. Res. Commun., 52, 648-655.
31. Vance, W.A. and Stumpf, P.K. (1978) Arch. Biochem. Biophys., 190, 210-220.
32. Brar, G.S. and Thies, W. (1979) Proceedings from the 5th International Rapeseed Conf., Malmö, Sweden (in the press).
33. Gurr, M.I., Robinson, M.P. and James, A.T. (1969) Eur. J. Biochem., 9, 70-78.
34. Stymne, S. and Appelqvist, L-Å. (1978) Eur. J. Biochem., 90, 223-229.
35. Barron, E.J. and Mooney, L.A. (1968) Anal. Chem., 40, 1742-1744.
36. Nichols, B.W. and Safford, R. (1973) Chem. Phys. Lipids, 11, 222-227.
37. Howling, D., Morris, L.J., Gurr, M.I. and James, A.T. (1972) Biochim. Biophys. Acta, 260, 10-19.
38. Pugh, E.L. and Kates, M. (1975) Biochim. Biophys. Acta, 380, 442-453.
39. Nichols, B.W., James, A.T. and Breuer, J (1967) Biochem. J.,104, 486-496.
40. Nichols, B.W. (1968) Lipids, 3, 354-360.
41. Nichols, B.W. and Moorhouse, R. (1969) Lipids, 4, 311-316.
42. Appleby, R.S., Safford, R. and Nichols, B.W. (1971) Biochim. Biophys. Acta, 248, 205-211.
43. Willemot, C. and Verret, G. (1973) Lipids, 8, 588-591.
44. Roughan, P.G. (1975) Lipids, 10, 609-614.
45. Bartels, C.T., James, A.T. and Nichols, B.W. (1967) Eur. J. Biochem., 3, 7-10.
46. Slack, C.R. and Roughan, P.G. (1975) Biochem. J., 152, 217-228.
47. Bedord, C.J., McMahon, V. and Adams, B. (1978) Arch. Biochem. Biophys., 185, 15-20.
48. Slack, C.R., Roughan, P.G. and Balasingham, N. (1978) Biochem. J., 170, 421-433.
49. Gurr, M.I., Blades, J., Appleby, R.S., Smith, C.G., Robinson, M.P. and Nichols, B.W. (1974) Eur. J. Biochem., 43, 281-290.
50. Slack, C.R. and Roughan, P.G.(1978) Biochem. J., 170, 437-439.
51. Dybing, C.D. and Craig, B.M. (1970) Lipids, 5, 422-429.
52. Wilson, R.F. and Rinne, R.W. (1976) Plant Physiol., 57, 556-559.
53. Roughan, P.G., Slack, C.R. and Holland, R. (1978) Lipids, 13, 497-503.
54. Stumpf, P.K. (1976) in Plant Biochemistry, Third Edition, Bonner, J. and Varner, J.E., eds. Academic Press, New York, pp. 428-460.
55. Weaire, P.J. and Kekwick, R.G.O. (1975) Biochem. J., 146, 425-437.

SOME ASPECTS OF GALACTOLIPID FORMATION IN SPINACH CHLOROPLASTS

AERNOUT VAN BESOUW, GERARD BÖGEMANN, and J.F.G.M. WINTERMANS
Department of botany, Catholic University of Nijmegen, 6525 ED Nijmegen
the Netherlands

ABSTRACT

Galactosylation of diacylglycerols occurs in two separate steps; the first of which involves the incorporation of galactose from UDPGal into the membrane lipids as MGDG. The second step is the interlipid exchange of galactose.

INTRODUCTION

In their model of 1961 for the formation of galactolipids, Ferrari and Benson[1] use the following scheme:

$$\text{diacylglycerol} + \text{UDPGal} \longrightarrow \text{MGDG} + \text{UDP} \qquad (I)$$

$$\text{MGDG} + \text{UDPGal} \longrightarrow \text{DGDG} + \text{UDP} \qquad (II)$$

In our experiments with isolated chloroplast envelopes we could confirm the first step of the reaction by NMR, but for reasons we shall elucidate further below, we had to exclude the second reaction step. Instead we proposed[2] an interlipid galactosyltransferase activity according to the scheme:

$$2 \text{ MGDG} \longrightarrow \text{DGDG} + \text{diacylglycerol} \qquad (III)$$

MATERIALS AND METHODS

Materials. Radioactive UDPGal was obtained from the Radiochemical Centre, Amersham; ^{3}H labelled: batch 4 specific activity 16,3 Ci/mmol, ^{14}C labelled batch 27 specific activity 0.337 Ci/mmol. UDPGal (Sigma grade) was obtained from the Sigma Chemical co. Nucleotides, uridine, uracil, galactose and galactose-1-phosphate were obtained from Boehringer, Mannheim. All other reagents were, analytical grade , from Merck, Darmstadt.

Isolation procedure. Chloroplast envelopes were isolated according to Joyard and Douce[3] with only minor differences. Contamination of chloroplast envelopes with thylakoid material was measured with a Beckman spectrophotometer model 25 after extraction with ethanol[4].

Reaction Mixtures. The complete reaction mixture contained 100μg/ml of envelope proteins, measured according to Lowry et al[5].,0.2 mM UDPGal, 25·10^5 cpm if ^{3}H labelled and 10^5 cpm if ^{14}C labelled, 0.1 M tricine/KOH buffer pH 7.2

and 10 mM $MgCl_2$, all final concentrations. The reaction was started by addition of envelope material and was carried out in a waterbath at $30°$ C with continuous shaking. The reaction was stopped by addition of a triple volume of methanol and mixing on a Vortex shaker.

Chromatography and quntification of label. Lipids were extracted according to Bligh and Dyer[6], the chloroform layer being washed twice and taken to dryness with a stream of air. Lipoluma (Lumac) was added and the vials were counted in a Philips PW 4540 L.S.A.. For separation of lipids, drying was performed with N_2, the residue was dissolved in 40 μl $CHCl_3/CH_3OH$ 2:1 v/v and brought on a thin layer plate (Kieselgel 60, Merck) with a 50 μl Hamilton syringe. After elution with $CHCl_3/CH_3OH/H_2O$ 65:25:4 v/v, radioactivity was scanned with a Desaga thin layer scanner model 12-2, or autoradiographed with Kodirex X-ray film (Kodak). Spots with activity were scraped off and counted as mentioned.

RESULTS AND DISCUSSION

We have found several effectors with different effects on the two reaction steps given above, thus supporting our hypothesis of two separate reactions. For instance, uridine nucleotides show competetive inhibition on reaction(I) as shown in fig.1, but have no effect on the percentage label in MGDG after prolonged incubations. UDP is the strongest inhibitor, 0.13 mM being necessary for an inhibition of 50%. UMP is a weaker inhibitor, achieving 50% inhibition with a concentration of 2 mM.

Fig. 1. Lineweaver Burk plot of monogalactolipid synthesis

The reaction is inhibited by UDP (□) in a competitive way, compared with a reference incubation without UDP (▲). The mixture was incubated for 5 minutes at $30°C$ and pH 7.2.

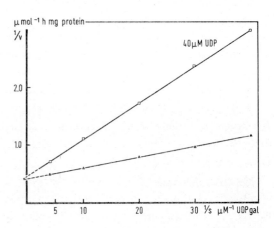

UTP inhibits the reaction 50% at a concentration of 2.5 mM. Since the preparation contains about 4% of UDP, according to the manufacturer, this means that also a concentration of about 0.1 mM UDP is present. Therefore most of the inhibition is due to UDP and UTP is only a very weak inhibitor.

Gal-1-P and Gal have no effect at all, which leads to the conclusion that the UDPGal molecule is recognised by the enzyme in the UDP part. On the other hand, it is obvious from the lack of effect on the second reaction step, that DGDG and higher homologues are synthesised without direct use of UDPGal.

Temperature dependence is also a parameter with marked differences for both reactions. An Arrhenius plot of the rates of incorporation of galactose into the the lipid fraction shows two straight lines with a breakpoint at about $17^{\circ}C$ (fig.2). This breakpoint did not correspond with a phase transition of the membrane lipids, as was checked in an experiment in which differential calorimetric scans were made of envelope membranes. No phase transition could be detected in this experiment between $0^{\circ}C$ and $60^{\circ}C$. The break point in the plot appeared to be independent of the age of the leaves. Both envelopes isolated

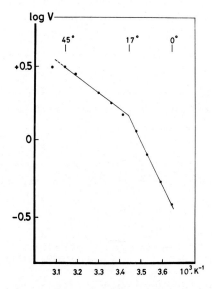

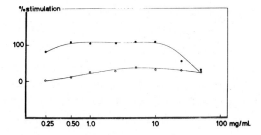

Fig. 3. The effect of albumin on galactolipid synthesis

On the abscissa is plotted the albumin concentration in mg/ml, on the ordinate the stimulation compared with a blanc without albumin (=0%).
a) Stimulation of MGDG synthesis in a 5 min. incubation (●).
b) % of label in higher homologues after one hour of incubation (o).

Fig. 2. Arrhenius plot of MGDG synthesis.
On the ordinate is plotted the logarithm of the reaction velocity in μmol/mg h, on the abscissa 1000/T in Kelvin. The enzyme appears to be stable up to $45^{\circ}C$.

from market spinach as well as those isolated from young leaves from spinach grown in a greenhouse showed the same breakpoint at 17°C. It is interesting, that above this same temperature the activity of the interlipid galactosyl transfer (reaction III) can be observed, as shown in table 1.

TABLE 1
Label in galactolipids after one hour of incubation with ^{14}C-UDPGal at various temperatures.

Temp.	MGDG		DGDG		T(e)GDG	
	cpm	%	cpm	%	cpm	%
0°C	1625	100	0	0	0	0
15°C	2860	94	150	5	35	1
30°C	2800	75	810	22	130	3
45°C	2860	60	1240	26	640	14

Below 17°C over 90% of the label is found in MGDG, whilst the proportion of the higher homologues rapidly increases with increasing temperature. This clearly indicates that both mechanisms are different involving different enzymes.

A third interesting effector is albumin. From fig.3 it appears, that both reactions are enhanced in a different way. The enhancement of MGDG synthesis is not specific, as can be concluded from the wide range of albumin concentrations that have the same effect. This stimulation is probably due to the free fatty acid binding capacity of albumin. For indeed we found, that free fatty acids are potent uncompetitive inhibitors of the galactosylation reaction of diglycerides, and that isolated envelopes contain free fatty acids. We do not yet understand the effect of albumin on the intergalactolipid galactosyl transfer, but it is likely to be a more specific effect, although much weaker, since it has a more definite optimum concentration (5 mg/ml). This optimum concentration is 20 fold higher than the lowest concentration of albumin, which enhances the first reaction optimally.

REFERENCES

1. Ferrari, R. and Benson, A. (1961) Arch. Biochem. Biophys. 93, 185-192
2. van Besouw, A. and Wintermans, J. (1978) Biochim. Biophys. Acta 529, 44-53
3. Joyard, J. and Douce, R. (1976) Physiol. Vég. 14, 31-48
4. Wintermans, J. and de Mots, A. (1965) Biochim. Biophys. Acta 109, 448-453
5. Lowry, O., Rosebrough, N., Farr, A. and Randall, R. (1951) J. Biol. Chem. 193, 265-275
6. Bligh, E. and Dyer, W. (1959) Can. J. Biochem. Physiol. 37, 911- 917

CHANGES IN THE PHOTODYNAMIC PROPERTIES OF THE CHLOROPHYLL(IDE) DURING THE EARLY STAGES OF GREENING

LENNART AXELSSON AND EVA SELSTAM

Dept of Plant Physiology, University of Göteborg, Göteborg (Sweden)

ABSTRACT

Dark grown plants were greened in weak light (white fluorescent, 4 W m^{-2}). At different stages in the greening, different absorption forms of the chlorophyll(ide) predominated. After photoreduction of the protochlorophyllide, the chlorophyllide 684-form was the main pigment form during the first few min of greening (stage I). After 30-45 min of greening all 684-form had been transformed into the Chlorophyll 673-form. The Chlorophyll 673-form remained stable for about 30 min (stage II), but then started to transform into pigment form(s) absorbing at 676-8 nm (stage III). Short lasting (min) high intensity irradiations (red light, 1000 W m^{-2}) given to the leaves at certain stages in the greening, caused a decomposition of the chlorophyll(ide) form(s) present. Depending on the greening stage, this irradiation could also result in a certain decrease in the carotenoid content and in the content of galactolipids. In some cases the high intensity irradiation also caused an inhibition of further *synthesis* of both carotenoids and chlorophyll. The described effects of the high intensity irradiation was interpreted as photodynamic effects sensitized by the pigment form(s) present. Thus the *Chlorophyllide 684-form* sensitized *in vivo* photodecomposition of carotenoids and of the pigment form itself, as well as photoinhibition of chlorophyll and carotenoid synthesis. The chlorophyllide present in isolated etioplasts did not sensitize photodecomposition of the linolenic acid of monogalactosyl diglyceride. The *chlorophyll 673-form* sensitized photodecomposition of the linolenic acid of monogalactosyl diglyceride in isolated chloroplasts, and of the pigment form itself *in vivo*, but did not sensitize photodecomposition of carotenoids or photoinhibition of pigment synthesis. The *chlorophyll forms present after 180 min* of greening (stage III) sensitized the same processes as did the chlorophyllide 684-form, but the 'sensitizing capacity' of these pigment form(s) was less pronounced. The fact that different absorption forms of the chlorophyll(ide) sensitized different processes was thought to reflect a different localization, or molecular organization, of the pigment forms in question.

INTRODUCTION

Some of the early stages in the chlorophyll biosynthesis (the Chlide 684-form and the chlorophyll 673-form) are easily decomposed by light. During chlorophyll photo-decomposition *in vitro*, singlet oxygen functions as an intermediate energy carrier. This is probably also the case when chlorophyll is photodecomposed *in vivo*. Since the singlet oxygen molecule is very reactive, it is not surprising that in some cases, only a slight bleaching of chlorophyll *in vivo* may lead to a complete inhibition of further chlorophyll biosynthesis. This is the case e.g. in leaves treated with δ-aminolevulinic acid[1]. But in other cases a photodecomposition of considerable amounts of the Chl(ide) may have a very small effect on furhter chlorophyll biosynthesis. In such cases the pigment form being photodecomposed must either be compartmentalized from structures sensitive to singlet oxygen (by e.g. membranes containing carotenoids), or the reaction sequence for the photodecomposition may not involve singlet oxygen.

The present investigation deals with effects on chlorophyll and carotenoid biosynthesis, caused by photodecomposition of different absorption forms of chlorophyll(ide). Studies of Chl(ide) sensitized photodecomposition of carotenoids and of monogalactosyl diglycerides has also been included. The results indicates that during greening at low temperatures, when the light sensitive pigment forms exist over a longer period of time, even moderate intensities may have a destructive effect on the enzyme system for chlorophyll biosynthesis. The degree of damage on a certain process, caused by photodecomposition of a certain pigment form, gives informations about the localization of this pigment form in relation to the sensitive component of the process (singlet oxygen must diffuse from the pigment form to the sensitive component).

<u>Abbreviations</u>. PChlide, protochlorophyllide; Chlide, chlorophyllide; Chl(ide), chlorophyll and/or chlorophyllide (the wavelengths of the absorption maxima are denoted for the different forms); MGDG, monogalactosyl diglyceride; HIR, high intensity red light (irradiation); LI, low intensity (irradiation).

MATERIAL AND METHODS

Plant material

For the *in vivo* experiments, six-day-old dark grown leaves of wheat[2] (*Triticum-aestivum* L. cv. Weibull's Starke) were used. Etioplast were prepared from five-day-old dark grown barley seedlings[3] (*Hordeum disticum* L. cv. Weibull's Ingrid).

Etioplast preparation

Etioplasts were isolated according to Griffiths[4], except that the pH of the isolation medium was changed to 7.6. Five cm sections of the barley leaves were used. The leaf sections were cut 2 cm below the tip.

Pigment determination

In vivo determination of Chl(ide) was done by absorption spectrophotometry. Pieces of leaves, 2.5 cm long, were cut 2.0 cm below the tip. Samples of three leaf-pieces were inserted in a specially designed sample cell, and the absorption spectra was measured on the midmost (one cm) parts of the leaves. Absorption spectra of etioplasts were measured in a similar sample cell, a small glass tube (3x15 mm), containing 200 µl of etioplast suspension. The spectrophotometer was a Shimadzu MPS 50 L, with the measuring range set at 1-2 absorbance units.

The height of the absorption peak in the red region was taken as a measure of the Chl(ide) content in the sample[2].

For *in vitro* determinations of carotenoids and chlorophyll, the leaf pieces were extracted by aceton. The absorption spectra of the extracts was recorded between 700 and 350 nm on a Shimadzu UV-200 spectrophotometer, thus making possible baseline corrections due to scattering. The absorption at 663 nm was used as a relative measure of the content of the Chl(ide) and the absorption at 480 nm as a relative measure of the carotenoid content. The absorption at 480 nm was corrected for the increase in absorption at this wavelength due to the photodecomposition of the Chl(ide).

Lipid analysis

MGDG was extracted from the etioplasts and purified by thin layer chromatography[3], and was determined quantitatively as the amount of linolenic acid methyl ester derived from the MGDG after transesterification with BF_3 in methanol[3]. The amount of linolenic acid methyl ester was determined by gas chromatography (Perkin-Elmer model 900)[3].

Irradiation equipment

High intensity red light irradiation (HIR). The light from a 1000 W projector lamp (Philips 6286C/05 G17Q) was filtered through 10 cm of water, and focused on the sample *in situ* in the spectrophotometer). The light was filtered through a heat reflecting filter and a red glass cut off filter (Baltzer's Calflex C and RG1). The intensity at the front of the sample cell was adjusted by a neutral grey filter to 1000 W m^{-2} and the irradiation covered the wavelengths between 610-740 nm. In case of leaf pieces only the midmost 1 cm parts were irradiated.

Low intensity irradiation (LI) was obtained from four flourescent tubes (Philips TL De Luxe), at a distance of about 60 cm from the samples. The intensity at the level of the sample was 4.0 W m^{-2}.

Experimental

The leaves (or the etioplasts) were treated in such a way that a certain form of the Chl(ide) predominated. This pigment form was then partly bleached by HIR. During the bleaching the temperature of the sample cell containing the leaves was kept at -1 °C, in order to avoid light independent transformations of the pigment form(s) present. The temperatur of the leaves increased to +2 °C during the first four min of HIR, but was then constant. The etioplasts were irradiated at 20 °C. Some processes or compounds affected by the HIR were studied as follows:

Chlorophyll and Carotenoid formation after HIR. The leaf pieces were 'greened' in Petri dishes in weak light for 24 hours. Damages on the chlorophyll synthesis then became clearly visible, as the middle part of the leaf pieces remained yellow. The Chl(ide) content of the midmost (1 cm) parts of the leaves (= sample) was then compared to the Chl(ide) content of such parts of the leaves which had not received HIR (= reference). These parts were 0.5 cm long, situated on both sides of the bleached parts, and separated from them by one mm. Greening ability was defined as: $100 \cdot Chl(ide)_{sample}/Chl(ide)_{reference}$, after 24 hours LI. The very small amount of Chl(ide) present before greening could be overlooked.

The ability to form carotenoids was studied in a similar way, but in this case the amount of carotenoids present before the greening had to be considered.

Chlorophyll and Carotenoid decomposition during HIR. Decrease in the content of chlorophyll(ide) was calculated from the absorption spectra of a sample before and after HIR (both for leaves and etioplast preparations). The decrease in carotenoid content was evaluated by comparing the carotenoid content in acetone extracts from irradiated parts of leaf pieces (sample) with the carotenoid content in acetone extracts from non-irradiated parts of the same leaf pieces (reference).

Decomposition of Galactolipids during HIR. The content of MGDG in samples of HIR irradiated etioplasts was compared with the MGDG content in samples of non irradiated etioplasts from the same preparation.

If not otherwise stated, all measurements are mean values of six measurements.

RESULTS

Effects of HIR of leaves. When dark grown leaves received LI, the ability of the Chl(ide) to sensitize its own photodecomposition in HIR decreased steadily with time. The 673-form (present after 60 min LI) needed about twice as long

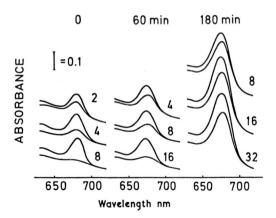

Fig. 1. Absorption spectra of samples of greening dark grown leaves showing the Chl(ide) content before and after a HIR. The time of HIR (min) is indicated at each pair of curves, and the preirradiation (LI) of the leaves is indicated at the top of the Figure.

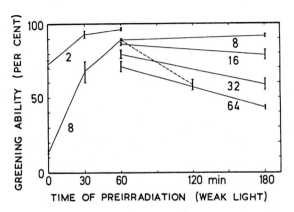

Fig. 2. Ability to resume chlorophyll synthesis (greening ability) in dark grown leaves, which after different time of preirradiation (x-axis) have received a certain amount of HIR (indicated in min at each curve). The vertical bars indicate standard error. The broken line shows the effect of returning the leaves into darkness.

irradiation time as the 684-form for a certain decrease in absorption (Fig. 1). The pigment form(s) existing after 180 min of LI needed about twice as long irradiation time as the 673-form for a certain decrease in absorption. The ability of the pigment to sensitize photoinhibition of further chlorophyll synthesis at first also decreased with the time of LI (Fig. 2). Eight min of HIR, which caused an almost total decomposition of the 684-form (Fig. 1), also resulted in a very high reduction in greening ability (Fig. 2), while irradiation nearly completely bleaching the 673-form (16 min HIR), had a very small effect on the greening ability. But from 60 to 180 min of LI, the ability of the Chl(ide) to sensitize photoinhibition of chlorophyll synthesis increased (Fig. 2). If the leaves after 180 min of LI received 16 min of HIR, the decrease in pigment absorption was less than when the 673-form was irradiated, but the reduction in greening ability was more pronounced.

The ability of the Chl(ide) to sensitize both photodecomposition of

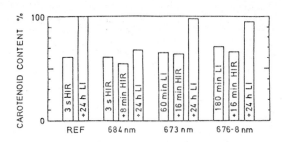

Fig. 3. The carotenoid content of dark grown leaves after pre-irradiation (3 s HIR, 60 min LI or 180 min LI), after a following bleaching of the Chl(ide) (8 or 16 min HIR, cf. Fig. 1), and after greening of the bleached leaves. The carotenoid content of greened dark grown leaves was used as reference. The wavelength of the absorption peak of the Chl(ide) before bleaching is indicated.

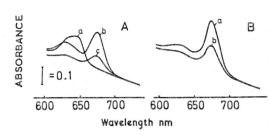

Fig. 4. A. Absorption spectra of etioplasts from dark grown barley (a), after photoreduction of the PChlide (b), and after partial bleaching of the Chlide (18 min. HIR; c). B. Absorption spectra of etioplasts isolated from dark grown barley which have received 20 min LI (a), and after partial bleaching of the chlorophyll (18 min HIR; b).

carotenoids and photoinhibition of carotenoid synthesis (Fig. 3) varied in the same way as described for the ability to sensitize photoinhibition of chlorophyll synthesis (Fig. 2).

<u>Effects of HIR of etioplasts.</u> The Chlide, which is formed by photoreduction of the PChlide in etioplasts (2 s HIR, 4 min dark, 2 s HIR) was partly photodecomposed by 18 min HIR (Fig. 4 A). HIR of this Chlide form caused a very little decrease in the content of galactolipids (1.7 $\pm$ 3.3 %). Irradiation (18 min HIR) of etioplast isolated from irradiated (20 min LI) seedlings also caused a decrease in the content of Chl(ide) (Fig. 4 B). In this case the Chl(ide) photodecomposition was accompanied by a pronounced decrease in galactolipid content (13.0 $\pm$ 3.5 %).

This investigation was supported by the Swedish Natural Science Research Council.

REFERENCES
1. Axelsson, L. (1974) Physiol. Plant., 31, 77-85.
2. Axelsson, L. (1976) Physiol. Plant., 38, 327-332.
3. Selstam, E. (1978) Thesis, Dept of Plant Physiology, Univ of Göteborg, III.
4. Griffiths, W.T. (1975) Biochem. J., 146, 17-24.

ON THE COMPARTMENTATION OF THE BIOSYNTHESIS OF AROMATIC AMINO ACIDS AND PRENYLQUINONES IN HIGHER PLANTS.

H. BICKEL, BARBARA BUCHHOLZ, G. SCHULTZ
Botanisches Institut and Institut für Tierernährung, Arbeitsgruppe für Phytochemie und Futtermittelkunde, Tierärztliche Hochschule D 3000 Hannover (FRG)

ABSTRACT

Aromatic amino acids are synthesized by shikimate pathway in spinach chloroplasts to some extent whereas peroxisomes seemed to be involved in forming the quinone moiety of the prenylquinones PQ and αT from Tyr.

INTRODUCTION

Two pathways are combined in the biosynthesis of prenylquinones: the mevalonate pathway for synthesizing the isoprenoid moiety of prenylquinones and the SKA pathway for aromatic moiety. Goodwin[1] could demonstrate that an intraplastidic and an extraplastidic site of mevalonate synthesis exist in higher plants because the chloroplast envelope is more or less a barrier for mevalonate and its pyrophosphate. Thus, carotene and the isoprenoid moiety of PQ and perhaps of tocols are coming from the intraplastidic mevalonate pool whereas the sterols and the isoprenoid moiety of ubiquinones originates from the extraplastidic one. Threlfall et al.[2] revealed the biosynthesis of the aromatic moiety of prenylquinones. SKA as well as Tyr is incorporated via HPP and homogentisate. Aim of present work was to investigate the compartmentation of SKA pathway and the subsequent conversion of Tyr to homogentisate in prenylquinone synthesis.

MATERIAL AND METHODS

In the present studies unpurified[3] and purified[4] spinach chloroplast, both fully intact, were compared. Beside chloroplasts

---

Abbreviations: Chlp - chloroplast; HPP - p-hydroxyphenylpyruvate; KDAHP - 2-keto-3-deoxyarabinoheptulonic-acid-7-P; αKGA - α-ketoglutuarte; PQ - plastoquinone-45; SKA - shikimate; αT - α-tocopherol.

suspensions of unpurified chloroplasts contained more or less large portions of peroxisomes, mitochondria, ER, cytosol etc. in part as multiorganelle complexes[4] whereas suspensions of purified chloroplasts were free from other organelles. This was tested by NADH-hydroxypyruvate reductase[5], cytochrome c oxidase[6] nonreversible nonphosphorylating NADP-linked glyceraldehyde-3-phosphate-dehydrogenase[7] as marker enzymes for peroxisomes, mitochondria and cytosol, respectively. Leaf peroxisomes from spinach were isolated as described in[8]. Homogentisate-/α-^{14}C/ was synthesized according to[9]. Chloroplast suspensions were illuminated with white light (1 x 10^6 erg/cm^2 sec) in 1 ml Medium C[3] containing 10 mM $NaHCO_3$ and the labelled substance for 30 min at 20 $\pm$ 2°C. For identification of amino acids as dansylated compounds and of prenylquinones see[10] and [11,12], respectively. Identification of HPP as 2,4-dinitrophenylhydrazone was performed as follows: the effluate(pH 2)[10] from the cation exchange column (Dowex 50 W x 8 in H^+-form) was extracted thrice with 2 vol $(Et)_2O$. The residue of evaporated $(Et)_2O$-phase was dissolved in 40 ml dry $(Et)_2O$ + 20 mg 2,4-dinitrophenylhydrazine + 20 mg $AlCl_3$ and stirred for 20 min (-15°C; N_2). After evaporation the products were dissolved in 2 ml EtOH. TLC on HPTC-Silicagel (Merck) with $CHCl_3$ - MeOH - NH_3 25 % 7 : 2 : 0,2 as solvent. R_f of HPP-dinitrophenylhydrazone was ca 0,16; homogentisate was also converted.

RESULTS

Synthesis of aromatic amino acids in unpurified chloroplasts. Intact chloroplasts if not purified[3] synthesize amino acids from photosynthetically fixed CO_2. These are mainly Ala, Gly, Ser and some dicarboxylic amino acids. As Buchholz et al.[8] could demonstrate, not only multiorganelle complexes but also chloroplasts, peroxisomes and mitochondria present in these suspensions co-operate completely in synthesizing the above amino acids.

Aromatic amino acids could not be detected before rechromatographed on system II in ref.[10]. Compared to aliphatic amino acids unpurified spinach chloroplasts synthesize small but definated amounts of aromatic amino acids[8,10]. The following experiments were performed to decide wether aromatic amino acids are synthesized in the chloroplasts themselves or in organelles accompanying these as in the case of aliphatic amino acids[13].

Indications for synthesis of aromatic amino acids in chloroplasts. To prove this, experiments were performed under two points of view.

(i) Considering primary products of photosynthesis mainly consisting of triosephosphates are translocated by shuttle mechanism of phosphate translocator from the isolated chloroplasts into the medium[14], the substrate concentrations for operating of the SKA pathway might be too low. Consequently, it was argued that it might be possible to preserve substrates within the chloroplasts by lowering the export as it was achieved by omitting inorganic phosphate from the medium[14]. Doing so, it should effect a better supply of pathway providing complete glycolysis is present in chloroplasts.

(ii) Similar results but on another way are to be expected by levelling the gradient of substrate concentration between organelle and medium by the method of "enriching" the medium with substrates[15].

The results by omitting inorganic phosphate (i) and in "enriched" medium (ii) confirm this hypothesis (Table 1). Synthesis of aromatic acids and prenylquinones was twice to thrice as high as in the control.

TABLE 1
^{14}C-INCORPORATION OF $^{14}CO_2$ INTO AROMATIC AMINO ACIDS AND PRENYLQUINONES OF UNPURIFIED CHLOROPLASTS IN MEDIUM C[3], THE SAME WITHOUT INORGANIC PHOSPHATE[14] AND IN "ENRICHED" MEDIUM[15]

	Medium C	without P_i	"enriched"
	% of photosynthetically fixed CO_2/mg chlorophyll		
Phe + Tyr + Trp	0,0378	0,0740	0,0860
PQ + αT	0,0012	0,0026	0,0022

Synthesis of aromatic amino acids and prenylquinones in purified chloroplasts compared to unpurified ones. From the results in Table 1 it could be assumed synthesis of aromatic amino acids in purified chloroplasts might be in the same range as in unpurified ones. Applying SKA-/1,6-^{14}C/ as substrate (Table 2) this could be confirmed. Under these conditions it can be assumed SKA is translocated across the chloroplast envelope.

On the other hand, only unpurified chloroplasts synthesize prenylquinones (Table 2). In the following more direct precurors of prenylquinone synthesis were used to decide wether (i) the purified chloroplasts are damaged by isolation procedure or (ii) other organelles are involved in this synthesis.

TABLE 2
^{14}C-INCORPORATION OF SKA-/1,6-^{14}C/ INTO AROMATIC AMINO ACIDS AND PRENYLQUINONES OF CHLOROPLASTS

/dpm / mg chlorophyll/	unpurified Chlp	purified Chlp
Added Shikimate	3 320 000 (100)	20 340 000 (100)
Phe	13 500)	50 640)
Tyr	3 100) (0,81)	77 300) (2,16)
Trp	10 200)	26 700)
PQ	2 570) (0,33)	247) (0,04)
αT	8 220)	1 400)

As in the case of labelled SKA also Tyr-/β-^{14}C/ was incorporated into prenylquinones only by unpurified chloroplasts (Table 3).

TABLE 3
^{14}C-INCORPORATION OF TYROSINE-/β-^{14}C/ INTO PRENYLQUINONES OF CHLOROPLASTS

/dpm / mg chlorophyll /	unpurified Chlp	purified Chlp
Added Tyrosine	9 030 000 (100)	13 270 000 (100)
PQ	167 000 (1,8)	630 (0,006)
αT	133 000 (1,5)	1 050 (0,008)

Formation of HPP and homogentisate by chloroplasts and peroxisomes. ^{14}C-Incorporation from Tyr-/β-^{14}C/ into HPP and homogentisate took place only by unpurified chloroplasts (Table 4). This indicates that Tyr is converted in organelles other than chloroplasts.

TABLE 4
^{14}C-INCORPORATION OF TYROSINE-/β-^{14}C/ INTO HPP AND HOMOGENTISATE OF CHLOROPLASTS

/ dpm / mg chlorophyll /	unpurified Chlp	purified Chlp
Added Tyrosine	1 970 000 (100)	3 080 000 (100)
HPP	2 270 (1,2)	80 (0,002)
Homogentisate	1 290 (0,6)	390 (0,01)

Because of this, leaf peroxisomes and purified chloroplasts were compared under the same conditions as in Table 4, except that 1 mM α KGA as NH_2-acceptor for the aminotransfer-reaction was added. By this only leaf peroxisomes (Table 5) but not the purified chloroplasts (Table 6) exhibits a strong enhancement of amino transfer from Tyr to HPP and also of oxidation of the ketoacid to homogentisate. Thus, an active aminotransferase specific for Tyr and a HPP-oxidase might be present in peroxisomes whereas only low activities of amino acid oxidase and no transferase could be found in chloroplasts (see also[16]).

Studying the distribution, HPP and homogentisate formed by unpurified chloroplasts from Tyr-/β-^{14}C/ sedimented largely with the organelles. Only a slow transfer to the medium occured.

TABLE 5
^{14}C-INCORPORATION OF TYROSINE-/β-^{14}C/ INTO HPP AND HOMOGENTISATE OF LEAF PEROXISOMES IN MEDIUM C^3 ADDING OR OMITTING α KGA.

/ dpm /	+ 1 mM α KGA	without α KGA
Added Tyrosine	9 500 000 (100)	6 080 000 (100)
HPP	18 000 (0,2)	183 (0,003)
Homogentisate	10 900 (0,11)	123 (0,002)

TABLE 6
^{14}C-INCORPORATION OF TYROSINE-/β-^{14}C/ INTO HPP AND HOMOGENTISATE OF PURIFIED CHLOROPLASTS IN MEDIUM C^3 ADDING OR OMITTING α KGA.

/ dpm / mg chlorophyll /	+ 1 mM α KGA	without α KGA
Added Tyrosine	5 750 000 (100)	11 105 000 (100)
HPP	3 770 (0,06)	6 310 (0,06)
Homogentisate	157 (0,003)	376 (0,003)

In vivo a more or less tight connection between plastids and peroxisomes is supposed. This might be a prerequisit for some of their functions. From this is might be understood that in contrast to whole plants[9,10] homogentisate-(α-^{14}C) was only poorly incorporated into prenylquinones of purified as well as of unpurified chloroplasts (ca 0,1 % of substrate uptaken by chloroplasts).

CONCLUSIONS

Beside in the extraplastidic site[8,17] the shikimate pathway acts in chloroplasts too (Figure 1). In these, E4P and PEP as substrates for the KDAHP-synthesis might come from Calvin cycle and glycolysis, respectively (for restrictions of this assumption see[18]). For extraplastidic syntheses E4P might be translocated from plastids[14] as well as it is formed by pentose-P-pathway. Aromatic amino acids are transfered in considerable amounts across chloroplast envelope[14]. Conversion of Tyr to HPP and Homogentisate seems to occur in peroxisomes.

Fig. 1. Proposed scheme for compartmentation of the synthesis of aromatic amino acids and prenylquinones PQ and αT in spinach.

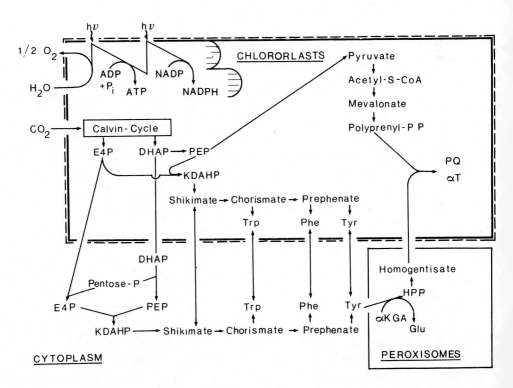

ACKNOWLEDGEMENT

Financial support by the Deutsche Forschungsgemeinschaft is gratefully acknowledged.

REFERENCES

1. Goodwin, T.W. (1965) in Biosynthetic Pathways in Higher Plants, Pridham, J.B. and Swain, T. eds., Academic Press, London, pp. 57-71.
2. Threlfall, D.R. and Whistance, G.R. (1971) in Aspects of Terpenoid Chemistry and Biochemistry, Goodwin, T.W. ed., Academic Press, London, pp. 335-404.
3. Jensen, R.G. and Bassham, J.A. (1966) Proc. Natl. Acad. Sci. U.S. $\underline{56}$, 1095.
4. Larsson, C., Andersson, B. and Roos, G. (1977) Plant Sci. Letters $\underline{8}$, 291-298.
5. Tolbert, N.E., Yamazaki, R.K. and Oeser, A. (1970) J. Biol. Chem. $\underline{245}$, 5129-5136.
6. Schnarrenberger, C., Oeser, A. and Tolbert, N.E. (1971) Plant Physiol. $\underline{48}$, 566-574.
7. Kelly, G.J. and Gibbs, M. (1973) Plant Physiol. $\underline{52}$, 111-118.
8. Buchholz, B., Reupke, B., Bickel, H. and Schultz, G., submitted to Phytochemistry.
9. Whistance, G.R. and Threlfall, D.R. (1970) Biochem. J. $\underline{117}$, 593-600.
10. Bickel, H., Palme, L. and Schultz, G. (1978) Phytochemistry $\underline{17}$, 119-124.
11. Schultz, G., Huchzermeyer, Y., Reupke, B. and Bickel, H. (1976) Phytochemistry $\underline{15}$, 1383-1386.
12. Bickel, H. and Schultz, G. (1976) Phytochemistry $\underline{15}$, 1253-1255.
13. Larsson, C. and Albertsson, P.A. (1974) in Proceedings of the Third International Congress on Photosynthesis, Avron, M. ed., Elsevier Scientific Publ. Comp., Amsterdam, pp. 1489-1498.
14. Heldt, H.W. (1976) in Topics in Photosynthesis, Vol. 1, Barber, J. ed., Elsevier / North Holland Biomedical Press, Amsterdam, pp. 215-234.
15. Walker, D.A. in Topics in Photosynthesis, Vol. 1, loc. cit., pp. 235-278.
16. Löffelhardt, W. (1977) Z. Naturforschg. $\underline{32c}$, 345-350.
17. Haslam, E. (1974) The Shikimate Pathway, Butterworths, London.

REGULATION OF PRENYLQUINONE SYNTHESIS BY SHIKIMATE PATHWAY IN HIGHER PLANTS.

H. BICKEL, G. SCHULTZ
Botanisches Institut and Institut für Tierernährung, Arbeitsgruppe für Phytochemie und Futtermittelkunde, Tierärztliche Hochschule D 3000 Hannover (FRG)

ABSTRACT

In spinach chloroplasts Phe and Tyr exerts regulation by feedback over their own rates of synthesis whereas Trp control shikimate pathway on a step between shikimate and chorismate synthesis. On the other hand, synthesis of prenylquinones, both PQ and αT, is regulated only by Tyr concentration.

INTRODUCTION

Until now regulation of shikimate pathway by feedback mechanisms has been known only in microorganisms (for reviews see[1,2]). Aim of the following work was to demonstrate the presence of this mechanisms in higher plants.

METHODS

Experiments were performed with suspensions of unpurified intact spinach chloroplasts isolated according to Jensen et al.[3]. Beside chloroplasts these suspensions contain more or less large portions of peroxisomes, mitochondria and cytosol, in part as multiorganelle complexes[4]. Recent results[5] show that peroxisomes are involved in converting Tyr to the quinone moiety of the prenylquinones PQ and αT.

RESULTS

$^{14}CO_2$ as substrate. The first experiment was carried out to study the ^{14}C-incorporation of $^{14}CO_2$ (via photosynthetical CO_2-fixation) into aromatic amino acids and the prenylquinones PQ and αT by

---

Abbreviations: PQ - plastoquinone 45; αT - α-tocopherol; Phe - phenylalanine; Tyr - tyrosine; Trp - tryptophan; KDAHP - 2-keto-3-deoxyarabinoheptulonic-acid-7-phosphate;

adding simultaneously Phe, Tyr and Trp, respectively (Table 1). By adding Phe and Tyr only the synthesis of the same amino acid was reduced whereas that of the competing one was increased. In both cases Trp synthesis was enhanced. On the other hand, adding Trp the synthesis of all three aromatic amino acids was decreased considerably.

The synthesis of prenylquinones, both PQ and αT, was regulated largely by the concentration of Tyr. Consequently, Trp controls the formation of prenylquinones over the rate of Tyr synthesis.

Basing on the known mechanism of shikimate pathway, this indicates an attack of Phe and Tyr in spinach on an enzymic step behind the prephenate synthesis whereas Trp does this on a step before chorismate.

Feedback in the synthesis of aromatic amino acids also effected an increase in the synthesis of alanine as non aromatic amino acids.

TABLE 1
^{14}C-INCORPORATION FROM $^{14}CO_2$ AND SHIKIMATE-/1,6-^{14}C/, RESPECTIVELY, IN THE LIGHT INTO AMINO ACIDS AND PRENYLQUININES OF CHLOROPLASTS BY ADDING PHENYLALANINE, TYROSINE AND TRYPTOPHAN, RESPECTIVELY. (Medium C according to^3, 30 min light, T = 20 $\pm$ 2°C)

	+ Phe	+ Tyr	+ Trp
	- each 5 mM -		
	^{14}C-incorporation in % of control		
Chloroplasts + $^{14}CO_2$			
Ala	254	112	161
Phe	82	163	18
Tyr	386	25	37
Trp	545	121	12
PQ	162	217	55
αT	220	160	70
Chloroplasts + shikimate-/1,6-^{14}C/			
Phe	5	152	7
Tyr	147	84	10
Trp	120	208	38

Shikimate-/1,6-^{14}C/ as substrate. To determine the point of attack of Trp more exactly, a second experiment was done using shikimate-/1,6-^{14}C/ as a more direct precursor (Table 1). Adding Phe and Tyr the same results were obtained as in the $^{14}CO_2$-experiment. The same was in the case of Trp. This indicates that Trp attacks a step between shikimate and chorismate and not at the step of KDAHP-synthesis.

Incorporating Tyr-/ß-^{14}C/ into prenylquinones no influence of Phe (1 mM) could be observed. It seems Phe does not influence the pathway via p-hydroxyphenylpyruvate and homogentisate.

From this the following scheme of feedback regulation of shikimate pathway in spinach is proposed (Figure 1).

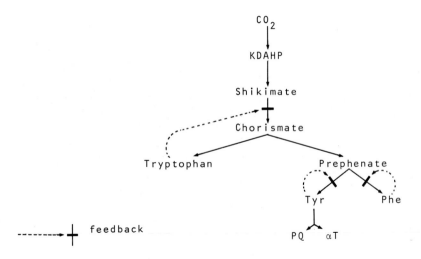

Fig. 1. Proposed scheme of feedback regulation of shikimate pathway in spinach.

DISCUSSION

Feedback regulation by Trp on a step other than the KDAHP synthesis seems to be not unusualy in higher plants. Also in cauliflower (Brassica oleracea ssp.) the KDAHP synthetase is not inhibited by Trp[6]. The enzyme is influenced neither by aromatic amino acids nor by chorismate and prephenate. Furthermore the regulation in other plants seems to be more complex than in spinach, e.g. in mung bean (Phaseolus aureus) the activity of the CM-1 form[7] but not the CM-2 form[8] of the chorismate mutase is increased by Trp but is decreased by Phe and Tyr.

ACKNOWLEDGEMENT

Financial support by the Deutsche Forschungsgemeinschaft is gratefully acknowledged.

REFERENCES
1. Haslam, E. (1974) The Shikimate Pathway, Butterworth, London.
2. Lingens, F. (1968) Angew. Chem. 80, 384-394.
3. Jensen, R.G. and Bassham, J.A. (1966) Proc. Natl. Acad. Sci. U.S. 56, 1095.
4. Larsson, C., Andersson, B. and Roos, G. (1977). Plant Sci. Letters 8, 291-298.
5. Bickel, H., Palme, L. and Schultz, G., (1978) Phytochemistry 17 119-124.
   Bickel, H., Buchholz, B. and Schultz, G., in preparation.
6. Huisman, O.C. and Kosuge, T.(1974) J. Biol. Chem. 249, 6842-6848.
7. Gilchrist, D.G. and Kosuge, T. (1974) Arch. Biochem. Biophys. 164, 95-105.
8. Gilchrist, D.G. and Kosuge, T. (1975) Arch. Biochem. Biophys. 36-42.

PHOSPHATIDYLGLYCEROL BIOSYNTHESIS BY ISOLATED CHLOROPLASTS OF *Euglena gracilis*.

ANTOINE CHAMMAI and RODOLPHE SCHANTZ
Laboratoire de Physiologie Végétale, Université Louis Pasteur, 28 rue Goethe, 67083 Strasbourg (France)

ABSTRACT

During chloroplast development, phosphatidylglycerol (PG) present in traces amount in etiolated cells, is synthesized after a lag phase similar to that of chlorophyll. The study of the subcellular distribution of the enzymes involved in the biosynthesis of PG shows that the first reaction step involved in the formation of CDP diglyceride from CTP and Phosphatidic acid proceeds at a higher rate with the microsomal fraction than with the chloroplast's one. In etiolated cells this activity was found in microsomes and mitochondria. However, the second step leading to the formation of PG occurs with a higher specific activity in purified plastids than in the microsomal fraction. *Euglena* chloroplasts are able to synthesize their own PG molecules as well as their galactolipids and thus appear to have a large autonomy concerning their specific lipid biosynthesis.

INTRODUCTION

In *Euglena* as well as in higher plants during light-induced differentiation an increase in specific lipids, galactolipids and phosphatidylglycerol (PG) occurs [1,2,3,4]. In dark-grown etiolated cells, galactolipids are present in small amounts (0.5 µg/10^6 cells) and only traces of PG can be detected. During greening galactolipids accumulate without the typical lag phase showed by chlorophylls and phosphatidylglycerol [5]. Recently we have found that in *Euglena* the biosynthesis of galactolipids occurs in the plastid envelope like in higher plants but the synthesis proceeds differently by the fact that digalactosyldiglyceride is more actively synthesized than monogalactosyldiglyceride [6]. However, little informations are available on the biosynthesis of PG. In Spinach leaves it was shown that most of the synthetic activity was associated with the 40 000 g pellet [7,8]. In earlier experiments we found that in *Euglena* PG does not derive from another phospholipid by a transphosphatidylation mechanism [9]. In contrast to higher plants, *Euglena* chloroplasts are able to incorporate CTP into CDP-diglycerides [10]. The aim of this investigation was to determine in

‡ On leave from the National Council for Scientific Research of Lebanon.

Euglena the second step in the biosynthetic pathway of PG: the condensation of CDP diglyceride and of glycerophosphate leading to the formation of PG *via* PGP.

MATERIAL and METHODS

Euglena Gracilis, strain Z 1225 5/25 was obtained from the University of Göttingen (Germany). Synchronous dark-grown cultures and greening experiments on resting medium were performed as previously described [11,12]. For cell labelling, $^{32}PO_4HNa_2$ was added to the resting medium (25 μCi/ml) at the onset of illumination. At different phases of chloroplast development, cells were harvested phospholipids extracted and separated by thin-layer chromatography on silica gel using the system described by Lepage [13]. The different phospholipids were identified by specific reactions and by their Rf values. Autoradiography was used for localization of labelled products and the silica gel area corresponding to each phospholipid was scraped off into a vial containing 10 ml of the scintillation mixture and counted.

Chloroplasts were isolated from green cells grown either autotrophically or photoheterotrophically with 5g/l Sodium acetate as carbon source and purified following the methods described elsewhere [10].

Enzyme assays: A typical reaction mixture contained tris-maleate buffer 0.1M pH 7.2, mannitol 0.3M, phosphatidic acid 2mM, CTP 2mM, $MgCl_2$ 40mM, ^{14}C glycerophosphate 4mM, chloroplast proteins (0.6-1mg) in a final volume of 0.2 ml. After incubation at 28°C lipids were extracted , separated and analysed as described previously. In some experiments, CDP diglyceride was added instead of phosphatidic acid and CTP.

RESULTS

Evolution of phospholipids during chloroplast development

When ^{32}P is added to the medium of dark synchronous growing cells, the radioactivity is incorporated mainly in the two major phospholipids,Phosphatidylethanolamine (PE) and phosphatidylcholine (PC). A following chase (figure1) during cell illumination shows that the radioactivity incorporated in PC is fairly constant whereas that incorporated in PE decreases more rapidly. In this experiment PG is synthesized but not labelled , indicating that this phospholipid does not derive from another phospholipid and has its own biosynthetic pathway. However, if ^{32}P is added at the beginning of the light period (figure2), the labelling appears first in PE and after a lag period in PG. PC remains all over the greening process mainly unlabelled. This result confirm that PC is very stable and seems to be exchanged from another cellular compartment to the thylakoïd membranes in formation. On the other hand, PG becomes labelled after

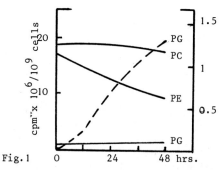 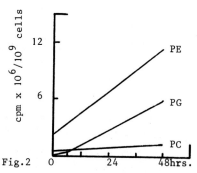

Evolution of radioactivity in the major Phospholipids during cell greening.
Fig.1 - Chase after labelling during one life-cycle in the dark. Dotted line: amount of PG. Full lines: radioactivity
Fig.2 - Labelling during cell illumination.

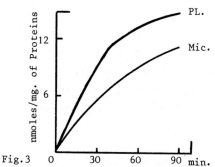 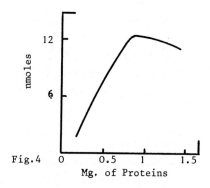

Biosynthesis of PG by microsomal and plastidial fractions as a function of time.

Effect of enzyme concentration (chloroplasts) on PG biosynthesis.

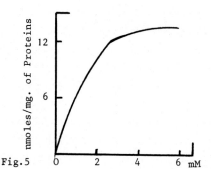 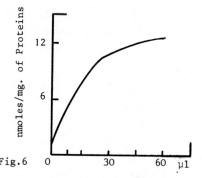

PG biosynthesis by chloroplasts in relation with GP concentration (Fig.5) and with CDP-DG concentration (Fig.6).

a lag period similar to that of chlorophyll indicating that these molecules are newly synthesized.

Incorporation of GP by isolated chloroplasts

Incubation for different periods of time of labelled glycerophosphate (GP) with the same concentration of proteins (plastids and microsomes) is shown in figure 3. The plastidial system incorporated GP into PG at higher rate than the microsomal one. The formation of PG was linear with time up to 1hour with both systems.

In figure 4 it can be seen the relation between PG biosynthesis and protein concentration. The incorporation of radioactivity proceeded linearly with increasing amounts of enzyme up to a protein amount of 1mg. Above this concentration, the PG synthesis tailed off.

The dependence of PG formation rate (nmoles synthesized / hour) on the amount of GP added is given in figure 5. The rate of incorporation was directly proportional to the concentration of GP up to 2mM.

The amount of PG synthesized was linearly proportional to the concentration of CDP diglyceride added as shown in figure 6. It is interesting to note that addition of exogenous CDP- diglyceride resulted in incorporation of radioactivity from GP exclusively into PG whereas addition of the system CTP + PA to the reaction mixture led to the formation of PG and labelled diglycerides.

DISCUSSION

The main lipids in chloroplasts are galactolipids. Besides them, phospholipids have been found, especially PC and PG, PE being considered to be absent in chloroplast membranes. During chloroplast development in *Euglena*, the total amount of the two major phospholipids, PC and PE, remains fairly constant while a marked synthesis of PG is observed. Addition of ^{32}P to the resting medium at the beginning of the greening process resulted in a fast labelling of PE, denoting a high turn-over rate of this non-plastidial phospholipid.
Evolution of radioactivity in PG shows the same time course than the amount accumulation of these molecules. One can assume that PC is exchanged during plastid differentiation from another membrane system i.e, from endoplasmic reticulum.
Chloroplasts, and particularly the plastid envelopes are the site of galactolipid synthesis[14,15]. Isolated plastids have the capacity of synthesizing most component fatty acids. However, the synthesis of phospholipids could not be obtained with isolated plastids, except the formation of phosphatidic acid[16]. Previous studies have described the biosynthesis of PG in Spinach leaves[7,8].

The highest specific activity was found to be associated with the microsomal fraction. It was suggested that in Spinach, PG is synthesized in microsomes and subsequently transferred to the plastids, but up to now, an exchange of PG between microsomes and chloroplasts could not be observed.

Using isolated plastids purified by several methods, we could demonstrate an enzymatic activity which catalyses the formation of CDP-diglycerides[10] although microsomal enzyme presented the highest specific activity.

The results presented in this study give further evidences of the capacity of the PG biosynthesis by isolated chloroplasts of *Euglena*. On specific activity basis, the plastids have a higher activity than the microsomal fraction.

Therefore the activity found in the plastids cannot be attributed to a contamination by endoplasmic reticulum membranes. Furthermore the activity of NADH-cytochrome C oxydoreductase used as microsomal marker was absent or negligible in our plastid preparations.

Thus, *Euglena* chloroplasts are able to synthesize their own PG molecules. The failure to found an activity in Spinach chloroplasts can be explained by the presence in these plastids of high levels of ATP[17], PA, Ca^{++}[18] or pyrophosphate[19] which are known to inhibit the first reaction step of CDP-diglycerides formation.

REFERENCES

1. Wintermans, J.F.G.M. (1960) Biochim. Biophys. Acta, 44, 49-54.
2. Benson, A.A. (1964) Ann. Rev. Plant Physiol. 15, 1-16.
3. Haverkate, F. and Van Deenen, L.L.M. (1965) Proc. Koninkl. Ned. Akad. Wetenshap. Ser. B, 68, 141-153.
4. Allen, C.F., Good,P., Davies,H.F., Chisum, P. and Fowler, S.D. (1966) Amer. Oil Chem. Soc. , 43, 223-231.
5. Schantz, R., Blee, E. and Duranton, H. (1976) Physiol. Veg. 14, 141-157.
6. Blee, E. and Schantz, R. (1978) Plant Sci. Let. in press.
7. Sastry, P.S. and Kates, M. (1966) Can. J. Biochem. 44, 459-467.
8. Marshall, M.O. and Kates, M. (1971) Biochim. Biophys. Acta, 260, 558-570.
9. Schantz, R.,Douce,R. and Duranton,H. (1972) Febs Lett. 20, 157-161.
10. Chammai,A. and Schantz,R. (1978) in Proc. of the Symposium on "Chloroplast development", Spetsaï, Elsevier/North-Holland Biomedical Press.
11. Schantz, R., Salaün,J.P., Schantz,M.L., and Duranton,H.(1972) Physiol. Veg. 10,133-151.
12. Schantz,R., Schantz, M.L., and Duranton, H. (1975) Plant Sci. Let. 5,313-324.
13. Lepage,M. (1967) Lipids 2, 244

14. Van Hummel,H.C., Hulsebos, Th.J.M. and Wintermans, J.F.G.M. (1975) Biochim. Biophys. Acta, 380, 219-226.
15. Joyard, J. and Douce,R. (1976) Biochim. Biophys. Acta, 424,125-131.
16. Joyard,J. and Douce,R. (1977) Biochim. Biophys. Acta, 486, 273-285.
17. Sribney, M., Dove, J.L. and Lyman, E.M. (1977) Biochem. Biophys. Res. Com. 79, 749-755.
18. Mc Caman, R.E., and Finnerty, W.R. (1968) J. Biol. Chem. 243,5074-5080.
19. Carter,J.R., and Kennedy, E.P. (1966) J. Lipid Res. 7, 678-683.

INCORPORATION OF $[^{14}C]$-ACETATE AND $[^{32}P]$-PHOSPHATE INTO PHOSPHOLIPIDS OF SYCAMORE CELL SUSPENSIONS

RICHARD BLIGNY, FABRICE REBEILLE-BORGELLA, ROLAND DOUCE

DRF/Biologie Végétale, CEN-G. et USM-G., 85 X, 38041 Grenoble Cédex (France)

ABSTRACT

- Suspension cultures of sycamore cells incorporated $[^{14}C]$-acetate very rapidly into the fatty acid moieties of phospholipids when incubated at 25°C for up to 20 h. The half-maximal rate was achieved in the presence of 2mM acetate. There was no rapid decrease in $[^{14}C]$-PC after pulse-chase experiments indicating that C 18:1 and C18:2 of PC were not transferred to other glycerolipids.

- In PC, $[^{14}C]$-C 16:0 and $[^{14}C]$-C 18:0 appeared specifically in position C-1 of the glycerol moiety, whereas $[^{14}C]$-C 18:1 was associated predominantly with position C-2 but appeared also in position C-1. The desaturation of C 18:1 to C 18:2 occured in both position C-1 and C-2.

- After a lag phase, sycamore cells incorporated $[^{32}P]$- phosphate very rapidly into phospholipids. The half-maximal rate was achieved in the presence of 0.5 mM phosphate. The most rapidly laballed $[^{32}P]$- lipids were PA and PG.

- During the exponential phase of growth, the maximal rate of phospholipid synthesis was roughly 5 $\mu g \cdot h^{-1} \cdot mg^{-1}$ phospholipid.

INTRODUCTION

Routine cultivation of plant cells in rapidly growing suspension cultures under controlled conditions is a useful system to study the lipid metabolism and membrane assembly [1-8].

This paper will describe experiments designed to examine the capacity of sycamore suspension cells to incorporate quantitatively exogeneously added $[^{14}C]$-acetate and $[^{32}P]$-phosphate into their polar lipids. It will also examine the distribution of ^{14}C among the fatty acids in PC.

MATERIALS AND METHODS

<u>Culture conditions</u>. Cells from *Acer pseudoplatanus* L. were cultivated in a phytostat for automatic mass culture of plant cells in liquid medium. This apparatus described previously[9] allowed the culture of 20 l cell suspensions

ABBREVIATIONS : PC, phosphatidylcholine ; PE phosphatidylethanolamine ; Pi phosphatidylinositol ; PG phosphatidylglycerol ; DPG, diphosphatidylglycerol ; PA, phosphatidic acid.

under batch conditions. The automatic recording of cell suspension growth was carried out by means of turbidity measurements since it was shown[9] that cell number and culture turbidity were closely correlated throughout the exponential phase of growth.

Uptake studies. Typical experiment started from 200 ml sycamore cell suspensions in exponential phase of growth (10^6 cell.ml^{-1}). At zero time 2 mCi [^{14}C]-sodium acetate and/or 10 mCi [^{32}P]-sodium phosphate were added and the suspension incubated with shaking at 25°C for the indicated time. For phosphate and acetate incorporation rate measurements a range of 0.1 to 50 mM phosphate or acetate was used. At stated intervals, cells (20 ml aliquots) were harvested by vacuum filtration (Fiberglass filter ; suction pressure, 0.2 bar ; 2 min) and lipids were extracted as described below.

For pulse-chase labelling experiments the cells (10^6 cells.ml^{-1}) were labelled for 1 h, rinced with fresh culture medium during 2 min and resuspended in culture medium containing 10 mM cold sodium phosphate and acetate. At stated intervals 20 ml aliquots were taken for lipid analysis.

Lipid analysis. The cells were rapidly fixed in boiling water during 2 min in order to destroy all the lipases (phospholipase D[10], lipolytic acylhydrolase[11,12] etc...) and the lipids were extracted by the method of Bligh and Dyer[13]. The total lipids were dissolved in chloroform and aliquots were taken for thin layer chromatography (TLC). Radioactivity was determined by counting aliquots in 10 ml of New England Nuclear Aquasol-2 cocktail with an Intertechnic model SL 4000 liquid scintillation spectrometer.

The chloroform-soluble products were applied to Silica gel G TLC (Merck) plates and chromatographed with the following solvents : chloroform-methanol-water (65 : 25 : 4) in the first dimension and chloroform-acetone-methanol-acetic acid-water (100 : 40 : 20 : 20 : 10) in the second dimension. The polar lipids were identified by co-chromatography with known standards and by the use of specific spray reagents[14]. Fatty acid methyl esters from individual lipids separated by TLC were prepared directly by adding 4 ml of methanol-sulfuric acid-benzene (100 : 5 : 5) to the Silica gel scraped from the plate and heating at 70°C for 2 h. The methyl esters were extracted with hexane and analyzed by gas chromatography on a column of 15% diethylene glycol succinate on chromosorb Varaport 30, 80/100 mesh at 175°C using a flame ionization detector (Hewlett-Packard Model 5750). The column was fitted with an effluent stream splitter and radioactive methyl esters were monitored continuously by a gas radioactivity counting system (Gas Proportional Counter Packard model 894).

TABLE 1

FATTY ACID COMPOSITION (PER CENT BY WEIGHT) of SYCAMORE CELL PHOSPHOLIPIDS

	C 14:0	C 16:0	C 18:0	C 18:1	C 18:2	C 18:3	unsaturated / saturated
Total lipids	tr	24	4	38	24	10	2.6
PC	tr	20	4	45	21	10	3.2
PE	tr	29	4	25	30	11	2.0
PI	1	43	4	22	21	9	1.1
PG	6	53	4	18	12	7	0.6
DPG	1	24	6	22	31	16	2.2
PA	5	38	8	20	21	8	0.9

RESULTS AND DISCUSSION

Sycamore cells harvested during their exponential phase of growth contained 72 ± 7 mg total lipid and 22 ± 3 mg phospholipids per g of dry matter. The phospholipid composition expressed in per cent by weight was : PC 45-48 ; PE 28-30; PI 15-17 ; PG 4-5 ; DPG 1.7-1.9 ; PA 0.8-1.0. The fatty acid composition of the sycamore cells (Table I) was characterized, for cells cultivated at 25°C in the 20 l batch conditions, by a high rate of monounsaturated fatty acids (up to 45% oleic acid).

Fig. 1 shows the time course of [^{14}C]-acetate and [^{32}P]-phosphate incorporation into sycamore cell phospholipids. The shape of the curve was linear with [^{14}C]-acetate and there was no deviation from linearity for several hours. In good agreement with previous results on soybean cell suspension[8] we have observed that phospholipids accounted for most of the ^{14}C radioactivity incorporated in the lipids at all the times. Of these, PC showed the most rapid labelling (results not shown). The half-maximal incorporation rate was achieved in the presence of 2 mM acetate.

In contrast, there was a lag in ^{32}P incorporation into the phospholipids. In this case the rate of incorporation increased during the 8 first hours. PA and PG showed the most rapid labelling whereas PE and PC were labelled slowly. The half-maximal rate was achieved in the presence of 0.5 mM phosphate.

The lag phase observed only with ^{32}P corresponds very likely to the fact that the labelled phosphate entering the cell was diluted by a large amount of free cellular phosphate.

Fig. 2 clearly shows that after a chase experiment, the radioactivity from [^{14}C] acetate incorporated into the cell phospholipids remained essentially constant up to 20 h. This result indicates that the fatty acids incorporated

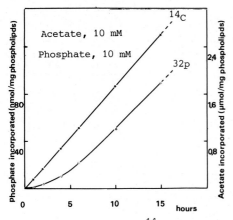 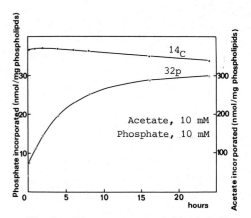

Fig. 1 Time course of [^{14}C]-acetate and [^{32}P]-phosphate incorporation into phospholipids of sycamore suspension culture cells. Cells were incubated with 10 mM acetate and phosphate as described in text.

Fig. 2 Time course evolution of the radioactivity incorporated into sycamore cell phospholipids during pulse-chase experiments carried out with [^{14}C]-acetate and [^{32}P]-phosphate. The chase started at time zero. For experimental conditions see text.

into polar lipids are not further metabolized.

In contrast, the radioactivity from [^{32}P]-phosphate incorporated into phospholipids was still rapid up to 10 hours, until all the pool of [^{32}P]-phosphate present in the cell was metabolized, and remained practically constant up to 20 h. The mathematical analysis of the [^{32}P] phosphate incorporation kinetics into phospholipids (see fig. 1 and 2) and the knowledge of the different cellular parameters[9] allowed indirectly the determination of the phosphate concentration in the cell soluble compartments that can be used for the phospholipid biosynthesis : values of 0.8 to 1 mM were found.

All these results together allowed the maximum incorporation rates of acetate and phosphate to be calculated. It appeared that intact sycamore cells incorporated [^{14}C]-acetate into phospholipids at rates up to 110 nmol.h^{-1}.mg^{-1} phospholipids when acetate concentrations were 1.000 fold higher than used previously[8]. Sycamore cells incorporated [^{32}P]-phosphate into phospholipids at rates up to 8 nmol.h^{-1}.mg^{-1} phospholipids.

Expressed in µg of newly synthesized phospholipids per h per mg phospholipids, 5 µg of phospholipids were synthesized by using acetate and 6 µg by using phosphate. These two values which are not significantly different correspond to the synthesis of new membranes during the cell cycle. It is interesting to note that the cell number doubling time in our assay conditions (about 90 h) was comparable to the time necessary to renew all the phospholipids (about 160 h). Consequently the rate of phospholipid synthesis gives an accurate reflection

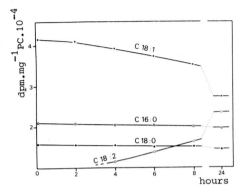

Fig. 3 Time course evolution of the radioactivity incorporated into PC fatty acids during pulse chase experiments carried out with [^{14}C]- and [^{32}P]-phosphate. The chase started at time zero. For experimental conditions see text. Note that [^{14}C]-labelled PC is a substrate for desaturation of C 18:1 to C 18:2.

of the expansion of the membrane systems inside the cell. For example we have found (results not shown) that the rates of phospholipid synthesis during the stationnary phase were at least 5 to 10 times lower than during the exponential phase of growth.

After a short time of labelling with [^{14}C]-acetate, PC contained a high proportion of labelled C 16:0, C 18:0 and C 18:1 whereas the mass peaks from gas liquid chromatography revealed that about 35% of the fatty acids in the polar lipids were in the form of C 18:2 and C 18:3 (Table I). With time more label appeared in unsaturated fatty acids and the labelling patterns showed a slow adjustment towards the mass patterns. In all experiments the acyl residues of the other phospholipids did not incorporate label to as high a specific activity as did the acyl residues of PC. Our results on the accumulation of radioactive C 18:1 in PC and the slow desaturation to C 18:2 (fig. 3) agree with previous observations[8,15], but do not support the proposed function of PC as donor of polyenoic acids in sycamore cells. In contrast with soybean cell suspensions[8] we were unable to show any rapid decrease of [^{14}C]- PC after pulse-chase experiments indicating that the C 18:1 and C 18:2 of PC are not transferred to other glycerolipids, in particular glycolipids.

Finally, after submitting the labelled PC to the action of the phospholipase A$_2$ (fig. 4) it appeared that [^{14}C]-C 16:0 and [^{14}C]-C 18:0 were associated specifically with C-1 position of the glycerol moiety whereas [^{14}C]-C 18:1 was associated predominantly with the C-2 position and a little with the C-1 position. As [^{14}C]-C 18:2 appeared simultaneously in position C-1 and C-2 it is very likely that the desaturation of C 18:1 to C 18:2 occured in both positions C-1 and C-2. Alternatively a deacylation-reacylation mechanism may bring about the specific fatty acid patterns of PC.

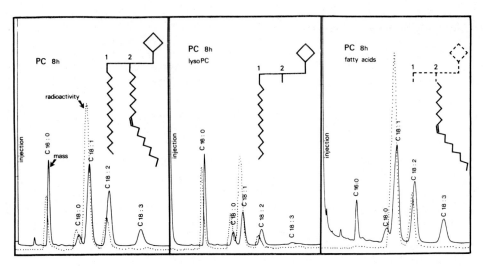

Fig. 4 Distribution of labelled fatty acids between C-1 and C-2 position of PC glycerol moiety in sycamore cells incubated with [^{14}C] acetate for 8 h. 0.5 mg of [^{14}C] labelled PC obtained by TLC (see text) was incubated under N_2 at room temprature with 10 µl of phospholipase A_2 from *Vipera russelli* (Sigma) in 250 µl phosphate buffer (20 mM, pH 6.5) and 1 ml ether. After 2 h the reaction products (free fatty acids from C-2 and lyso PC) were separated by TLC.

REFERENCES

1. Benveniste,P., Ourisson, G. and Hirth, L. (1970) Phytochemistry,9, 1073-1081.
2. Moore, T.S. and Beevers, H. (1974) Plant Physiol. 53, 261-265.
3. Radwan, S.S. and Mangold, U.K. (1976) Adv. Lipid Res., 14, 171-211.
4. Siebertz, H.P., Heinz, E. and Bergmann, L. (1978) Plant Science Letters, 12 119-126.
5. Stearns, E.M., Morton, J.R. and Morton W.T. (1975) Lipids, 10, 597-601.
6. Stumpf, P.K. and Weber, N. (1977) Lipids 12, 120-124.
7. Tattrie, N.H. and Veliky I.A. (1973) Can. J. Bot. 51, 513-516.
8. Wilson, A.C., Kates, M. and de la Roche, A.I. (1978) Lipids, 13, 504-510.
9. Bligny, R. (1977) Plant Physiol., 14, 499-515.
10. Douce, R., Faure, M. and Maréchal, J. (1966) C.R. Acad. Sci. Paris, 262, 1549-1552.
11. Galliard, T. (1975) *in* Recent Advances in the Chemistry and Biochemistry of Plant Lipids, Galliard, T. and Mercer, E.I. eds, Academic Press New-York, pp. 319-357.
12. Bligny, R. and Douce, R. (1978) Biochim. Biophys. Acta, 529, 419-428.
13. Bligh, E.G. and Dyer, W.J. (1959) Can. J. Biochem. Physiol.37, 911-917.
14. Kates, M. (1972) *in* Laboratory Techniques in Biochemistry and Molecular Biology, Work, T.S. and Work, E. eds., North Holland Publishing Co., Amsterdam and London, pp. 436-441.
15. Roughan, P.G. (1970) Biochem. J. 117, 1-8.

BIOSYNTHESIS OF THE SATURATED VERY LONG CHAIN FATTY ACIDS IN HIGHER PLANTS FROM EXOGENOUS SUBSTRATES.

Claude CASSAGNE and René LESSIRE

Département de Biochimie, CNRS ERA 403, 351 cours de la Libération, 33405 Talence (France)

ABSTRACT

The microsomal elongases evidenced in various plant tissues form C20 to C30 saturated fatty acids in the presence of malonylCoA and NADPH. But the question of the primer of the plant elongases is far from clear. Stearic acid could be the substrate in the case of germinating seeds, whereas in the microsomes of the developing leaf, stearoylCoA is far better accepted than stearic acid and stearoylACP. The endoplasmic reticulum seems to be the major site of synthesis and contains also a stearoylCoA synthetase able to furnish the substrate of the elongase.

INTRODUCTION

Saturated very long chain fatty acids in the C20-C30 range are well-known components of the wax layer of the higher plants. Recently their occurrence has been established in the plasma membranes from yeast [1], or higher plants.[2,3]. Numerous studies both *in vivo* and *in vitro* have been devoted to their biosynthesis.[1-18]. From these studies with developing leaves[11-14], germinating seeds[8] or aged potated slices [17], there is a general agreement that NADPH is the preferred reductant and that malonylCoA is the elongating agent [4-6], but the question of the primer of the elongase is still in debate; although *in vivo* studies have shown that substrates from C2 to C24 may be easily elongated into very long chain fatty acids,[4-7] in the cell free systems thus far examined very few data have been presented because most of the studied systems do not accept exogenous substrates.

We shall describe recent findings on the nature and on the synthesis of the primer of the plant elongase (s).

BIOSYNTHESIS OF SATURATED VERY LONG CHAIN FATTY ACIDS FROM EXOGENOUS SUBSTRATES

Developing leaves. The major site of synthesis is the epidermal layer of cells [10-11]. In Pea epidermal cells no exogenous substrate was elongated[1,1] but the microsomal fraction from Leek epidermal cells elongates stearoyl-CoA into very long chain fatty acids in the presence of malonyl CoA and NADPH[2,3].From 20 µM (1-14C)-stearoyl-CoA about 5 nmol/mg protein/hour were synthesized. The replacement of stearoyl-CoA by free stearic acid, free stearic acid + CoA + ATP, stearic acid + ACP + ATP or by stearoyl-ACP led to a decrease of the radioactivity found in the very long chain fatty acids (Table 1).

TABLE 1

VERY LONG CHAIN FATTY ACID SYNTHESIS FROM VARIOUS SUBSTRATES

Biosynthesis of the saturated very long fatty acids by a microsomal fraction from leek epidermal cells in the presence of malonyl-CoA and NADPH. The relative value 100 % was given to the results obtained in the presence of (1-14C)-stearoyl-CoA.

substrate	VLCFA[a] synthesis (%)	substrate	VLCFA[a] synthesis (%)
StearoylCoA	100	Stearic acid	41
Stearoyl-CoA + ACP	47	Stearic acid + CoA + ATP	60
Stearoyl-ACP	37	Stearic acid + ACP[b] + ATP	9
Stearoyl-ACP + ACP	33		

[a] very long chain fatty acids.
[b] acyl carrier protein.

The effect on elongation rate of a CoA + ATP addition to stearic acid could be interpreted by an early formation of an acyl-CoA which in turn could be the primer of the elongase (s)[2,14]. On the other hand, an ACP addition drastically reduced the stearate elongation; in the hypothesis of stearoyl-ACP as the primer of the elongase, that effect could be interpreted by a lack of stearoyl-ACP formation. But recent observations [19] established that a stearoyl-ACP synthetase is present in the microsomal pellet of leek epidermal cells as in bacteria[20]; the acyl ACP synthetase required ACP, ATP and Mg^{2+} for full activity; the specific activity was about 1.5 nmol/mg protein/15 min. From these results it is assumed that in the microsomes of leek epidermal cells, stearoyl-CoA and not stearic acid or stearoyl-ACP is the primer of the elongase. The study of the label distribution in the C20 to C32 fatty acids formed from C18 in presence of various cofactors revealed a different behaviour of the icosanoic acid and of the C22-C30 fatty acids : a) icosanoic acid formation was not stimulated by a (CoA + ATP) addition, which therefore doubled the C22-C30 fatty acid synthesis. b) a (CoA + ACP + ATP) addition diminished by 50 % the icosanoic acid formation, but had no effect on the C22-C30 fatty acids. c) an ATP addition to stearoyl-CoA increased chiefly the C22 to C30 fatty acid synthesis, and not that of the icosanoic acid. Not to speak of the stearate formation, these results suggest that there are at least two elongases the first one leading from C18 to C20 and the second one from C22 to C30, as was suspected from previous studies *in vivo*[10,12].

Germinating seeds. The germinating pea seed synthesizes fatty acids in the C16-C26 range from labelled malonyl-CoA[8,17]. The addition of free stearic acid markedly increased the C20 to C24 fatty acid formation[17], whereas a stearoyl-CoA addition resulted in some incorporation of radioactivity in the icosanoic acid. Stearoyl-ACP was not elongated at all [17]. The CoA had an opposite effect to that observed with leek microsomes. The very long chain fatty acids were only found as free fatty acids [17].

BIOSYNTHESIS OF STEAROYL-COENZYME A

Because of its possible role as the primer of the leek elongase, the stearoyl CoA has been further investigated. The presence of a stearoyl-CoA synthetase has been demonstrated in various subfractions from leek epidermal cells [21] (Table 2)

TABLE 2
SUBCELLULAR LOCALIZATION OF STEAROYL CoA SYNTHETASE

Fraction	activity (n.mol/30 min/mg protein)
Mitochondria	3.6
Plasma Membrane	10.2
Endoplasmic Reticulum	9.1
Cytosol	0.01

TABLE 3
COFACTOR REQUIREMENTS OF THE MICROSOMAL STEAROYLCoA SYNTHETASE

Assay	remaining activity (%)
Complete[a]	100
- CoA	6.3
- ATP	14
- Mg^{2+}	21.8
- microsomes	0.9
time zero	0.1

[a]The complete assay contained microsomes, stearate, CoA, ATP and Mg^{2+}

As seen from Table 2, the purified membrane fractions, especially plasma membrane and endoplasmic reticulum exhibit the highest specific activity. Since one can assume that acylthioesterases are present in the membrane fractions [22,23], the stearoyl-CoA formation reflects the balance between acylCoA esterases and acylCoA synthetase activities. Replacement of ATP by CTP, GTP, UTP, CMP or ADP led to a complete loss of acylCoA synthesis.

SITE OF SYNTHESIS OF SATURATED VERY LONG CHAIN FATTY ACIDS

There is a general agreement that the synthesis occurs chiefly in the crude 100000xg pellet of "microsomes", in the various plant systems so far studied, and the other subcellular fractions synthesized little amounts of very long chain fatty acids. Further fractionation of the microsomes from leek epidermal cells was achieved [24]. The fraction enriched in endoplasmic reticulum was the most active : from 125 µM (1-14C)-stearoyl-CoA the very long chain fatty acid formation was about 13-15 nmol/mg protein/h. In the same conditions the plasma membrane enriched fraction formed 3-4 nmol/mg protein/h [14]. Thus the main site of synthesis in leek epidermal cells seems to be the endoplasmic reticulum (Table 4).

TABLE 4

STEAROYL-CoA ELONGATION BY PURIFIED MEMBRANE FRACTIONS

Fatty acid chain length	Fatty acid synthesis in plasma membrane	(nmol/mg protein / λ) in endoplasmic reticulum
20	0.5	2.5
21	0.3	0.9
22	0.24	1.71
23	0.2	0.18
24	0.54	2.0
25	0.24	0.9
26	0.09	2.0
27	0.06	1.2
28	0.03	1.8
29	n.d.[a]	1.6
30	0.12	0.6
31	n.d.	n.d.
32	n.d.	0.2

[a] n.d. = not detected

The label incroporation in the C20 to C30 fatty acids is obviously diminished in the plasma membrane and the general pattern of synthesis is different in the two studied fractions as the contribution of the odd-chain fatty acids is significantly increased in the plasma membrane.

The plasma membrane of the leek epidermal cells contains very long chain fatty acids which account for about 20 % of the total fatty acids of that fraction and differ markedly from the wax fatty acids in that odd-and even-chain fatty acids have approximately equal amounts. The possibility of an α-oxidation of the very long chain fatty acids, as demonstrated in yeast [25], has to be further investigated. At the present state the reason of the advantage of the presence of odd-chain fatty acids in the plasma membrane is totally unknown.

CONCLUSION

The elongases of the epidermal cells of the developing leaves play a great role in synthesizing the C20 to C32 fatty acids, which in turn are the precursors of the alkanes. These substances may reach the wax layer, and a steady state seems to be established not only between the wax and the internal lipids of the leaf but also between the different components of the wax [26]. Independently of the wax formation it is believable that the very long aliphatic chains play an important role on the structure (e.g. rigidity) and on the function (e.g. permeability) of the plant plasma membranes.

The elongation pathway occurs in the endoplasmic reticulum employing, as the substrate, the acylCoAs synthesized in that cell fraction. No role in that synthesis could be assigned to the acyl-ACP derivative. The nature of the products of the elongase (free fatty acids, neutral lipids, acyl-CoAs, acyl-ACP) is not well established in developing leaves. Further works are thus necessary before a definitive statement can be made upon the above mentioned points.

REFERENCES

1. Blanchardie, P. and Cassagne, C. (1977) Biol. Cellulaire 30, 127-136.
2. Cassagne, C., Lessire, R. and Blanchardie, P. (1976) Proc. Int. Cong. Biochem Hamburg, p. 620.
3. Cassagne, C., Lessire, R., Blanchardie, P., Darriet, D. and Larrouquère-Régnier, S. (1977) 11th Febs Meeting, Copenhagen, Abst. A 58-759.
4. Kolattukudy, P.E., Croteau, R. and Buckner, J.S. (1976) in Chemistry and Biochemistry of Natural Waxes, Kolattukudy, P.E. ed., Elsevier, Amsterdam, pp. 289-347.
5. Kolattukudy, P.E. (1975) in Recent advances in the Chemistry and Biochemistry of Plant Lipids, Galliard, T. and Mercer, E.T. eds., Academic Press, New York pp. 203-246.
6. Harwood, J.L. (1975)in Recent advances in the Chemistry and Biochemistry of Plant Lipids, Galliard, T. and Mercer, E.T. eds., Academic Press, New York, pp. 43-93.
7. Mazliak, P. (1963) Ph. D. Thesis, Univ. Paris, France.

8. Macey, M.J.K. and Stumpf, P.K. (1968) Pl. Physiol. 43, 1637-1647.
9. Harwood, J.L. and Stumpf P.K. (1970) Pl. Physiol. 46, 500-508.
10. Cassagne, C. (1970) Ph. D. Thesis, Univ. Bordeaux, France.
11. Kolattukudy, P.E. and Buckner, J.S. (1972) Biochem. Biophys. Res. Comm. 46, 801-807.
12. Cassagne, C. and Lessire, R. (1974) Physiol. Vég. 12, 149-163.
13. Cassagne, C. and Lessire, R. (1974) Arch. Biochem. Biophys. 165, 274-280.
14. Cassagne, C. and Lessire, R. (1978) Arch. Biochem. Biophys. 190, in press.
15. Mikkelsen, J.D. (1978) Carlsberg Res. Comm. 43, 15-35.
16. Bolton, P. and Harwood, J.L. (1976) Phytochemistry 15, 1501-1506.
17. Bolton, P. and Harwood, J.L. (1977) Biochem. J. 168, 261-269.
18. Harwood, J.L. (1977) Biochem. Soc. Trans. 5, 1259-1263.
19. Lessire, R. and Cassagne, C. (1978) Plant Sci. Letters, in press.
20. Ray, T.K. and Cronan, J.E. Jr. (1976) Proc. Natl. Acad. Sci. USA, 73, 4374-4378.
21. Lessire, R. (1975) C.R. Acad. Sci. Sér. D. 281, 1765-1768.
22. Jaworski, J.G. and Stumpf, P.K. (1974) Arch. Biochem. Biophys. 162, 158-165.
23. Stumpf, P.K. (1977) in Lipids and Lipid Polymers in Higher Plants, Tevini, M. and Lichtenthaler, H.K. eds., Springer-Verlag, Berlin, pp. 75-84.
24. Cassagne, C., Lessire, R. and Carde J.P. (1976) Plant Sci. Letters, 7, 127-135.
25. Blanchardie, P. and Cassagne, C. (1976) C.R. Acad. Sci. Sér. D. 282, 227-230.
26. Cassagne, C. and Lessire, R. (1975) Plant Sci. Letters 5, 261-268.

LIPID FORMATION IN OLIVE FRUIT (*Olea europea L.*)

ABDELKADER CHERIF, ABDELMAJID DRIRA AND BRAHIM MARZOUK

Laboratoire de Physiologie Végétale, Institut de Recherche Scientifique et Technique - Faculté des Sciences - Tunis (Tunisia) -

ABSTRACT

The lipid content of the olive seed was investigated. The neutral lipid fraction constituted the major class.

The fatty acids of the embryo and the endosperm were the same but their relative amounts in these two parts of the kernel were different. The latter observation was also true for other fatty acids of various molecular categories.

The oleic acid is the most important fatty acid, it increased from 15 to 70% (for total fatty acids) is the developing fruit.

After 6 to 8 weeks the relative proportions of the various fatty acids remained unchanged.

INTRODUCTION

Except some early work[1,2] dealing with the fatty acids of the kernel oils no sufficient investigation related to the metabolism of the lipids in the different parts of the olive seed were carried out.

The objectives of the study were :

1 - To study the fatty acid composition of the parts of the seed, i,e : the kernel as a whole, the embryo and the endosperm.

2 - To follow the accumulation of the lipids in the ripening fruit.

MATERIAL AND METHODS

Olive seeds of the widely grown tunisian cultivar " Chemlali " were used. After soaking the seeds for 24 hours in water, the embryo and the endosperm were separated.

The Bligh and Dyer method[3] was used to extract total lipids. The methyl esters fraction of the corresponding fatty acids were prepared for gas chromatography analysis according to the Metcalfe " et al. " method[4].

To separate the different lipid fraction we used thin layer chromatography following Gardner's method as modified by Grenier " et al "[5].

RESULTS AND DISCUSSION

1 - The fatty acids of the different parts of the olive fruit. Kernel, embryo, endosperm and pulp (mesocarp) have the same fatty acids, two are saturated (palmitic and stearic acids) and two are unsaturated (oleic and linoleic).

The fatty acids of the kernel are the same as those of the pulp; The major fatty acid is the oleic acid, while among the saturated ones the palmitic acid is the most important. However, the kernel is richer in linoleic acid than the pulp.

Fatty acids of the different parts of the kernel showed that the endosperm has a higher percentage of linoleic acid (20% of the total F.A) than the embryo (8%). More over, the endosperm is lower in oleic acid (66% of the total F.A) than the embryo (73%).

Investigation of the fatty acids in the phospholipids and galactolipids of the embryo and the endosperm showed a high level of palmitic acid while the level of oleic acid detected was much lower (Table 1).

TABLE 1

FATTY ACIDS OF DIFFERENT LIPIDS OF THE KERNEL, THE EMBRYO AND ENDOSPERM OF THE OLIVE FRUIT (%)

	$C_{16:0}$			$C_{18:0}$			$C_{18:1}$			$C_{18:2}$		
	A	B	C	A	B	C	A	B	C	A	B	C
TL	10,3	10,7	9,9	4,6	7,4	3,1	66,4	73,7	66,3	18,5	8,2	20,7
NL	11,9	9,8	11,7	3,3	4,9	3	64,8	72,9	61,1	20	12,4	24,2
MGDG	19,6	12,8	13,5	2,9	3,2	2,5	58,9	66	54,2	18,6	18	29,8
DGDG	30,1	22,2	45,6	11,5	6,8	7,5	47,6	61,4	40,3	10,8	9,6	6,6
PL	25,1	18,9	20,5	2,3	6,7	3,8	52,4	69,1	56,5	20,2	5,3	19,2

A:kernel B:embryo C:endosperm TL:total lipids NL:neutral lipids MGDG: monogalactosyldiacylglycerol DGDG:digalactosyldiacylglycerol PL:phospholipids.

The kernel showed a higher content (18%) of linoleic acid than the pulp (12%). This observation was also true in other fruits such the avocado the apples and the grapes in which the amounts of linoleic acid were higher in the seed than the pulp (Table 2).

TABLE 2

CONTENT IN LINOLEIC ACID OF THE SEED OF THE FRUIT OF SOME PLANTS (% of total fatty acids).

	Fruit	Seed
Olea europea L.	12	18
Persea americana L[6]	10,4 - 11,3	41,5
Pirus malus L[7]	53	60 - 70
Vitis vinifera L[8]	40	68

2 - <u>The lipids in the developing and ripening olive fruit</u>. The curve showing the accumulation of lipids in the olive fruit is of the sigmoid type with three distinctive phases:
- a first phase lasting about 12 weeks.
- a rapid phase of 14 weeks.
- a show down phase in the rate of lipid accumulation levelling off at ripening of fruit. Similar patterns of lipid build up were observed in other oleaginous seeds[9].

The sigmoid curve is true for each fatty acid taken separately including those present at low level (Fig. 1). From the first to the second phase, the unsaturated fatty acids and especially the oleic one increase dramatically leading to a high oil content. The oleic acid increased 300 fold in newly formed fruits.

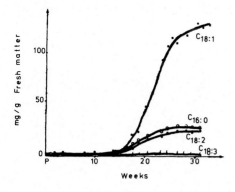

Fig. 1. Fatty acids content in the developing and the ripening olive fruit (P : pollinization).

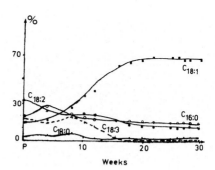

Fig. 2. Changes in fatty acids composition during the development and the maturation of the olive fruit.

The relative variations in the proportions of each fatty acid during the fruit developing and ripening showed (Fig. 2):
- A near stability in the lipid composition. During the first six weeks, the percentages of the different fatty acids remained unchanged.
- A drop in the palmitic, linoleic and linoleic acid percentages there after.
- A considerable increase of the oleic acid in the final stages (15% in newly formed fruits vs 70% at maturity).

The above variations were not observed beyond the first 6 to 8 weeks of the development of the fruit. The relative proportions of the various fatty acids remained constant there after.

REFERENCES

1 - Brandoniso V. (1935) in L'Olivicoltore, Palazzo Margherita ed , Roma , pp 16 - 18 .
2 - Demingo , M. and Romero , J.H. (1953) Rev. Real Acad. Cienc. Ex. Fis. Nat. (Madrid) , 47, 557 - 585.
3 - Bligh , E.G. et Deyer , W.J. (1959) Can. J. Biochem., 37 , 911 - 917.
4 - Metcalfe , L.D., Schmitz A.A. , and Pelka J.R. (1966) Anal. Chem. , 38 514 - 515.
5 - Grenier,G. , Trémolières , A., Therrien , H.P. and Willemot,C. (1972) Can. J.Bot. , 50 , 1681 - 1689.
6 - Mazliak,P. (1968) in le métabolisme des lipides dans les plantes supérieures, Masson et Cie eds , Paris, pp. 174 - 184.
7 - Mazliak,P. (1965) Fruits, 20 , 49 - 57 .
8 - Drira,A. (1978) Unpublished Results.
9 - Cherif,A. and Mazliak,P. (1978) Rev. Fr. Corps Gras , 25 , 15 - 20 .

LIPID SYNTHESIS IN IMBIBED FERN SPORES

ARMIN R. GEMMRICH
Abteilung Allgemeine Botanik, Universität Ulm, Oberer Eselsberg,
D-7900 Ulm (Donau)

ABSTRACT

Imbibing spores of the fern *Anemia phyllitidis* kept in total darkness incorporate ^{14}C-acetate into lipids which increase linearely with time. The incorporation rate is reduced to about one half by cordycepin and cycloheximide. ^{14}C-acetate was incorporated into palmitic, oleic, and linoleic acid. These newly synthesized fatty acids were distributed preferably among neutral lipids which constitute the majority of storage lipids in the spore.

INTRODUCTION

The spores of the fern *Anemia phyllitidis* (Schizaeaceae) require water and light for germination. They do not germinate in total darkness with one exception: gibberellic acid induces germination in darkness[1]. In germinating spores storage lipids are degraded to generate C-2 compounds for biosynthesis of membrane lipids in the protonema cell[2]. Thus, during germination catabolic as well as anabolic processes are active[3,4].

In view of the observation that enzymes for fatty acid synthesis are already present in the dry seeds of Pisum[5] it seemed interesting to investigate the capacity for lipid synthesis in germinating spores which are characterized by a dramatic lipid metabolism. In order to separate lipid degradation and synthesis which both proceed in germinating spores, the capacity for lipid synthesis and degradation in imbibing spores kept in total darkness was investigated. Since this treatment does not induce spores to germination, lipid degradation rather than synthesis might be expected. However, the contrary is true.

MATERIALS AND METHODS

The source of spores, culture conditions, extraction of lipids and GLC methods have been described previously[2].

Incorporation studies. Na(1-^{14}C) acetate (56mCi/nmol) was used at a concentration of 5 μCi in sterile culture medium. Inhibitors were applied at a concentration which inhibits germination, cycloheximide at 5×10^{-6} g/ml and cordycepin at 5×10^{-4} g/ml.

Lipid analysis. Lipids were separated by TLC on precoated silica Gel G plates using the following systems: (A) petroleum benzine 60-80°/diethyl ether/acetic acid (40:10:1; v/v), (B) chloroform/methanol/7 N ammonia (32:15:2; v/v/v), (C) acetone/acetic acid/water (50:1:0,5; v/v/v). Neutral lipids were chromatographed two-dimensionally with diethyl ether/benzene/ethanol/acetic acid (40:50:2:0,2; v/v/v/v) in the first dimension and with system (A) in the second dimension. Lipid bands were detected by spraying with specific reagents: molybdate/H_2SO_4 for phospholipids[6], periodate-Schiff stain for glycolipids[7]. For general detection spots were made visible by charring after spraying with sulfuric acid/acetic acid (1:1; v/v) or by iodine vapour. Lipids were identified by co-chromatography of authentic compounds in different solvent systems.

Radioactivity measurement. The radioactivity incorporated into lipids was recorded qualitatively and quantitatively by scanning thin layer plates with a Berthold thin layer scanner equipped with a multichannel analyser. The radioactive regions on two dimensional chromatograms were detected by a ß-camera (Berthold). Quantitative recovery of radioactivity in individual compounds was determined by scraping the appropriate areas from the plates and assaying radioactivity by standard scintillation techniques. The ^{14}C-fatty acid methyl esters from the GLC column effluent were trapped in cooled glass tubes. The entire contents of each tube were washed into a scintillation vial containing 5 ml Unisolve (Koch Light Laboratories) as scintillator fluid and counted in a Intertechnique Isocap 300 liquid scintillation counter.

All experiments were conducted at least twice and the results are representative of those replications.

RESULTS

Lipid degradation. Spores imbibed in total darkness do not germinate. Even after 21 days of imbibition in darkness neither a decrease in the lipid content nor an alteration in the proportions of lipid classes or fatty acids from the triglycerides and polar lipid fraction can be detected, while during this time in germinating spores the lipid content decreases to about 60 %. This clearly indicates that the enzymes or their cofactors involved in lipid degradation are under light control.

TABLE 1
EFFECT OF CORDYCEPIN (5×10^{-4} g/ml) AND CYCLOHEXIMIDE (5×10^{-6} g/ml) ON THE DISTRIBUTION OF RADIOACTIVITY FROM ^{14}C-ACETATE IN LIPID CLASSES OF SPORES IMBIBED IN DARKNESS

	no inhibitor	CD	CH
Polar lipids	29	33	35
1,2-Diglyceride	18	10	19
1,3-Diglyceride	1	1	3
Free fatty acids	1	1	3
Triglyceride	35	26	34
Carotene	14	28	4
Phosphatic acid	1	8	-
Phosphatidyl choline	6	4	8
Unknown	6	6	17
Phosphatidyl glycerol	8	4	14
Monogalactosyl diglyceride	3	-	-
Neutral lipids	75	78	61

CD = Cordycepin
CH = Cycloheximide
The R_F of the unknown lipid was lower than PG in acid and basic solvent systems

Incorporation of ^{14}C-acetate into lipid classes. The incorporation rate into lipids is linear during 48 h. Labelled lipids can already be detected after 1 h. As shown in Table 1 the major label was found in neutral lipids including carotene. To obtain information on the regulation of fatty acid synthesis, the effect of two inhibitors was studied; cycloheximide as an inhibitor of protein synthesis and cordycepin which inhibits RNA synthesis through incorporation into the growing RNA chain. Both inhibitors reduced the incorporation of label into lipids to about one half. Cordycepin caused a reduction in the proportion of label in diglycerides, triglycerides and in phosphatidyl glycerol and an enhancement of label in carotene. The accumulation of the key intermediate, phosphatidic acid, suggests that the subsequent steps in the synthetic pathway are inhibited by cordycepin. Cycloheximide inhibited incorporation of label into neutral lipids, particularly into carotene, and stimulated incorporation into phospholipids (Table 1).

Incorporation of ^{14}C-acetate into fatty acids. Alkaline hydrolysis of total lipid extract showed that label is located only in the fatty acid moiety. Most of the radioactivity accounts for those fatty acids representing also quantitatively the major fatty acids in the spore, i. e. palmitic, oleic, and linoleic acid. Cordycepin inhibited incorporation of acetate into long chain and unsaturated fatty acids.

DISCUSSION

From the results presented two aspects of spore germination of *Anemia phyllitidis* have become evident. Firstly, lipid synthesis is independent of the induction of the germination process by external factors. Secondly, the facts that the incorporation of acetate into lipids starts nearly without a lag phase and that the labelling pattern does not change significantly during 48 h of imbibition, indicate that the enzymes involved in the biosynthesis of fatty acids and acyl lipids are preexistent in the dry spore. As lipid synthesis rises linearly during the imbibition phase we conclude that the onset of lipid synthesis is only correlated to a specific water content in the spore. In this respect spores resemble the seeds of higher plants[5]. In support of this conclusion are the results obtained with inhibitors of protein and RNA synthesis. Despite strong evidence for the presence of stable mRNA in the dormant spore [8-11] it is suggested that the regulation of a limited number of enzymes involved in fatty acid and lipid synthesis is neither at the stage of mRNA synthesis nor at the stage of protein synthesis but that the preexistent enzymes are activated by the imbibed water. However, it appears likely that apart from preexisting enzymes other enzymes are also synthesized de novo during the early germination phase. This is concluded from the difference in labelling pattern caused by inhibitors.

Unexpected was the high incorporation of acetate into triglycerides since these are regarded as storage lipids being utilized during germination. Since this has also been described for another lipid storing fern spore [4] it seems likely that this phenomenon is typical for the early phase of spore germination. We therefore suggest that in imbibing spores as well as during the early stages of germination the system involved in storage lipid synthesis is still operative. This capacity might represent the legacy of the drying spore terminating its development. The enzymatic system required for the synthesis of lipids typical of vegetative tissue is initiated after the spore has been induced to germination. Most likely these enzymes are synthesized de novo.

ACKNOWLEDGEMENTS

The author thanks Prof. Dr. H. Schraudolf for stimulating discussion, Dr. R. Grill for revising the English text and Mrs. M. Schirrmacher for excellent technical assistance.

REFERENCES

1. Schraudolf, H. (1962) Biol. Zentralbl., 81, 731-740.
2. Gemmrich, A. R. (1977) Plant Science Letters, 9, 301-307.
3. Towill, L. and Ikuma, H. (1975) Plant Physiol., 56, 468-473.
4. Robinson, P. M., Smith, D. L., Safford, R. and Nichols, B.W. (1973) Phytochemistry, 12, 1377-1381.
5. Harwood, J. L. and Stumpf, P. K. (1970) Plant Physiol., 46, 500-508.
6. Dittmer, J. C. and Lester, R. L. (1964) J. Lipid Research, 5, 126-127.
7. Kates, M. (1972) Techniques of Lipidology, North-Holland, Amsterdam, 428-446.
8. Schraudolf, H. (1967) Planta, 74, 123-147.
9. Raghavan, V. (1970) Exp. Cell Res., 63, 341-352.
10. Towill, L. and Ikuma, H. (1975) Plant Physiol., 55, 803-808.
11. Raghavan, V. (1977) J. Cell Sci., 23, 85-100.

ULTRASTRUCTURAL SITES INVOLVED IN PETROSELINIC ACID ($C18:1\Delta^6$) BIOSYNTHESIS DURING IVY SEED (*Hedera helix* L.) DEVELOPMENT

M. GROSBOIS and P. MAZLIAK

Laboratoire de Physiologie Cellulaire, ERA 323, Université de Paris, 12 rue Cuvier, 75005 Paris (France)

ABSTRACT

A composite subcellular fraction containing both mitochondria and microsomes was the most effective one for petroselinic acid biosynthesis from ^{14}C-acetate.

Treatment of this fraction by concentrated sucrose solution highly depressed petroselinic acid synthesis. The activity could be restored by the addition of untreated microsomes.

In addition, the complex cellular fraction directly conditionned incorporation of labelled precursor into the fat fraction itself.

INTRODUCTION

Despite the large number of results now available concerning fatty acid anabolism in plants[1], the problem of desaturation, especially the specific problem of "unusual" isomer formation, is not yet completely elucidated[2].

Petroselinic acid is the major fatty acid elaborated during Ivy fruit maturation and accounts for 70 to 80% of the total fatty acids in the ripe seed. Our work is an attempt to answer the question of the mode of formation of petroselinic acid. First of all, we tried to elucidate the ultrastructural site of its biosynthesis.

MATERIAL AND METHODS

Common Ivy plants were grown in a botanical garden. Fruit development and maturation occurred during the cold season (September to March).

The following subcellular fractions were prepared by centrifuging homogenates of Ivy albumen tissues[3] :
1) cellular debris were eliminated by pelleting at 3 000 g x 10 mn.
2) a composite fraction, called F_{II}, was sedimented at 2 000 g x 30 mn.
3) a microsomal fraction was sedimented at 100 000 g x 45 mn.
4) the floating layer corresponding to the oil body fraction.

The experimental conditions used for incubation of these fractions with ^{14}C-acetate have been extensively described elswhere[3].

RESULTS

The first approach was to experiment with *in vivo* systems consisting of seed slices fed with ^{14}C-acetate in aqueous solution.

The fatty acid labelling patterns obtained in the various subcellular fractions defined above are shown in figure 1.

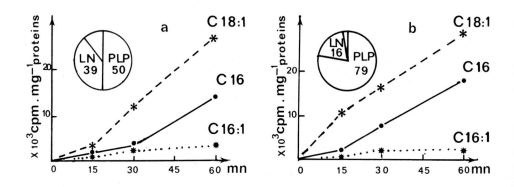

Fig. 1. *In vivo* ^{14}C-acetate incorporation into fatty acids.
1.a Incorporation into the fatty acids of the composite F_{II} fraction.
1.b Incorporation into the fatty acids of microsomes.
The percentages of the total lipid radioactivity recovered in neutral lipids (LN) and phospholipids (PLP) are indicated in the circular diagrams.

In F_{II} fraction and microsomes, octadecenoïc acids were the most heavily labelled fatty acids. The fat fraction incorporated ^{14}C-acetate very actively and almost exclusively into petroselinic acid. The diagrams on Figure 1 giving the percentage of radioactivity found in each category of complex lipids, show that phospholipids are the most intensively labelled lipids *in vivo* in F_{II} and microsomal fractions. On the contrary, in the oil droplets fraction neutral lipids account for 83% of the total radioactivity incorporated into complex lipids.

In vitro experiments with the same subcellular fractions as defined above showed that the F_{II} composite fraction was the most effective for the synthesis of $C18:1\Delta^6$ from ^{14}C-acetate. F_{II} fraction incorporated ^{14}C predominantly into monounsaturated fatty acids and mostly into Δ^6 isomers. Petroselinic acid was, at any time of incubation tested between 15 and 60 mn, the most labelled fatty acid (Figure 2).

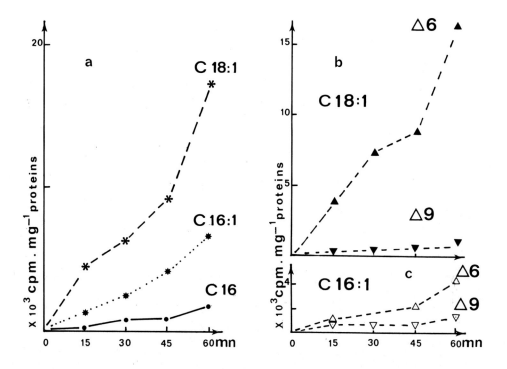

Fig. 2. *In vitro* ^{14}C-acetate incorporation into fatty acids
2.a Incorporation into C16 and C18 fatty acids.
2.b Incorporation into octadecenoïc positional isomers.
2.c Incorporation into hexadecenoïc positional isomers.

The optimal pH was near 8.5, oxygen was required for $C18:1\Delta^6$ synthesis. Absence of ATP or CoASH in the medium decreased dramatically acetate incorporation. Only 2% of the initial activity was maintained without ATP and no more than 7% when CoASH was missing.

Significant amount of the newly synthetized petroselinic acid was recovered in phospholipids although this acid is not there major in term of mass.

We determined that treating F_{II} fraction, in a way similar to that involved in sucrose gradient centrifugation partition, disturbed greatly the fatty acid labelling pattern obtained with the original F_{II} fraction. After treatment by a concentrated sucrose solution, we observed a significant increase of ^{14}C-acetate incorporation into palmitic acid which became the most labelled fatty acid, whereas C18:1 radioactivity was lowered twice (Figure 3a).

Addition of microsomes to the F_{II} fraction, treated as described previously, restored completely the ^{14}C incorporation into monoenes and enhanced the C18:1 activity approximately to its original value. TLC controls have shown that C18:1Δ^6 was directly implicated in that increase. This result underlines the role of microsomes in monoenes formation, especially C18:1Δ^6 (Figure 3b).

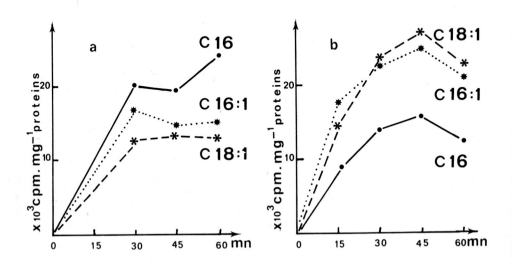

Fig. 3. ^{14}C-Acetate incorporation into fatty acids by treated F_{II} fraction (3.a) and by treated F_{II} fraction in the presence of untreated microsomes (3.b). For treatment, organelles were suspended 2 mn in a 1 M sucrose solution. The suspension is then diluted to 0.4 M and the organelles pelleted by centrifugation.

An other approach has been developped which gave similar results. In an attempt to recreate the characteristic membrane structure association occurring in F_{II} fraction, microsomes have been added to isolated mitochondria.

Mitochondria isolated from Ivy seed albumen showed the classical predominant labelling of C16, and some incorporation of ^{14}C into monoenes[4]. Microsomes *in vitro* always exhibited a very poor incorporation of acetate and no labelling in unsaturated acids.

Addition of microsomes to a mitochondrial fraction actually restored the original distribution of radioactivity, formerly observed amongst fatty acids with the untreated F_{II} fraction (Figure 4).

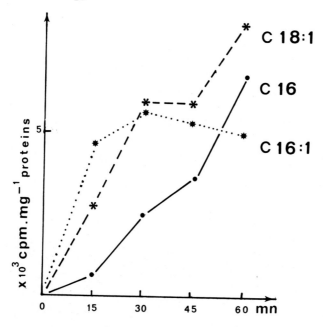

Fig. 4. ^{14}C-Acetate incorporation into fatty acids by F_{II} fraction reconstituted by addition of mitochondria to microsomes.

Several authors have reported synthesis of fatty acids and triacylglycerols associated with the fat fraction itself[2]. Oleosomes isolated from Ivy seed albumen did not show any capacity to synthetize petroselinic acid from acetate *in vitro*, but addition of the F_{II} fraction to these oleosomes led to a significant labelling of petroselinic acid.

CONCLUSION

Cooperation between distinct cell compartments (microsomes and mitochondria) is evidenced from our experiments ; this can be paralleled with the cooperation between different cell compartments previously shown to occur in certain plant tissues[5].

The role of phospholipids in petroselinic acid biosynthesis although not yet clearly elucidated should be underlined.

REFERENCES

1. Harwood, J.L. (1975) in Recent Advances in the Chemistry and Biochemistry of Plant Lipids, Galliard, T. and Mercer, E.I. eds., Academic Press, London, N.Y., S.F., pp 44-88.
2. Applequist, L.Å. (1975) in Recent Advances in the Chemistry and Biochemistry of Plant Lipids, Galliard, T. and Mercer, E.I. eds., Academic Press, London, N.Y., S.F., pp 244-283.
3. Grosbois, M. (1978) Thèse de Doctorat d'Etat, Université P. et M. Curie, Paris.
4. Mazliak, P., Oursel, A., Ben Abdelkader, A. and Grosbois, M. (1972) Eur. J. Biochem., 28, 399-411.
5. Ben Abdelkader, A. and Mazliak, P. (1970) Eur. J. Biochem., 15, 250-262.

GLYCEROLIPID SYNTHESIS IN THE LEAVES OF VICIA FABA AND HORDEUM VULGARE TREATED WITH SUBSTITUTED PYRIDAZINONES (San 9785, San 9774 and San 6706)

M. KHAN, D.J. CHAPMAN, N.W. LEM, K.R. CHANDORKAR AND J.P. WILLIAMS
Department of Botany, University of Toronto, Toronto, Ontario M5S 1A1 (Canada)

ABSTRACT

The major effect of San 9785 was a depression of the linolenic acid (18:3) content of the glycerolipids in *V. faba* and *H. vulgare*, particularly in the galactolipids. San 9785 inhibited the formation of trans-Δ^3-hexadecenoic acid (16:1) in phosphatidyl glycerol (PG) at higher concentrations.

San 6706 was found to inhibit the formation of 16:1 in PG at low concentrations. It also decreased the level of galactolipids in the leaves of *V. faba* but did not affect their fatty acid composition. In *H. vulgare*, this herbicide decreased the 18:3 content of the galactolipids (MGDG and DGDG).

Further studies revealed that the pyridazinones affected the normal development of chloroplasts, chlorophyll synthesis and photosynthesis.

INTRODUCTION

Substituted pyridazinones have previously been found to alter the fatty acid compositions of galactolipids in wheat[1], cotton[2] and the roots of winter wheat[3]. The effect of these herbicides was observed particularly in the reduction of linolenic acid (18:3) content which was accompanied by an increase in the level of linoleic acid (18:2) and other saturated fatty acids. These compounds have also been reported to affect the chlorophyll and carotenoid contents of leaves of treated plants[4]. Willemot has suggested that these pyridazinones have a direct effect on the linoleate desaturase[3]. Recently, the results of Feierabend and Schubert[5] have indicated that these compounds effect a reduction in the level of carotenoids leading to the photodestruction of chloroplast ribosomes. Photosynthetic CO_2 fixation and the Hill reaction in isolated chloroplasts were observed to be inhibited by these pyridazinones[6].

The present studies were undertaken to elucidate the effects of three substituted pyridazinones (San 9785, San 9774 and San 6706) at three different concentrations on the distribution of fatty acids in leaf glycerolipids, with special emphasis on monogalactosyl diacylglycerol (MGDG), digalactosyl diacylglycerol (DGDG) and phosphatidyl glycerol (PG). Ultrastructural changes occurring in the chloroplasts of *V. faba* leaf tissues treated with these herbicides were also investigated. In addition, the effects of these herbicides on photosynthetic CO_2 fixation and on the rate of Hill reaction in isolated chloroplasts were investigated.

MATERIALS AND METHODS

Seedings of *Vicia faba* and *Hordeum vulgare* were treated with three different concentrations of San 9785, San 9774 and San 6706 immediately after germination (see Willemot[3]).

Lipid analysis. Lipids were extracted from the leaves of three-week old treated and control plants, separated by TLC, methanolysed and the fatty acid methyl esters were analysed by GLC[7,8].

Chloroplast preparation. Chloroplast thylakoid preparations for oxygen evolution and ferricyanide reduction were isolated from leaves by homogenization in a medium containing no osmoticum. The reaction medium (3 ml) was 0.74 mg $KFeCN_6$ in the isolation medium with 120-195 µg of chlorophyll and 0.03 ml of herbicide in acetone.

Electron microscopy. Pieces of *V. faba* leaf tissue were fixed conventionally, embedded in Epon and examined after staining with lead citrate.

RESULTS AND DISCUSSION

The effects of these substituted pyridazinones were detected primarily in the fatty acid compositions of MGDG, DGDG and PG from both plants. The major effect of San 9785 was on the level of 18:3 in most glycerolipids and particularly in the galactolipids (Table I). In both MGDG and DGDG, the quantity (µmole/g. fresh wt.), as well as the mole % of 18:3 decreased, while the levels of other fatty acids remained essentially unaffected. Increasing concentrations of San 9785 further reduced the concentration of 18:3 in the galactolipids but did not stimulate the accumulation of 18:2. Since 18:3 is the major fatty acid of the galactolipids the reduction in the level of this fatty acid resulted in a drop in the quantity of MGDG and DGDG. *H. vulgare* appeared to be more susceptible than *V. faba* showing a greater reduction in the level of 18:3 even at the lowest concentration of the herbicide used in this experiment (0.1 mM).

The fatty acid composition of PG in both plants was also found to be affected by San 9785. As in the galactolipids, the level of 18:3 in PG decreased. In *V. faba*, the level of 16:1 decreased only at higher concentrations (0.5 and 1.0 mM). The decrease in the level of 16:1 with corresponding increase in the level of 16:0 suggests the possible inhibitory action of this herbicide on palmitate desaturase. In *H. vulgare*, however, San 9785 appeared to stimulate the production of 16:1 in PG.

Table I

FATTY ACID COMPOSITIONS AND QUANTITIES OF LIPIDS OBTAINED FROM THE LEAVES OF V. FABA AND H. VULGARE TREATED WITH DIFFERENT CONCENTRATIONS OF San 9785

	V. faba							H. vulgare						
	Quant.	16:0[a]	16:1	18:0	18:1	18:2	18:3	Quant.	16:0	16:1	18:0	18:1	18:2	18:3
	μmole/gfw	mole %						μmole/gfw	mole %					
MGDG														
0.0mM	4.2	1	-	tr[b]	1	9	89	3.2	3	-	1	1	6	89
0.1mM	5.9	1	-	tr	2	13	84	4.2	2	-	tr	tr	40	58
0.5mM	2.9	1	-	tr	1	23	73	3.0	1	-	tr	tr	56	41
1.0mM	1.5	5	-	1	2	42	51	2.8	2	-	1	tr	60	38
DGDG														
0.0mM	3.7	8	-	2	1	7	82	2.8	11	-	1	1	5	83
0.1mM	3.4	8	-	1	1	14	76	2.8	8	-	1	1	41	50
0.5mM	2.1	8	-	2	1	29	60	2.3	9	-	1	tr	55	35
1.0mM	1.2	6	-	2	1	42	49	1.5	2	-	2	tr	58	34
PG														
0.0mM	0.8	23	32	1	3	15	26	1.0	15	21	2	2	8	53
0.1mM	1.2	16	37	1	2	25	18	0.9	12	30	1	1	29	26
0.5mM	0.8	29	23	1	3	24	20	0.9	12	31	2	1	37	18
1.0mM	0.6	42	3	2	4	35	13	1.0	16	26	2	2	35	19

a Number of carbon atoms : number of double bonds
b mole % less than 0.5

The major effect of San 6706 in V. faba was on the fatty acid composition of PG and unlike San 9785 it had little effect on the galactolipid fatty acid composition although the quantities of galactolipids decreased considerably (Table II). In both galactolipids, the level of 18:3 remained as high as the level in the control plants. The synthesis of PG 16:1 in both plants was inhibited even at 0.1 mM San 6706. At higher concentrations, the leaves from treated plants showed almost a complete absence of this fatty acid from PG. The seedlings of V. faba did not survive the treatment of 1.0 mM San 6706. As with San 9785 in H. vulgare, the treatment of San 6706 was found to decrease the level of 18:3 in both galactolipids and PG. While increasing concentrations of San 6706 decreased the quantity of MGDG, DGDG and PG in both plants, increasing concentrations of San 6706 stimulated the synthesis of phosphatidyl choline (data not shown). San 9774 had very little effect on the quantity, as well as on the fatty acid compositions of glycerolipids.

Table II

FATTY ACID COMPOSITIONS AND QUANTITIES OF LIPIDS OBTAINED FROM THE LEAVES OF V. FABA AND H. VULGARE TREATED WITH DIFFERENT CONCENTRATIONS OF San 6706

	V. faba							H. vulgare						
	Quant.	16:0[a]	16:1	18:0	18:1	18:2	18:3	Quant.	16:0	16:1	18:0	18:1	18:2	18:3
	µmole/gfw	mole %						µmole/gfw	mole %					
MGDG														
0.0mM	4.2	1	-	tr[b]	1	9	89	3.2	3	-	1	1	6	89
0.1mM	0.5	5	-	1	2	6	87	0.5	4	-	2	3	46	46
0.5mM	0.6	3	-	1	1	3	93	0.5	4	-	2	3	49	41
1.0mM	...[c]							0.7	4	-	2	4	42	48
DGDG														
0.0mM	3.7	8	-	2	1	7	82	2.8	11	-	1	1	5	83
0.1mM	0.4	7	-	3	1	3	86	0.7	9	-	1	2	52	37
0.5mM	0.5	7	-	1	3	3	86	0.6	8	-	1	2	58	30
1.0mM	...[c]							0.8	9	-	3	3	47	39
PG														
0.0mM	0.8	23	32	1	3	15	26	1.0	15	21	2	2	8	53
0.1mM	0.4	31	3	1	4	18	43	0.4	42	tr	3	5	34	17
0.5mM	0.4	37	2	2	5	17	37	0.3	41	1	3	5	36	14
1.0mM	...[c]							0.7	41	tr	2	5	35	17

a Number of carbon atoms : number of double bonds
b mole % less than 0.5
c V. faba seedlings did not survive at this concentration

Infrared gas analysis results (unpublished) indicated that the herbicides have similar effects on photosynthetic CO_2 uptake. The CO_2 uptake of leaves which had absorbed herbicide through the petiole for 2 hrs was almost completely inhibited at a concentration of 1.0 mM. It has been reported that San 6706 is more effective than San 9785 and San 9774 in inhibiting Hill reaction in isolated chloroplasts[6]. Our results indicated that all three herbicides inhibited Hill reaction by 50% at approximately the same concentration (Table III). V. faba appeared to be very susceptible to these herbicides, particularly San 9785. Similar results were obtained for V. faba from the oxygen electrode analysis. Experiments with isolated chloroplasts preincubated with these herbicides did not give significantly different results. The data indicated that San 9774 had similar toxicity in inhibiting the Hill reaction and oxygen evolution in isolated chloroplasts but had less effect on chlorophyll and lipid biosynthesis.

These results suggest that the change in fatty acid desaturation was not caused by the inhibition of Hill reaction and subsequent lowering of the oxygen tension in the leaf.

Table III

THE EFFECTS OF San 9785, San 9774 AND San 6706 ON THE HILL REACTION OF ISOLATED CHLOROPLASTS OF V. FABA AND H. VULGARE

Reaction	Species	Herbicide giving 50% inhibition		
		San 9785	San 9774	San 6706
		µmole/µmole of chlorophyll		
$FeCN_6$ reduction	V. faba	0.36	0.34	0.41
	H. vulgare	0.61	0.57	0.92
O_2 evolution	V. faba	0.25	0.42	0.50

Major ultrastructural changes occurring with the treatment of San 9785 and San 6706 are shown in Figs. 1 & 2 respectively. San 6706 prevented grana formation. Swellings of the loculi of granal lamellae were observed with the treatment of San 9785. The ultrastructural differences may be a direct consequence of the effect of these herbicides on lipid and fatty acid biosynthesis.

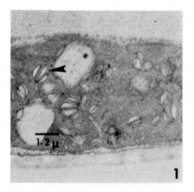
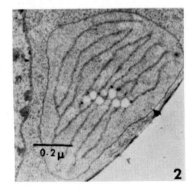

Fig. 1. Section of chloroplast from V. faba treated with San 9785.

Fig. 2. Chloroplast from V. faba treated with San 6706.

Our results do not explain the action of the herbicides on lipid biosynthesis. They do, however, show a clear difference in the susceptibility of the two species and a difference in the site of action of San 6706 and San 9785. While both herbicide inhibit desaturation reactions at high concentrations, at lower concentrations San 6706 appeared to affect palmitate desaturase while San 9785 affected linoleate desaturase. This specificity is difficult to explain in terms of a general inhibition of carotenoid and ribosome synthesis. The differences in and poor development of plastid structure may be a result of the inhibition of different fatty acid desaturase systems. We believe that these herbicides could be used to investigate the biosynthesis of glycerolipids in plants and hence to study the biogenesis of chloroplast membrane components. More details of the present investigation will be published later.

ACKNOWLEDGEMENTS

This work was supported by a National Research Council of Canada operating grant to J.P. Williams and postgraduate scholarship to N.W. Lem. We thank the Sandoz Co. Ltd., Switzerland, for their generous gift of substituted pyridazinones.

REFERENCES

1. St. John, J.B. (1976) Plant Physiol., 57, 38-40.
2. St. John, J.B. and Christiansen, M.N. (1976) Plant Physiol., 57, 257-259.
3. Willemot, C. (1977) Plant Physiol., 60, 1-4.
4. Bartels, P.G. and Hyde, A. (1970) Plant Physiol., 45, 807-810.
5. Feierabend, J. and Schubert, B. (1978) Plant Physiol., 61, 1017-1022.
6. Hilton, J.L., Scharen, A.L., St. John, J.B., Moreland, D.E. and Norris, K.H. (1969) Weed Sci., 17, 541-547.
7. Williams, J.P. and Merrilees, P.A. (1970) Lipids 5, 367-370.
8. Khan, M. and Williams, J.P. (1977) J. Chromatogr., 140, 179-185.

CALCIUM INHIBITION OF PHOSPHOLIPID BIOSYNTHESIS IN A CALCICOLOUS PLANT (*Vicia faba*) ; COMPARISON WITH A CALCIFUGE ONE (*Lupinus luteus*).

A. OURSEL

Laboratoire de Physiologie Cellulaire, ERA 323, 12 rue Cuvier, 75005 Paris (France).

ABSTRACT

^{14}C-acetate, ^{14}C-glycerol-3 phosphate and ^{32}P-phosphate were incorporated into the lipids of entire roots from plants of both species ; when the plants were grown on media enriched in calcium, some modifications were observed in the lipid metabolism of *Vicia faba* roots only.

CDP-choline- and CDP-ethanolamine-diacylglycerol-phosphotransferases present in *Vicia faba* microsomal membranes were inhibited either *in vivo* or *in vitro* by high concentrations of calcium in the medium.

INTRODUCTION

It has been shown previously (Oursel "et al."[1, 2], Lamant and Heller[3]) that when the amount of calcium increased in the growing medium of *Lupinus luteus* plants, the lipid composition of their membranes was not modified. On the contrary, the phospholipid content of the membranes of *Vicia faba* roots decreased as calcium increased in the medium. These facts can be offered several explanations. On one hand, it is well known that phospholipases are generally activated by calcium (Gatt and Barenholz[4]). On the other hand, we explore in this paper the direct action of calcium ions on the biosynthesis of lipid molecules. We have investigated the action of calcium on the activities of two enzymes involved in the biosynthesis of membrane phospholipids : CDP-choline-diacylglycerol-choline-phosphotransferase and CDP-ethanolamine-diacylglycerol-phosphotransferase. For this purpose, two kinds of experiments were realized :

1°) *In vivo* incubations of entire root slices with ^{14}C-acetate, ^{14}C-glycerol-3-phosphate and ^{32}P-phosphate ;

2°) *In vitro* incubations of microsomal fractions prepared from *Vicia faba* roots with CDP-^{14}C-choline and CDP-^{14}C-ethanolamine.

MATERIAL AND METHODS

Vicia faba and *Lupinus luteus* seedlings were grown for 6 days on a nutritive medium, as previously described (Oursel "et al."[1]). The calcium concentrations used were 0.1 mM in the normal medium and 50 mM in the calcium-enriched medium. Sodium-^{14}C-acetate (45 mCi/mmol), sodium-^{32}P-phosphate (10 mCi/mmol) were obtained from C.E.A. (France) ; [U-^{14}C] glycerol-3-phosphate (170 mCi/mmol) was from the Radiochemical Center (Amersham) ; ^{14}C-methyl-CDP-choline (49 mCi/mmol) was obtained from New England Nuclear and [2-^{14}C]-CDP-ethanolamine (28 mCi/mmol) was from the Radiochemical Center (Amersham).

In vivo incubations were performed during 16 hr with root slices plunged in aqueous solutions of radioactive precursors (10 µCi/ml). *In vitro* incubations, lipid analyses and radioactivity counting were performed as described previously (Oursel "et al."[5]).

RESULTS

I - *In vivo* labelling of root lipids. After 16 hr incubations, the following results were obtained (table I) ;

1) ^{14}C-acetate. With roots of normal plants (*Vicia faba*, F_N and *Lupinus luteus*, L_N), phospholipids were the most radioactive lipids in both species representing 76% (F_N) to 82% (L_N) of the total lipid radioactivity. When plants were grown on media enriched in calcium, the percentage of the total radioactivity found in the phospholipids was modified only in *Vicia faba* roots (F_{Ca}) ; the neutral lipid radioactivity represented 32% of the total against only 62% for the phospholipids.

2) [U-^{14}C]-glycerol-3-phosphate. For both species, phospholipids were the most radioactive lipids (84 to 88% of total lipid radioactivity). No modification of these percentages was observed when plants were grown on a medium enriched in calcium but, in that case, and for *Vicia faba* only (F_{Ca}), the total incorporation of the precursor was only 70% of that observed on a normal medium.

3) ^{32}P-phosphate. Two phospholipids, phosphatidylcholine and phosphatidylethanolamine, were highly labelled when *Vicia faba* or *Lupinus luteus* roots were incubated with ^{32}P-phosphate. 52% (F_N) to 66% (L_N) of the total lipid radioactivity was incorporated in those two phospholipids. High concentrations of calcium in the medium apparently neither modified these percentages nor changed the intensities of ^{32}P incorporations into root lipids.

TABLE 1

IN VIVO LABELLING OF THE LIPIDS OF VICIA FABA AND LUPINUS LUTEUS ROOTS
(16 Hr INCUBATIONS)

	Vicia faba		Lupinus luteus	
	0.1mM $CaCl_2$ medium	50 mM $CaCl_2$ medium	0.1mM $CaCl_2$ medium	50 mM $CaCl_2$ medium
	% total lipid radioactivity after labelling from ^{14}C acetate			
Phospholipids	76	62	82	84
Neutral lipids	16	32	8	12
Free fatty acids	6	6	10	4
	% total lipid radioactivity after labelling from ^{14}C-glycerol-3-phosphate			
Phospholipids	84	84	88	86
Neutral lipids	14	12	6	6
Free fatty acids	2	4	6	6
	% total phospholipid radioactivity after labelling from ^{32}P-phosphate			
Phosphatidylcholine	46	44	62	62
Phosphatidyl-éthanolamine	6	10	4	4

II - <u>Inhibition by calcium of the CDP-choline- and CDP-ethanolamine-phosphotransferases of microsomes isolated from Vicia faba roots</u>. Figure 1 shows that the phosphotransferase activities of microsomal preparations from plants (F_N) grown on a medium poor in Calcium (0.1 mM $CaCl_2$) were more important than those from plants (F_{Ca}). grown on a medium enriched in calcium (50 mM $CaCl_2$). This was particularly obvious for the CDP-choline phosphotransferase which was 100 times more active in plants grown on a normal medium. In these experiments, the incubation media did not contain any calcium ; we can thus suppose that it is the calcium fixed in vivo on the microsomal membranes of plants which inhibited the phospholipid biosynthesis.

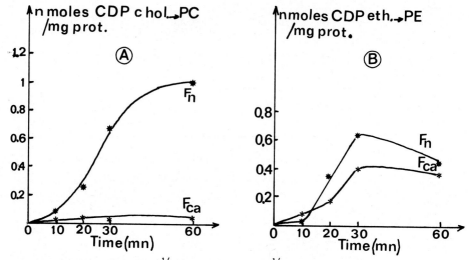

Fig. 1. Incorporation of ^{14}C-choline (A) or ^{14}C-ethanolamine (B) into the phospholipids (PC or PE) of *Vicia faba* root microsomes.
F_n = microsomes from roots grown on a normal medium (0,1 mM $CaCl_2$)
F_{Ca} = microsomes from roots grown on a 50 mM $CaCl_2$ medium

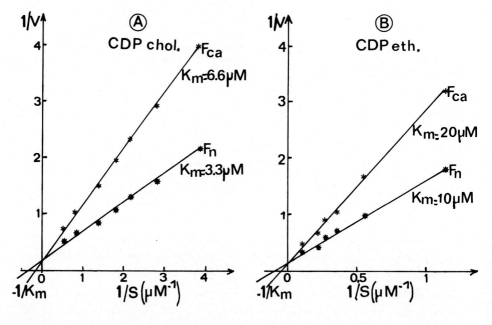

Fig. 2. Lineweaver and Burk plots corresponding to the CDP-choline-transferase (A) and CDP-ethanolamine-transferase (B) activities of *Vicia faba* root microsomes (F_n, F_{Ca} as in Fig. 1).

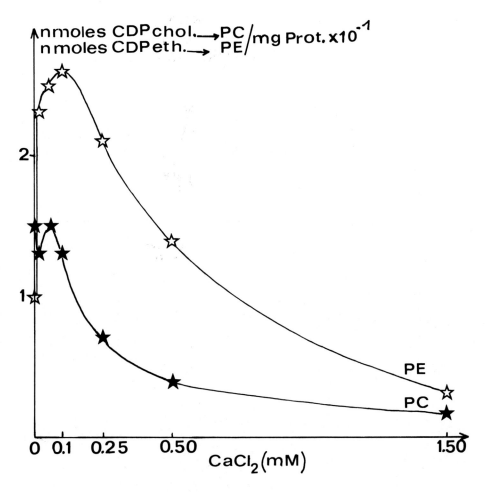

Fig. 3. Influence of calcium ions added to the incubation medium on the activities of the CDP-choline (★—★ PC) and the CDP-ethanolamine-transferases (☆—☆ PE) of *Vicia faba* root microsomes.

III - <u>Calculation of kinetic constants (apparent K_M and V_{MAX}) for the two enzymes in presence or absence of calcium.</u> Figure 2 shows that in plants grown on a normal medium (F_N), the affinities of both enzymes for their substrates were twice greater than those of enzymes from plants grown on a calcium-enriched medium (F_{Ca}).

IV - <u>Effect of calcium, *in vitro*, on the biosynthesis of phosphatidylcholine and phosphatidylethanolamine by microsomes of plants (F_N) grown on a normal medium.</u> When microsomes of plants grown on a normal medium were incubated in the presence of increasing amounts of $CaCl_2$ (0 to 1.50 mM), inhibition of the enzymatic activities could be observed (figure 3) for $CaCl_2$ concentrations greater than 0.1 mM. The inhibition was virtually complete with 1.50 mM.

CONCLUSION

The preceding results strongly suggest that the lower quantity of phospholipids observed in the membranes of *Vicia faba* roots grown on a medium enriched in calcium is the result of a direct inhibition of the phospholipid synthetizing enzymes by calcium ions.

REFERENCES

1. Oursel, A., Lamant, A., Salsac, L. and Mazliak, P. (1973) Phytochemistry, 12, 1865-1874.
2. Oursel, A., Lamant, A., Mazliak, P. and Heller, R. (1974) First Meeting of the International Association of Plant Physiology, Würzburg, Abstract.
3. Lamant, A. and Heller, R. (1975) Physiol. Vég., 13, 685-700.
4. Gatt, S. and Barenholz, Y. (1973) Ann. Rev. Biochem., 42, 61-90.
5. Oursel, A., Tremolieres, A. and Mazliak, P. (1977) Physiol. Vég., 15, 377-385.

CONTROL OF FATTY ACID AND LIPID FORMATION IN BALTIC MARINE ALGAE BY ENVIRONMENTAL FACTORS

P.POHL and F.ZURHEIDE
Institute of Pharmaceutical Biology, Grasweg 9, 2300 Kiel, Germany

ABSTRACT

The influence of several environmental factors on the formation of fatty acids and lipids was studied in the two Baltic marine algae, Phycodrys sinuosa (Rhodophyceae) and Fucus vesiculosus (Phaeophyceae). The main factors appeared to be a) the water temperature and b) the contents of nitrogen ($NH_4^+ + NO_2^- + NO_3^-$) and of phosphate in the seawater.

INTRODUCTION

The biosynthesis of fatty acids and lipids in microscopic freshwater algae can be influenced by various environmental factors such as light and light intensity[1-4] and temperature[5,6]. The composition of the nutrient media seems to be important, too: At higher concentrations of nitrogen (NO_3^-, NH_4^+), the algae produce larger proportions of polyunsaturated fatty acids and of polar lipids (mostly glyco- and phospholipids); low levels of nitrogen favour the formation of saturated and mono-unsaturated fatty acids and of unpolar lipids[4,7]. Similar results have been obtained with varying concentrations of Mn^{++} (Euglena gracilis[8]).

We have now extended our previous investigations[4,7] to macroscopic marine algae from the Baltic Sea. As the Baltic Sea is polluted by waste waters containing high amounts of nitrogen and phosphate, there appeared to be a good opportunity to study the influence of these and other environmental factors in a natural eco-system.

MATERIALS AND METHODS

Phycodrys sinuosa KÜTZ 1843 and Fucus vesiculosus L. were obtained from distinct localities in the western part of the Baltic Sea (Kiel Bight). P.sinuosa grew in the sublittoral zone, 1 km off shore at a depth of 8 m, while F.vesiculosus inhabited the

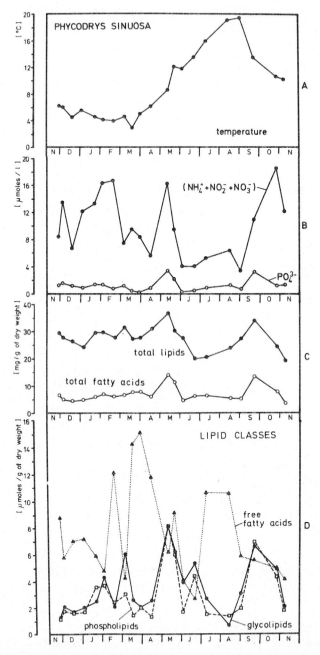

Fig.1: Phycodrys sinuosa; A: Water temperature; B: Contents of nitrogen and phosphate in the seawater; C: Total lipids and fatty acids; D: Lipid classes

uppermost intertidal zone close to the shore (depth 0.5 - 1 m).

RESULTS AND DISCUSSION

Phycodrys sinuosa: The temperature and the contents of total nitrogen and phosphate in the seawater at the location of this alga are shown in fig.1/A and 1/B. During the cold season, the nutrient (N and PO_4) levels were high. Normally (without pollution) these values decrease due to consumption of the nutrients by algae, and increase again in September/October when the growth of the algae ceases because the water temperature and the daily incident radiation become low. However, an additional abundance of nitrogen and phosphate was observed in April/May caused by pollution. The water temperature reached high values (15-20°) in July/August.

As for the alga: The main growth period of Phycodrys sinuosa starts in February/March. Fig.1/C shows that parallel to the maxima of N and PO_4 (fig.1/B) there were maxima in the formation of the total fatty acids and lipids. Within the total lipids, the formation of polar lipids (glyco- and phospholipids) reached maxima at the periods of high nitrogen and phosphate levels, whereas the unpolar lipids (mainly free fatty acids) were produced predominantly at periods of low levels of nitrogen and phosphate in the seawater (fig.1/D).

Total fatty acids (without fig.): With increasing supply of nitrogen and phosphate and at temperatures below 12°, there was an increase in the percentage of polyunsaturated fatty acids (mainly 20:4; 40-50%) and a decrease in the percentage of saturated and mono-unsaturated fatty acids (16:0, 16:1, 18:1; 40-30%). These data indicate an additional influence of the temperature.

The salinity and the pH of the seawater had no or only a minimal effect (without fig.).

Fucus vesiculosus: This alga was collected from two locations with similar water conditions. As Fucus vesiculosus grows close to the shore (3-5 m) and in rather shallow water (0.5-1 m), these locations were necessarily more exposed to waste waters and to changing currents. Consequently, the nutrient supply in the seawater changed more rapidly as at the location of Phycodrys sinuosa. Moreover, shallow water gets warm more easily. So the water temperature had a still greater influence (see fig.2): During the

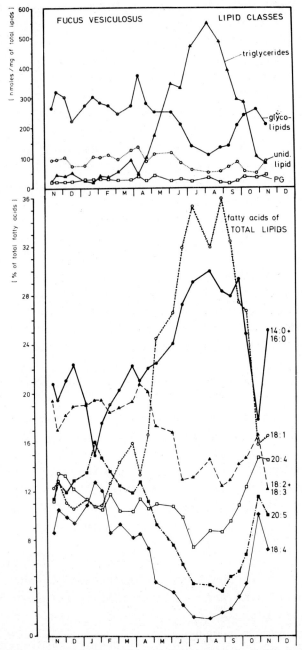

Fig.2: Fucus vesiculosus: Lipid classes and fatty acids of the total lipids

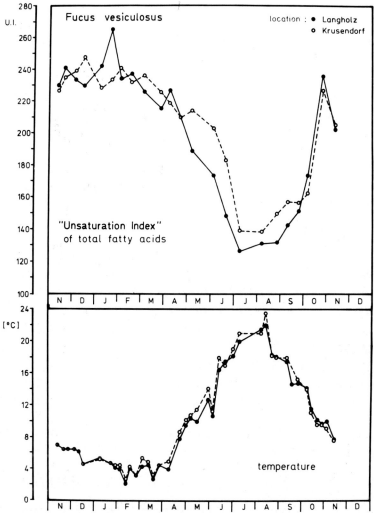

Fig.3: Fucus vesiculosus; Water temperature and unsaturation index

warm season (June-September) the formation of polar lipids (mainly triglycerides) and of saturated and mono-unsaturated fatty acids (14:0, 16:0, 18:1) increased, and the formation of polar lipids (mainly glycolipids, an unidentified lipid and phosphatidyl glycerol (PG)) and of polyunsaturated fatty acids (18:2, 18:3, 18:4, 20:4, 20:5) decreased.

In fig.3 a molar "unsaturation index" was calculated according to the method of Williams and coworkers[9]. This index is used to describe the total number of double bonds in a group of fatty acids and is proportional to the classical "iodine number". The curve of the unsaturation index is clearly symmetrical to the water temperature indicating that with regard to this alga and at these locations the water temperature is a major factor.

The contents of nitrogen and phosphate in the seawater had a similar influence on Fucus vesiculosus as on Phycodrys sinuosa. Due to the above dominating effect of the water temperature, this influence was lower marked (without fig.).

As with Phycodrys sinuosa, the salinity and the pH of the seawater had no or only little influence.

The results show that the temperature and the contents of nitrogen and phosphate of the water strongly influence the formation of lipids and fatty acids in marine algae throughout the year. According to each location and to each alga, however, the efficiency of these factors varies.

REFERENCES
1. Rosenberg, A. and Pecker, M. (1964) Biochemistry, 3, 254-258
2. Nichols, B.W. (1965) Biochim.Biophys.Acta, 106, 274-279
3. Constantopoulos, G. and Bloch, K. (1967) J.biol.Chemistry, 242, 3538-3542
4. Pohl, P. and Wagner, H. (1972) Z.Naturforsch., 27b, 53-61
5. Ackman, R.G. and Tocher, C.S. (1968) J.Fish.Res.Bd.Canada, 25, 1603-1620
6. Kleinschmidt, M.G. and McMahon, V.A. (1970) Plant Physiol., 46, 286-289 and 290-293
7. Pohl, P., Passig, T. and Wagner, H. (1971) Phytochemistry, 10, 1505-1513
8. Constantopoulos, G. (1970) Plant Physiol., 45, 76-80
9. Williams, M.A., Katyare, S. and Packer, L. (1975) Arch.Biochem.Biophys., 170, 353-359

ON THE FATTY ACID BIOSYNTHESIS IN CARROT ROOT. AN INDIRECT APPROACH

JUHANI SOIMAJÄRVI
Department of Biology, University of Jyväskylä, Vapaudenkatu 4, SF-40100 Jyväskylä, Finland

ABSTRACT

Fatty acids from carrot root lipids have been studied during growth. The total lipid content decreases and the relative distribution of major lipid fractions (NL, GL, PL) varies slightly, evidently irrespective of the wide changes in the fatty acid composition during the same period. The fatty acid composition is similar in each lipid fraction, 18:2 and 16:0 predominating. The percentage of 18:3 increases in the autumn (especially in the GL) but it is quite low during summer. Most monoenoic, minor acids are $\Delta 9$ isomers. The polyenoates present are n-6 dienoic acids ranging from C_{16} to C_{20} and the n-3 trienoic acids 16:3 and 18:3.

INTRODUCTION

Since the publication of some early reports on lipids from carrot root, for example, by Dalgarno and Birt[1], interest in this plant as a model of physiological study has recently increased and the lipid composition in root tissue cultures has been presented[2] and has also been compared with major lipids from root material[3]. Because of its long season of rapid growth and the absence of a reproductive phase, the carrot has been used as a rewarding object of growth measurements[4]. Consequently, analysing the fatty acid composition during growth is expected to present benefits when following environmental effects on lipids.

The present investigation was made in order to compare the changes in the fatty acid composition of carrot root lipids during growth with the development of major lipid classes. Experiments were also performed to determine the chemical structures in a detailed fatty acid composition.

MATERIALS AND METHODS

Plant material. Carrots (*Daucus carota*, cv. Feonia Hunderup) were grown in a field experiment, after sowing in late April. The root samples were harvested at intervals during the period from late June to September and roots without short green upper sections were used for analysis.

Extraction and fractionation of lipids. Lipids were extracted from the roots by homogenizing twice in a mixture of chloroform and methyl alcohol (2:1, v/v)

and purified from non-lipids[5]. Total lipids were separated into major fractions - neutral lipids (NL), glycolipids (GL) and phospholipids (PL) - by silicic acid column chromatography[5].

Derivatives and fractionation of fatty acids. Methyl esters were prepared by the BF_3 method[6]. Before a detailed GLC analysis the methyl esters were fractionated as mercuriacetate derivatives by TLC[7]. Pyrrolidides of fatty acids were prepared for mass spectral analysis[8].

Gas chromatography and mass spectrometry. The fatty acid composition of the growth samples was determined on a packed 3 % EGSS-X column and FID. To resolve isomeric esters a 35 m x 0.3 mm column coated with FFAP phase was used. The pyrrolidides were separated on a SE-30 phase capillary column.

Mass spectra of methyl esters and (monoenoic) pyrrolidides of fatty acids were obtained using respectively the FFAP and the SE-30 columns, coupled to a LKB 9000 gas chromatograph-mass spectrometer.

RESULTS AND DISCUSSION

During growth the total lipid content of the carrot root decreased gradually and it was about 250 mg/100 g fr. wt. in the autumn (Fig. 1, upper part). The NL formed the major part of total lipids but the comparatively low and stable NL content indicated only minute reserve function. In the young roots the GL were predominant polar lipids, but later the relative amount of the PL increased (Fig. 1). To discuss the changes in the relative distribution of major lipids the contents may be expressed in relation to growth of a single root. The root growth fits quite well a skew sigmoid growth function[9]. The net accumulation of lipid fractions in a single root is in allometric relation to weight and it fits also sigmoid growth equations, or even slightly better polynomial functions of third or fourth order. The growth equations may be transformed to the growth rate and net accumulation rates of lipids per unit time, as depicted in the lower part of Fig. 1. The different position of the maximum in each curve indicates an alternation of synthesis and possible degradation of lipids.

The fatty acid composition of lipids changed considerably during growth (Fig. 2). In each analysed lipid fraction it was nearly similar and there were nearly simultaneous analogous changes. Of the main acids, linoleic acid was predominant and it commenced its most prominent accumulation at the same time, in late summer, as palmitic acid started to decrease. Even before this highest relative amount of oleic acid was observed. Only a trace of linolenic acid was present during summer, but later this percentage increased, especially in the GL. This development may be caused by environmental factors (temperature) controlling generally unsaturation, and it may also indicate specific changes in plastids,

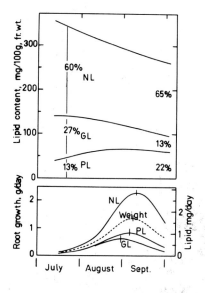
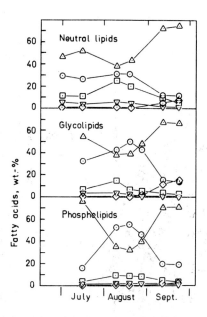

Fig. 1. The content (upper part) and net accumulation (lower part) of lipids in carrot root during growth.

Fig. 2. Fatty acid composition of lipids from carrot roots during growth. 16:0 ○, 18:0 ▽, 18:1 □, 18:2 △, 18:3 ◇.

since trienoic acids of carrot root lipids accumulate in plastid envelope[10].

In addition to the five most abundant acids introduced in Fig. 2, many others were present in small quantities. In the most detailed fatty acid composition obtained after mercuration TLC, several minor components were detected and the monoenoic fraction was resolved to certain positional isomers by capillary GLC.

The presence of most carrot fatty acids, presented in a simplified scheme of Fig. 3, is consistent with common desaturation and elongation mechanisms described in the literature. The first double bond is generally introduced by Δ9 desaturases predominant in higher plants. In carrot lipids this principle can explain the origin of 9-16:1 and 9-17:1 in addition to oleic acid. Palmitoleic acid and its shorter homoloque 15:1(n-8) are Δ7 acids which need another mechanism of synthesis. In addition, one Δ5 acid (15:1, n-10) is present.

The polyenes include common 18:2 and 18:3 together with 8,11-eicosadienoic acid as an apparent elongation product. The synthetic path of 16:2(n-6) and 16:3(n-3) accumulating in minor quantities is not proclaimed by the arrows pointing at them in Fig. 3, but it is noted that no shorter polyunsaturated homologues were present at the detection level of 0.01 % of total fatty acids.

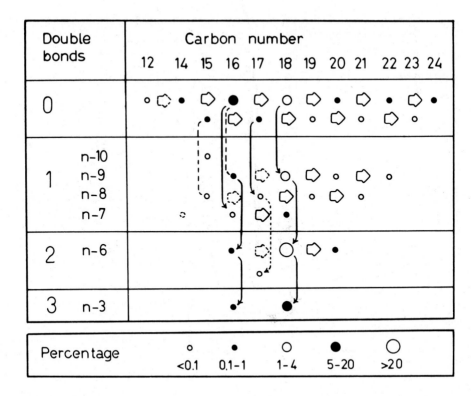

Fig. 3. Scheme of biological origin of fatty acids from carrot root lipids, based on compositional analysis.

REFERENCES

1. Dalgarno, L. and Birt, L.M. (1963) Biochem. J. 87, 586-596.
2. Kleinig, H. and Kopp, C. (1978) Planta 139, 61-65.
3. Gregor, H.-D. (1977) Chem. Phys. Lipids 20, 77-85; Phytochemistry 16, 953-955.
4. Stanhill, G. (1977) Ann. Bot. 41, 533-540; 541-552.
5. Soimajärvi, J. and Linko, R.R. (1973) Acta Chem. Scand. 27, 1053-1058.
6. Metcalfe, L.D., Schmitz, A.A. and Pelka, J.R. (1966) Anal. Chem. 38, 514-515.
7. Mangold, H.K. and Kammereck, R. (1961) Chem. & Ind. 1961, 1032.
8. Anderson, B.Å. and Holman, R.T. (1974) Lipids 9, 185-190.
9. Richards, F.J. (1959) J. Exp. Bot. 10, 290-300.
10. Tevini, M. (1978) personal comm.

OLEYL CoA AND LINOLEYL CoA DESATURASE ACTIVITIES AND α-LINOLENIC ACID BIOSYNTHESIS IN SUB-CELLULAR FRACTIONS FROM YOUNG PEA LEAVES.

A. TREMOLIERES, J.P. DUBACQ, M. MULLER, D. DRAPIER and P. MAZLIAK
Laboratoire de Physiologie Cellulaire, E.R.A. 323, 12 rue Cuvier, 75005 Paris (France)

ABSTRACT

The chloroplast of adult leaves is only able to synthetize saturated and mono-unsaturated fatty acids but oleyl CoA and linoleyl CoA desaturase activities are detectable in various sub-cellular fractions of young Pea leaves. The hypothesis of a necessary cooperation between chloroplast and other organelles for linolenic acid biosynthesis is discussed.

INTRODUCTION

The problem of the biosynthesis of α-linolenic acid is now an old problem in plant biochemistry ; 14 years ago Harris and James[1] first proposed (mainly from experiments of ^{14}C-acetate and ^{14}C-oleate incorporation into Chlorella lipids) a successive desaturation of oleate into linoleate and linolenate ; more recently Kannangara and Stumpf[2] have demonstrated an elongation of $C_{16:3}$ into $C_{18:3}$ in chloroplast of adult spinach leaves. The pathway of fig. 1 is now largely accepted as the major one and has been completely demonstrated by "*in vivo*" incorporation of radioactive acetate, oleate or linoleate into green tissue lipids[3,4]. The direct conversion of $1-^{14}$C-linoleate into $1-^{14}$C-linolenate has been proved by the analysis of the radioactivity distribution between the different carbon atoms of the linolenic acid formed from ^{14}C-linoleate.

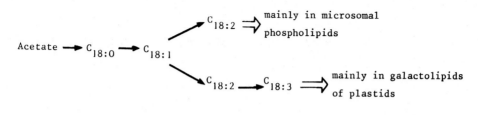

Fig. 1. Biosynthetic pathway of α-linolenic acid in developing leaves.

This desaturation pathway has now been evidenced in many growing organs including leaves, roots and seeds of higher plants and in some lower plants and algae[4]. But the demonstration of this pathway at the sub-cellular level remains an unsolved problem. In the present paper new results are given concerning linolenic acid biosynthesis at the sub-cellular level.

MATERIALS AND METHODS

Pea seeds (*Pisum sativum*, variety alaska) were germinated over a 7-day period under long photoperiods (16 h of light each day), at 22°C, with a relative humidity of 70%. Adult spinach leaves were purchased at a local market.

^{14}C-acetate (45 mCi/nmole), ^{14}C-oleate (51,7 mCi/nmole) and ^{14}C-linoleate (50 mCi/nmole) were obtained from C.E.A. (France) and their ammonium salts were prepared. 1-^{14}C-oleyl CoA and 1-^{14}C-linoleyl CoA were chemically prepared[5].

Microdroplets of radioactive precursors were put on intact leaves[6] and sub-cellular fractions were prepared as previously described[7].

Lipids and fatty acids were analyzed by radio-GLC and radio-TLC[6].

RESULTS

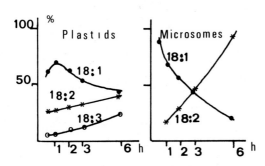 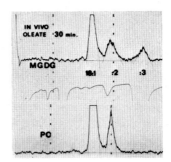

Fig. 2. "*in vivo*" desaturation and incorporation of 1-^{14}C-oleate into two sub-cellular fractions of Pea leaf.

Fig. 3. Radioactivity of MGDG and PC fatty acids in the plastids after 30 min. incorporation of 1-^{14}C-oleate.

In vivo incorporation of 1-^{14}C-oleate : Fig. 2 and fig. 3 show that, after short time incorporations of oleate into young Pea leaf lipids followed by a subsequent preparation of sub-cellular fractions, 30 minutes were sufficient to let 1-^{14}C-linolenic acid appear almost exclusively into the plastidial fraction. In this fraction the radioactive linolenic acid was very actively and preferentially incorporated into monogalactosyldiacylglycerol (MGDG) ; no radioactive linolenic acid was found in phosphatidylcholine (PC). The fact that after short time "*in vivo*" incorporations not only linolenic but also oleic and linoleic acids were found in MGDG shows that the peculiar composition of the galactolipid molecules (with a high content of linolenic acid) does not result from the specificity of the acylases.

<u>In vitro incorporations by the intact chloroplast of adult spinach leaves</u> :

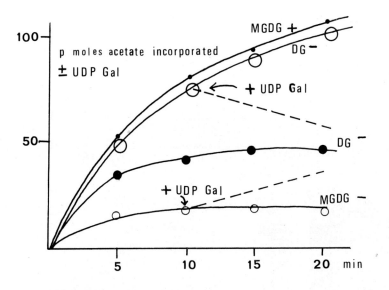

Fig. 4. Incorporation of 1-^{14}C-acetate into lipids by class A chloroplast of adult spinach leaves.

When 1-^{14}C-acetate was given to class A chloroplasts from adult spinach leaves, MGDG was synthetized only when UDP galactose was added to the medium ; in the absence of UDP galactose, radioactive diacylglycerol accumulates ; when UDP galactose was added radioactivity decreased in the diacylglycerol and increased in MGDG (fig. 4).

Intact chloroplasts synthetized palmitic, oleic and very few linoleic acid.

In vitro incorporation by the sub-cellular fractions of young Pea leaves :
An oleyl CoA desaturase activity was found in cell-free systems from young Pea leaves and required O_2 and $NADH_2$. The specific activity of this desaturase was highest in the microsomal fraction but remained important even in purified plastid fraction (with extremely low NADH-cyt c reductase antimycin insensitive activity). The linoleic acid formed in this reaction was always very actively incorporated into phosphatidylcholine.

A linoleyl CoA desaturase activity was found in the plastid fraction (fig. 5). If UDP galactose was added to the medium, the linolenic acid formed from linoleyl CoA was almost exclusively incorporated into MGDG. In the absence of UDP galactose desaturation occurred but very low incorporation into MGDG was observed. Then it appears that desaturation of precursor may occur before integration of the acyl CoA into galactolipid molecules.

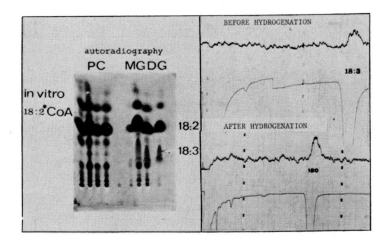

Fig. 5. Study of the linoleyl CoA desaturase activity of a crude plastid fraction of young Pea leaves. (TLC on SiO_2-NO_3Ag (left) and radio GLC (right) of fatty acid methyl esters)

Thus if formation of linolenate and integration of this acid into MGDG is now demonstrated in a cell-free system it remains difficult, in the actual state of young leaf cell fractionation, to decide between a necessary cooperation of microsomes and plastids for the biosynthesis of linolenic acid or a localization of the biosynthetic pathway entirely within plastid.

Are there lipid exchanges between plastids and microsomes or liposomes ?

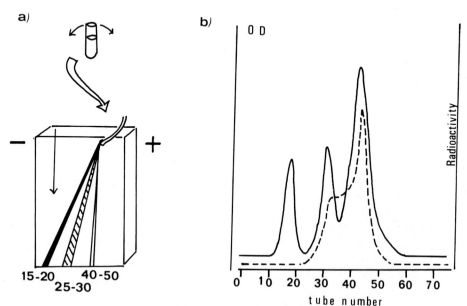

a) Scheme of a free-flow electrophoresis separation.

b) Analysis of tube content :
15-20 intact plastids
20-25 broken plastids
40-50 microsomes

Fig. 6. Analysis by free-flow electrophoresis of the plastids of young Pea leaves mixed with labelled Pea microsomes.

We have tested directly the possibility of a cooperative process between different organelles by means of free-flow electrophoresis. Microsomes labelled "*in vivo*" from ^{14}C-oleate or linoleate in oleyl or linoleyl-phosphatidylcholine or liposomes labelled with ^{32}P-phosphatidylcholine were mixed with the plastids of young or adult leaves, and then injected in the free-flow electrophoresis chamber. After separation the result clearly indicate that no radioactivity appears in the band of intact chloroplasts (even when the phospholipid exchange protein of Ricinus was added[8]). One can see that after free-flow electrophoresis separation, only the fractions of broken chloroplasts contained some radioactivity in their lipids.

Despite of these negative results, we think that the cooperation hypothesis must not be already forsaken because it remains possible that some specific leaf exchange protein is absolutely necessary for these exchanges or that the cooperation implicates some entirely different process such as membrane fusion.

REFERENCES
1. Harris, V. and James, A.T. (1964) Biochim. Biophys. Acta, 106, 456.
2. Kannangara, C.G. and Stumpf, P.K. (1972) Arch. Biochem. Biophys., 148, 414.
3. Tremolieres, A. and Mazliak, P. (1974) Plant Science Letters, 2, 193.
4. Cherif, A., Dubacq, J.P., Mach, R., Oursel, A. and Tremolieres, A. (1975) Phytochemistry, 14, 703.
5. Bourre, J.M. and Daudu, O. (1978) Neuroscience letters, 7, 225.
6. Tremolieres, A. (1972) Phytochemistry, 11, 3453.
7. Dubacq, J.P., Mazliak, P. and Tremolieres, A. (1976) F.E.B.S. letters, 66, 2, 183.
8. Kader, J.C. unpublished result.

PURIFICATION OF CHOLINE KINASE FROM SOYA BEAN

J. WHARFE and J.L. HARWOOD
Department of Biochemistry, University College, Cardiff CF1 1XL, (United Kingdom)

ABSTRACT

Choline kinase has been purified from germinating soya bean seeds. The steps involved ammonium sulphate fractionation, DEAE-cellulose chromatography and affinity chromatography. The purified enzyme had a molecular weight of 36,000 and was unstable. Sulphydryl reagents, glycerol and storage in a freeze-dried form all failed to stabilise the enzymatic activity. The loss of activity may have been due to protease action. The purified enzyme would also catalyse the phosphorylation of ethanolamine but at low rates.

INTRODUCTION

Phosphatidylcholine is the major phospholipid in almost every plant tissue which has been examined. In non-photosynthetic tissues phosphatidylethanolamine is the second most prevalent phospholipid, while in green leaves, phosphatidylglycerol is as equally important as phosphatidylcholine. Not only are these phospholipids important as structural components of cellular membranes, but they may also serve metabolic functions. For example, phosphatidylcholine itself has been implicated as a possible substrate for fatty acid desaturation in plants and other organisms[2].

Two major pathways are known for the biosynthesis of phosphatidylcholine de novo. These are the CDP-base pathway[3] and the methylation pathway from phosphatidylethanolamine[4]. Individual enzymes from the CDP-base pathway have been studied in a number of plant tissues and there is some indirect evidence that the methylation pathway may also be, in certain cases, of importance[5]. However, to date none of the individual enzymes concerned have been purified.

Since phosphatidylcholine is such an important component of higher plants we wished to clarify details of its biosynthesis with respect to (a) the relative importance of the various pathways and (b) differences from enzymes involved in forming phosphatidylethanolamine and other phospholipids. The soya bean was chosen because of its commercial importance and because it has been shown to have high rates of phospholipid synthesis during germination[6]. Experiments with the above in vivo system had indicated that, firstly, the methylation pathway was inactive during the first 24h of germination and, secondly, that the base kinase[5] and the glyceride transferase enzymes[7] were probably different for the synthesis of phosphatidylcholine and phosphatidylethanolamine. Because the base kinase enzymes are soluble we have concentrated our initial attention towards their purification. We have succeeded in isolating two base kinase enzymes from germinating soya bean and now report details of the purification of choline kinase.

MATERIALS AND METHODS

Preparation of fractions. Soya bean seeds var. Fiskeby V were surface sterilised and germinated at room temperature for 24h. They were then homogenised in 0.32M sucrose-5mM Tris-HCl pH 7.4. After filtration, the cell-free homogenate was centrifuged at 3000g X 5min, 18,000g X 20 min and 105,000g X 60 min. In each case the supernatant was used for the next spin. The final particle-free supernatant contained over 95% of the choline kinase activity of the homogenate.

For sub-cellular fractionation studies, the filtered homogenate was centrifuged at 400g X 5min to yield the cell debris pellet, 800g X 10min to yield a nuclear fraction, 23,500g X 20min to yield a mitochondrial fraction and 105,000g X 60min to give a microsomal pellet and a supernatant fraction. The floating fat fraction was also isolated during the above centrifugation[8]. The subcellular fractions were characterised by the use of enzyme markers[8].

Assay of enzymatic activity. Choline and ethanolamine kinase activities were assayed by a paper chromatographic method[9]. The optimal conditions were measured for the various reactants in the assay and the determinations were carried out in the following mixture: Tris buffer pH 8.5(33mM), ATP (10mM), $MgCl_2$ (10mM), dithiothreitol (1mM) and ^{14}C-base (5mM). The enzyme was incubated in the above system for 2h at $30°$ before aliquots were taken for chromatography. Zero time and no enzyme blanks were used and it was found important to always run standards during paper chromatography.

Purification of choline kinase. The final conditions which were used for purification were as follows. The particle-free supernatant was subjected to ammonium sulphate fractionation. The protein fraction precipitating in the 50%-saturated cut was sedimented at 10,000g X 20min. The protein pellet was resuspended in 0.32M sucrose-20mM Tris-HCl pH 7.4 and dialysed against 20mM K phosphate pH 7.0.

The protein was concentrated by ultrafiltration using the hollow fibre technique and applied to a column of DEAE-cellulose (DE-32, Whatman) of dimensions 50 x 1.4cm. The column was eluted with 200ml of 20mM K phosphate pH 7.0

followed by a gradient of 0.1-0.4M KCl in 20mM K phosphate pH 7.0. The active fractions were pooled and concentrated. They were then subjected to affinity chromatography.

Affinity chromatography was carried out on choline which had been coupled to epoxy-activated Sepharose 6B[10]. The column was eluted with 0.1mM Tris-HCl pH 9.0-0.5M NaCl-10mM dithiothreitol followed by 50mM Tris-HCl pH 7.5-10mM dithiothreitol-0.1mM choline[10]. The fractions were dialysed where necessary to remove choline and assayed as above.

Ethanolamine kinase was purified simultaneous to the choline kinase. After DE-32 chromatography the active fractions were run down a column of ethanolamine coupled to Sepharose 6B. The active fractions were then subjected to gel filtration on Sephadex G-100 which was eluted with 10mM Tris-HCl pH 9.0.

Protein estimation. Columns were initially monitored by using the $E_{260}/_{280}$ ratio. Individual fractions were then estimated by the Lowry method. Polyacrylamide gels were run on 10% gels at pH 8.5 and on 7.5% gels-1% sodium dodecyl sulphate at pH 7.0.

RESULTS

Subcellular fractionation of germinating soya bean seeds revealed, as expected, that the base kinase activities were soluble (97% for choline kinase and 99% for ethanolamine kinase). Ammonium sulphate fractionation of the soluble fraction gave percentage recoveries for choline kinase of 10.4%, 6.7%, 12.9% and 68.5% in the 0-30%, 30-40%, 40-50% and 50%-saturated ammonium sulphate cuts respectively.

In preliminary experiments we tried gel filtration, ion-exchange chromatography and affinity chromatography in different orders. As a result of these experiments we decided to use ion-exchange chromatography before affinity chromatography since, under such conditions, we were able to separate two kinases more easily. It should also be noted that in animal tissues the use of electrophoresis allowed the isolation of separate choline and ethanolamine kinases whereas affinity chromatography, when used immediately, did not[10].

Chromatography on DE-32 columns resulted in the rapid elution of a peak of ethanolamine kinase activity. In contrast, almost all of the choline kinase was eluted as a single peak at about 0.25M KCl in the salt gradient. The peak of choline kinase activity was associated with a second peak of ethanolamine kinase activity though the proportion of the latter enzyme in the two peaks varied from preparation to preparation.

Coupling of choline to Sepharose 6B columns permitted good purification of choline kinase. Almost all of the ethanolamine kinase activity which was also present in the fraction loaded was rapidly eluted from the column. In contrast choline kinase was firmly bound to the column and was only eluted by the addition of choline to the eluting buffer. After dialysing the excess choline away, the fractions were assayed and it was found that the purified protein had a ratio of choline: ethanolamine kinase activity of at least 3:1, in contrast to the peak from DEAE-cellulose where it was about 1:1. By comparison, the first peak of ethanolamine kinase activity which was eluted from the DE-32 column and further purified had extremely low choline kinase activity. It appears, therefore, that it is possible to separate choline and ethanolamine kinases from soya bean such as has been recently reported for rat liver[11]. A summary of the purification procedure is given in Table 1.

TABLE 1

SUMMARY OF THE PURIFICATION STAGES FOR CHOLINE KINASE

Stage	Recovery(% for each step)	Notes
$6 \times 10^6 g$ min supernatant	97	Ratio E/C kinase = 3
$(NH_4)_2SO_4$ precipitate	101	
DEAE-column	43	Ratio E/C kinase = 1
Affinity chromatography	40	Ratio E/C kinase = 0.3

The final purified enzyme showed one major and three minor bands by SDS-polyacrylamide gel electrophoresis. The major band had a molecular weight of 36,000 and was presumed to be choline kinase. Unfortunately, the purified choline kinase was rather unstable, a property which has been reported for purified kinases from animal tissues[10,12]. We tried a number of reagents in an attempt to stabilise the enzyme and the results of such experiments are shown in Table 2.

TABLE 2

EFFECT OF REAGENTS ON THE DECLINE OF CHOLINE KINASE ACTIVITY DURING STORAGE

Addition (conc.)	Activity (% control)	
	4^0 storage	-20^0 storage
None	61	39
Freeze-dried	90	48
Glycerol (20%)	54	73
Mercaptoethanol (5mM)	66	67
EDTA (5mM)+ protease inhibitor(0.5mg/ml)	110	99

Samples were stored for 13 days at the two temperatures.

It will be seen from the results that the only successful way of preserving activity appeared to be by the addition of a chelating agent, EDTA, and a soya bean trypsin inhibitor. This implied that, at least, part of the loss in activity was due to protease action. It is also interesting that storage of the freeze-dried and untreated preparations at 4° resulted in a smaller loss in activity than storage at -20°. This agrees with previous observations with the rat liver enzyme[10].

SUMMARY

Previous results obtained in vivo with germinating soya bean seeds indicated that separate kinases were present for the synthesis of phosphatidylcholine and phosphatidylethanolamine by the CDP-base pathway. These data have now been confirmed by the isolation of separate enzymes. The choline kinase was found to have a molecular weight of 36,000 and to have a small amount of ethanolamine kinase activity.

ACKNOWLEDGEMENT

We are grateful to the A.R.C. for financial support.

REFERENCES

1. Galliard, T. (1973) in Form and Function of Phospholipids, Ansell, G.B., Dawson, R.M.C. and Hawthorne, J.N. eds., Elsevier, Amsterdam, pp. 253-288.
2. Harwood, J.L. (1977) Biochem. Soc. Trans. 5, 1259-1263.
3. Kennedy, E.P. (1962) Harvey Lect. series, 57, 143-171.
4. Bremer, J. and Greenberg, D.M. (1961) Biochim. Biophys. Acta, 46, 205-216.
5. Dykes, C.W., Kay, J. and Harwood, J.L. (1976) Biochem. J., 158, 575-581.
6. Harwood, J.L. (1975) Phytochemistry, 14, 1985-1990.
7. Harwood, J.L. (1976) Phytochemistry, 15, 1459-1464.
8. Harwood, J.L. and Stumpf, P.K. (1972) Lipids, 7, 8-19.
9. Ulane, R.E., Stephenson, L.L. and Farrell, P.M. (1977) Analyt. Chem. 79, 526-534.
10. Brophy, P.J. and Vance, D.E. (1976) FEBS Lett., 62, 123-125.
11. Brophy, P.J., Choy, P.C., Toone, J.R. and Vance, D.E. (1977) Europ. J. Biochem. 78, 491-495.
12. Weinhold, P.A. and Rethy, V.B. (1974) Biochemistry 13, 5135-5141.

SYNTHESIS AND TURNOVER OF MOLECULAR SPECIES OF GALACTOLIPIDS OF VICIA FABA LEAVES.

J.P. WILLIAMS AND S.P.K. LEUNG
Department of Botany, University of Toronto, Toronto, Ontario M5S 1A1 (Canada).

ABSTRACT

The distribution of radioactivity in the galactose of chloroplast monogalactosyl diacylglycerol (MGDG) and digalactosyl diacylglycerol (DGDG) was examined after $^{14}CO_2$ feeding of *V. faba* leaves. The distribution of radioactivity in the molecular species of the lipids suggest that a large proportion of the fatty acids are desaturated prior to incorporation into the lipids. Further desaturation steps appear to occur after galactolipid formation, primarily 18:2 desaturation. The molecular species 18:3/16:0 of MGDG appears to be galactosylated to DGDG more rapidly than the 18:3/18:3 species, thus accounting for the high level of palmitic acid (16:0) in DGDG.

INTRODUCTION

The predominant lipids in plastids are MGDG and DGDG and they contain unusually high levels of unsaturated fatty acids, especially 18:3. It is generally accepted that MGDG is formed by galactosylation of a diglyceride and DGDG, by galactosylation of MGDG.

Previous work in our laboratory[1,2] has shown that the diglyceride precursors contain fatty acids that are largely desaturated prior to incorporation into MGDG and DGDG. The problems remain, however, as to where and when this occurs and how the fatty acid composition of MGDG and DGDG can differ if DGDG is formed from MGDG. The data presented here attempt to answer these problems by examining the labelling of galactoses of the molecular species of the two lipids after $^{14}CO_2$ feeding of *V. faba* leaves.

MATERIALS AND METHODS

Vicia faba (broad bean) plants were grown at 20 C in 16 hr light/8 hr dark regime under 1100 ft - c illumination for 3 to 4 weeks.

$^{14}CO_2$ feeding. Detached leaves or leaf discs were offered $^{14}CO_2$ in a closed plastic chamber illuminated with fluorescent lights (1100 ft-c) at 20 C.

Lipid separation. Lipids were extracted from leaves using the method of Williams and Merrilees[3] and separated by the TLC method of Khan and Williams[4].

Molecular species were separated by argentation TLC using 20% silver nitrate-silica gel G plates. The galactolipids were methanolysed and the fatty acids extracted with hexane. The amount of radioactivity in the galactose-glycerol (polar) fraction was determined in the remaining methanol-water phase.

Fatty acids were separated by GLC on 10% EGSS - X or 10% Silar 10 C. An internal standard of methyl pentadecanoate was used to determine the fatty acids and lipids quantitatively.

The two galactoses of DGDG were separated according to the methylation procedure of Williams et al[5].

RESULTS AND DISCUSSION

Analyses of the molecular species of MGDG and DGDG (Table 1) show the major similarities and differences between the two lipids. The major molecular species in both MGDG and DGDG is 3/3. In DGDG, however, there are significant quantities of 3/0 and 2/0 reflecting the high levels of palmitic acid (16:0) in DGDG.

Table 1

DISTRIBUTION OF MOLECULAR SPECIES OF GALACTOLIPIDS FROM VICIA FABA LEAVES

Lipid	Molecular species						
	3/3	3/2	3/1	2/2	3/0	2/0	1/0, 0/0
	% distribution						
MGDG	90	4	tr*	2	2	1	tr
DGDG	69	5	1	tr	16	8	tr

* less than 1%

Leaves were fed $^{14}CO_2$, the lipids were extracted and separated into molecular species and the radioactivity of the galactose-glycerol moiety was estimated. Previous determinations have shown that ~95% of the radioactivity is found in the galactose portion of the lipid. Therefore it is concluded that the radioactivity is indicative of galactose radioactivity. The results obtained shortly (10 min) after $^{14}CO_2$ feeding reflect the molecular species composition of the precursors of MGDG (i.e. a diglyceride) and DGDG (i.e. MGDG). If the fatty acids of the galactolipids are desaturated before being incorporated into the lipids by galactosylation then the pattern of radioactivity in the galactose moiety added should mirror the molecular species composition of the lipids in Table 1. If not, then the radioactivity should be found in molecular species containing more saturated fatty acids. The fatty acids in this case would be desaturated after galactolipid formation.

Table 2

DISTRIBUTION OF RADIOACTIVITY IN MOLECULAR SPECIES OF MONOGALACTOLIPID (GALACTOSE) FROM VICIA FABA LEAVES AT DIFFERENT PERIODS FOLLOWING $^{14}CO_2$ FEEDING

Time	Molecular species*					
	3/3	3/2	3/1	3/0, 2/2	2/0	1/0, 0/0
hours			% distribution			
0	26	35	7	23	4	4
1	31	27	8	19	8	6
2	62	15	2	12	5	4
8	66	14	5	9	5	2
24	72	8	4	9	4	4

* combinations of unsaturated bonds in fatty acids

The data in Table 2 illustrate that the initial galactose (^{14}C) incorporation (0 hour) into MGDG is not into the major molecular species 3/3, but is mainly divided between 3/3, 3/2, 2/2 and 3/0 species. In MGDG this would appear to be an anomaly as this does not reflect the molecular species composition of MGDG, especially as MGDG has little of the 3/0 molecular species. However, the precursors do appear to contain predominantly unsaturated fatty acids (18:2 and 18:3). In DGDG, the labelling of the galactose of the molecular species resembled more closely the species composition of this lipid.

The labelling patterns of the galactose of MGDG was examined over a period of time to determine whether further desaturation of fatty acids occurs after galactosylation (Table 2). While the initial incorporation of radioactivity is into the galactose of several molecular species, over a period of time the level of radioactivity increases in the 3/3 while decreasing in 3/2, 2/2 and 3/0. It would appear that desaturation of linoleic acid (18:2) in the 3/2 and 2/2 molecular species is occurring and the radioactivity is then found later in the 3/3 molecular species. It seems unlikely that desaturation of palmitic acid is occurring in the 3/0 molecular species, therefore, it is possible that this molecular species is converted more rapidly into DGDG than the others. The relative radioactivity of the two galactoses of DGDG were determined to see whether any of the molecular species of newly formed MGDG were turned over (to DGDG) more rapidly than others (Table 3). The data in Table 3 compares the labelling of the inner (DG I) and outer (DG II) galactoses of DGDG.

Table 3

SPECIFIC RADIOACTIVITIES OF THE TWO GALACTOSES FROM TWO MOLECULAR SPECIES OF DIGALACTOLIPID AT PERIODS FOLLOWING $^{14}CO_2$ FEEDING OF <u>VICIA FABA</u> LEAVES

Time	Molecular species					
	3/3			3/0 (2/2)		
	DG II	DG I	DG II/DG I	DG II	DG I	DG II/DG I
hours	dpm/µmole x 10^{-3}			dpm/µmole x 10^{-3}		
0	7.7	0.8	9.7	6.1	1.7	3.6
1	9.8	1.9	5.2	5.1	3.4	1.5
2	11.7	3.3	3.6	6.8	4.0	1.7
8	-	-	-	4.7	4.3	1.1
16	9.5	6.3	1.5	5.9	6.0	0.98

A high level of radioactivity in DG I may be interpreted as showing that the MGDG precursor of a particular species is rapidly turned over to DGDG after formation; this would result in a low ratio of DG II/DG I. The results indicate, therefore, a greater rate of turnover of 3/0 (2/2) molecular species of MGDG to DGDG and would thus help to account for the decrease in 3/0 (2/2) species labelling shown in Table 2.

SUMMARY

From the data in Table 2 it would appear that the diglyceride precursor of MGDG consists predominantly of molecular species containing 6 (3/3), 5 (3/2), 4 (2/2) and 3 (3/0) double bonds. This would indicate that desaturation to these stages had occurred before galactosylation to MGDG. The subsequent conversion to mainly 3/3 species would suggest further desaturation steps may occur after MGDG formation. Our data, therefore, indicate that the majority of the desaturation occurs prior to galactosylation and that a final "tuning" occurs afterwards.

The higher level of palmitic acid in DGDG than in MGDG may be accounted for by a higher turnover rate of 3/0 molecular species of MGDG to DGDG. These data will be published in greater detail at a later time.

ACKNOWLEDGEMENTS

This work was supported by a National Research Council of Canada operating grant to J.P. Williams.

REFERENCES
1. Williams, J.P., Watson, G.R., Khan, M. and Leung, S. (1975) Plant Physiol., 55, 1038-1042.
2. Williams, J.P., Watson, G.R. and Leung, S.P.K. (1976) Plant Physiol., 57, 179-184.
3. Williams, J.P. and Merrilees, P.A. (1970) Lipids, 5, 367-370.
4. Khan, M. and Williams, J.P. (1977) J. Chromatogr., 140, 179-185.
5. Williams, J.P., Khan, M. and Leung, S. (1975) J. Lipid Res., 16, 61-66.

AUTHOR INDEX

M.T. Alsasua 251
B.Å. Andersson 257, 263
L.-Å. Appelqvist 343
L. Axelsson 363

R. Bazier 151
C. Bengston 269
E. Bergeron 159
H. Beringer 133
M. Bertrams 139
A. van Besouw 359
P.Å. Biacs 275
H. Bickel 369, 377
R. Bligny 387
G. Blosczyk 287
G. Bögemann 359
J. Boldingh 231
P. Bolton 219
M.E. Breimer 281
C.H. Brieskorn 287
G. Britton 165
B. Buchholz 369
C. Buschmann 145

C. Cassagne 393
A. Chammai 381
K.R. Chandorkar 415
D.J. Chapman 415
A. Cherif 399
M. Chuzel 181
C. Costes 151

G.R. Daleo 313
A.O. Davies 219
R. Douce 79, 99, 181, 387
R. Douillard 159

D. Drapier 437
A. Drira 399
J.P. Dubacq 437

R. Frey 225
A. Frimer 193

T. Galliard 121
F. Gemeinhardt 293
A.R. Gemmrich 403
T. W. Goodwin 165
M. Grosbois 409
S. Grossman 193
K.H. Grumbach 165
T. Guillot-Salomon 169

M.F. Hallais 169
G.C. Hansson 281
G. Harnischfeger 175
J.L. Harwood 219, 443
E. Heinz 99, 139
K.-P. Heise 175
B. Herslöf 301
H.E. Hopp 313

A.T. James 219
R. Jeffcoat 219
J. Joyard 79, 99, 181

O. Kane 211
K.-A. Karlsson 281
M. Kates 329
J.J. Katz 37
J. Kesselmeier 187
M. Khan 415
K. Knobloch 293

G. Larson   281
K. Larsson   27
S. Larsson   269
H. Leffler   281
N.W. Lem   415
Y.Y. Leshem   193
R. Lessire   393
S.P.K. Leung   449
H.K. Lichtenthaler   57, 145, 319
C. Liljenberg   199, 269, 307
M. Linscheid   99
M. de Lubac   169
L. Lundgren   263

R.O. Mackender   205
B. Marzouk   399
P. Mazliak   211, 409, 437
V. McMahon   215
M. Muller   437

F. Nothdurft   133

A. Oursel   421

E. Palacios-Alaiz   251
I. Pascher   281
M.P. Percival   219
W. Pimlott   281
P. Pohl   427
R. Pont Lezica   313
U. Prenzel   319
S. Puang-Ngern   27

F. Rebeille-Borgella

A.I. de la Roche   329
P.A. Romero   313
H.G. Ruppel   187

B.E. Samuelsson   281
A.S. Sandelius   307
R. Schantz   381
G. Schultz   369, 377
E. Selstam   199, 307, 363
H.P. Siebertz   99
M. Signol   169
J. Soimajärvi   433
G. Stenhagen   263
S. Stymne   343

T. Tanaka   243
M. Tevini   225
A. Tremolieres   437
C. Tuquet   169

G.A. Veldink   231
J. Verhagen   231
J.F.G. Vliegenthart   231

M. Wettern   237
P. von Wettstein-Knowles   1
J. Wharfe   219, 443
J.P. Williams   415, 449
A.C. Wilson   329
J.F.G.M. Wintermans   359

M. Yamada   243

J. Ziv   193
F. Zurheide   427

SUBJECT INDEX

Acer pseudoplatanus 387
acyl-ACP 139
acylated glycolipid 104, 238, 435
  molecular species 117
  positional distribution 100, 102, 107
acyl-CoA 142
acyltransferase 104, 119, 123
alcohols 1, 270
  primary 1
  secondary 1
aldehydes 1
algae 427
alkanes 1, 270. 287
Anemia phyllitidis 403
argentation chromatography 108, 113, 118
aromatic amino acids 369
  synthesis 369
autoxidation 129
Avena sativa 187, 205, 269
avocado 133

barley 220, 364, 415
  chlorophyll-b less mutant 220
  normal 220
bimolecular membranes 27
blue-green algae 102, 103
broad bean 220

capillary gas chromatography - mass spectrometry 265
carotenoids 42, 61, 88, 145, 165, 206, 237, 319, 363
  labelling studies 165
  secondary 237

  synthesis 88
carrot 433
  root 433
castor bean 243
CDP-choline-diacylglycerol-phosphotransferase 421
CDP-ethanolamine-diacylglycerol-phosphotransferase 421
CDP-diglyceride 381
cell culture 329
  lipid composition 330
  lipid synthesis 329
cell-free fractions 293
  membranes 293
cell surface recognition 281
cellulose precursor 313
chilling injury 212
Chlorella pyrenoidosa 165
chlorophylls 27, 37, 59, 145, 151, 165, 200, 206, 238, 319, 363
  antenna pigment 37
  electrophilic properties 42
  esterifying alcohol 39
  geranylgeranyl-chlorophyll 59
  labelling studies 165
  - lipid interactions 37
  nuclear magnetic resonance spectroscopy 43
  photoreaction center 37
  phytyl-chlorophyll 59
chloroplast 151, 159, 165, 169, 175, 369, 377, 381, 438
  density gradient centrifugation of 159
  lamellae 159
  lyophilized 151

photosynthetic activity of 177
pigments 165
ultrastructure 169
chloroplast envelope 79, 103, 111,
 181, 359
 ATP translocator 90
 dicarboxylate translocator 90
 fatty acid composition 83
 galactolipid synthesis 181
 lipids 103, 111
 metabolite transport 89
 phosphate translocator 90
 pigments 86
 polar lipids 83, 92, 94
 proteins 89
 protein transport 91
choline kinase 443
cuticular transpiration 269
Cyanidium caldarium 215
cytokinin 193

diacylglycerol 102, 105, 113, 117
digalactosyldiacylglycerol 27, 46,
 151, 177, 181, 207, 307, 449
 lamellar liquid crystalline phase
  27
 molecular species 449
dihydroxyacetone phosphate 142
β-diketones 1
dolichols 313

elongase 393
 microsomal 393
elongation system 2
 elongation-decarboxylation 2
 elongation-reduction 2
endoplasmic reticulum 393
epicuticular wax 1, 269, 275

genetics 1
mutation 1
synthesis 1
ultrastructure 8
esters 1
Euglena gracilis 381
extraction, mild 151

fatty acid desaturases 215
fatty acid hydroperoxides 231
fern spores 403
 imbibition 403
free acids 1, 18, 257, 270, 275, 293
 329, 344, 389, 393, 428, 433, 435,
 437, 449
 composition 332, 389, 428, 435
 desaturation 339, 344, 437, 449
 isomers 433
 pyrrolidides 257
 synthesis 18, 358
 very long-chain 275, 393
free radical scavenging 193
Fritschiella tuberosa 237
Fucus vesiculosus 427

galactolipids 227, 359, 363, 399, 417
 fatty acids 399, 417
 synthesis 359
galactosylation 106, 115, 117, 119
galactosyltransferase 101, 113, 359
geranylgeraniol 200
germination 251
glycerol phosphate acyltransferase
 139
glycerol-3-phosphate 142, 181
glycolipids 281
granal stacking 219
grape 287

green algae 237, 313

Hedera helix 409
hemp 275
  chlorophyll deficient mutant 275
  normal 275
herbicides 61
high performance liquid chromatography 319
Hordeum vulgare 220, 364, 415
hydroperoxylinoleic acid 231

ivy 409

ketone 1

leek 393
linoleic acid 343, 437
  biosynthesis 343, 437
linolenic acid 200, 215, 343, 437
  biosynthesis 343, 437
lipid acyl hydrolase 122, 228
lipid class separation 303
lipid content 428
lipid synthesis 403
  effect of cordycepin and cycloheximide 403
lipid - water systems 27
lipoxygenase 126, 159, 193, 231
Lupinus luteus 421
lysophosphatidic acid 141

mango 211
mass spectrometry 102, 105, 257, 265, 281
membrane degradation 121
membrane fluidity 215
membrane turnover 129

microsomes 244, 381, 438
mitochondria 213, 244
  changes in fatty acid composition 213
  oxidative capacity 212
molecular associations 151
monogalactosyldiacylglycerol 27, 46, 151, 177, 181, 200, 207, 307, 449
  hexagonal liquid crystalline phase 27
  molecular species 449
monolayer 201

neutral lipids 177, 435

olive fruit 399
  embryo 399
  endosperm 399
  lipid accumulation 399
α-oxidation 125

palmitic acid 213
palmitoleic acid 213
pea 194, 393, 438
peroxisomes 369
petroselinic acid synthesis 409
  microsomes, role in 413
  mitochondria, role in 413
phosphatidylcholine 151, 217, 243
phosphatidylcholine - desaturase 346
phosphatidylglycerol 151, 169, 381
  fatty acids 417
  synthesis 381
phospholipase A 228
phospholipase D 124, 228
phospholipid 177, 227, 238, 251, 293, 399, 421, 435
  biosynthesis 421

fatty acids 399
   inhibition by Ca 421
phospholipid exchange protein 243
photodecomposition 363
photosynthetic activity 145
Phycodrys sinuosa 427
phylloquinone K_1 63, 145
phytohormones 145
phytol 200
Pinus pinea 251
plant development 275, 399
plant growth 433
plastids 206, 225, 226
   aging 225
   development 206
   pigment composition 207
   size 226
plastoquinone-9 62, 145
polarizing microscopy 28
polyprenylphosphate 313
potato 307
prenols 319
prenyllipids 57, 145
   biosynthesis 58
   in the chloroplast envelope 64
   in the thylakoid 66
   labelling degree of 70
   turnover 70
prenylquinones 319, 369, 377
   synthesis 369, 377
prenylvitamins 317
prolamellar bodies 187
protochlorophyll(ide) 145
Prototheca zopfii 313

Raphanus sativus 145
regreening 238
Rhodopseudomonas palustris 293

Rhodopseudomonas spheroides 293

safflower 348
saponins 187
senescence 193
shade plants 169
shikimate pathway 377
   feedback regulation 377
soya bean 231, 338, 348, 443
spinach 225, 359, 369, 377, 439
stearic acid 200
stearoyl-CoA 393
   synthetase 393
steroids 287
stratification 251
structural analysis 257, 281
substituted pyridazinones 415
   effects on lipid synthesis 415
sulphoglycolipid 281
sulphoquinovosyldiacylglycerol 43, 151, 217
sunflower 135
sun plants 169
surface film balance 200
suspension culture 387
   lipid synthesis 387

ternary system MGDG-DGDG-water 28
thin layer chromatography/flame ionization detector 301
thylakoid membranes 27, 206
tocochromanol 133
α-tocopherol 64, 133, 145, 194
α-tocoquinone 64
trans-3-hexadecenoic acid 169, 219, 417
triacylglycerol 104, 105, 238, 307
triterpenes 1, 287

UDP-galactose 108, 109, 113, 119, 181, 359

Vicia faba 136, 415, 421, 449
volatile substances 263
  analysis of 263

water stress 269
wax esters 287
wheat 151, 159, 364

x-ray analysis 28